T0202706

CHEMISTRY AND MOLECULAR ASPECTS OF DRUG DESIGN AND ACTION

CHEMISTRY AND MOLECULAR ASPECTS OF DRUG DESIGN AND ACTION

E. A. Rekka
P. N. Kourounakis

CRC Press
Taylor & Francis Group
Boca Raton London New York

CRC Press is an imprint of the
Taylor & Francis Group, an **informa** business

CRC Press
Taylor & Francis Group
6000 Broken Sound Parkway NW, Suite 300
Boca Raton, FL 33487-2742

First issued in paperback 2019

ISBN-13: 978-0-8493-9006-7 (hbk)
ISBN-13: 978-0-367-38736-5 (pbk)

Library of Congress Cataloging-in-Publication Data

Chemistry and molecular aspects of drug design and action / editors, E.A. Rekka and P.N. Kourounakis.
    p. ; cm.
"A CRC title."
Includes bibliographical references and index.
ISBN-13: 978-0-8493-9006-7 (hardcover : alk. paper)
ISBN-10: 0-8493-9006-0 (hardcover : alk. paper)
    1. Pharmaceutical chemistry. 2. Drugs--Design. I. Rekka, E. A. II. Kourounakis, P. N. III. Title.
    [DNLM: 1. Chemistry, Pharmaceutical. 2. Drug Design. 3. Pharmacologic Actions. QV 744 C5173 2007]

RS403.C424 2007
615'.19--dc22
                                                2007010012

Visit the Taylor & Francis Web site at
http://www.taylorandfrancis.com

and the CRC Press Web site at
http://www.crcpress.com

Contents

PART I

CHEMICAL, BIOCHEMICAL, AND BIOLOGICAL ASPECTS OF PATHOPHYSIOLOGICAL CONDITIONS

PART II
CLASSICAL MEDICINAL CHEMISTRY

PART III
DRUG DESIGN, CHEMICAL AND MOLECULAR
ASPECTS OF DRUG ACTION

PART IV

DRUG – XENOBIOTIC METABOLISM

PART V

PHYSICAL ORGANIC AND THEORETICAL MEDICINAL CHEMISTRY

Editors

Eleni A. Rekka is an associate professor in pharmacy, Department of Pharmaceutical Chemistry, School of Pharmacy, Aristotelian University of Thessaloniki, Greece. She holds a first degree in pharmacy (1977) and a Ph.D. in medicinal chemistry (1986), both from the School of Pharmacy, Aristotelian University of Thessaloniki. Since 1978 her research activities in the department have been in the fields of design and synthesis of novel biologically active compounds, including *in vivo* and *in vitro* investigations, aimed at the elucidation of the molecular mechanism of action and the establishment of relationships between chemical structure and biological activity. Special areas of scientific and research interest are those of free radicals, and oxidative stress and antioxidant agents, in relation to the design of bioactive compounds for the treatment of conditions such as inflammation and atheromatosis. An important part of her activity covers xenobiotic metabolism. For the last 30 years, she has participated in the teaching, research, and administrative activities of the department. She has written a number of publications in scientific journals, and has given many presentations at international conferences.

She has conducted research as a visiting scientist on topics in her specialty in the following European research institutes: Department of Pharmacochemistry, Free University of Amsterdam (1987–1988); Janssen Research Foundation, Beerse, Belgium (1989); School of Biological Sciences, University of Surrey, Guildford, UK (1992); Department of Pharmacology and Therapeutics, University of Sheffield, UK (1994); Department of Pharmacology and Toxicology, Philipps University, Marburg, Germany (1995); University of Mainz, Germany (1997); School of Pharmacy, Catholic University of Louvain, Brussels, Belgium (1999); Université Rene Descartes (Paris V), Unité mixte des recherché NRS, Paris, France (2001).

Panos N. Kourounakis is professor emeritus in pharmacy, Department of Pharmaceutical Chemistry, School of Pharmacy, Aristotelian University of Thessaloniki, Greece. He graduated from the Department of Pharmacy, University of Athens, and obtained his Ph.D. in pharmaceutical–medicinal chemistry from Chelsea College of Science and Technology, University of London. He also did postdoctoral research in drug metabolism at Chelsea College. He served as a research associate in medicinal chemistry, Faculty of Pharmacy, University of Montreal, where he became head of the biochemical laboratory and assistant professor, Institute of Experimental Medicine and Surgery, Faculty of Medicine. He became professor of pharmaceutical chemistry, School of Pharmacy, Aristotelian University of Thessaloniki, Greece, then head and deputy head. For about 20 years he was director of the department. Since 2003 he has been professor emeritus.

Dr. Kourounakis is a member of the council of the National Drug Organisation, and has been both president and vice president of the scientific committee for the approval of medicines in Greece. He was a member of the Supreme Health Council in Athens. He has guided research leading to a Ph.D. or master's degree of 22 graduate students, and has been invited to lecture in a number of institutions in the United States, Canada, Germany, Poland, Czechoslovakia, and Cyprus. He has been honored by the Deutsche Pharmazeutische Gesellschaft with the Herman-Thoms Medaille for his contribution to the European Pharmaceutical Sciences. He is one of the founding members of the European Federation for Pharmaceutical Sciences (EUFEPS). He has published 177 research papers and has written four scientific books, also editing with Dr. E.A. Rekka the book *Advanced Drug Design and Development. A Medicinal Chemistry Approach* (Horwood Ltd., Hemel-Hempstead, Hertfordshire, U.K. 1994).

His research interests include the following: drug design (synthesis, physical organic chemical studies); molecular–biochemical mechanisms of drug action; drug metabolism; biologic and oxidative stress interrelationships; inflammatory and neurodegenerative diseases, and the design of related drugs.

Contributors

Vildan Alptuzun
Faculty of Pharmacy
Ege-University
Bornova-Izmir, Turkey

Lucilla Angeli
Università degli Studi di Siena
Siena, Italy

J. Balzarini
Katholieke Universiteit Leuven
Brussels, Belgium

Thomas Beach
Sun Health Research Institute
Sun City, Arizona

F. Bitter
National Magnet Laboratory
Massachusetts Institute of Technology
Cambridge, Massachusetts

Maurizio Botta
Università degli Studi di Siena
Siena, Italy

Longchuan Chen
VA Medical Center
Long Beach, California
  and University of California, Irvine
Irvine, California

E.A. Coats
Research Center Borstel
Leibniz Center for Medicine and
  Biosciences
Borstel, Germany

E. De Clerq
Katholieke Universiteit
Leuven, Belgium

Xiaoming Deng
VA Medical Center
Long Beach, California
  and University of California, Irvine
Irvine, California

P.-T. Doulias
University of Ioannina
Ioannina, Greece

Bernard Faller
Novartis Pharma AG
Basel, Switzerland

Gerd Folkers
Institute of Pharmaceutical Sciences
ETH Zurich
Zurich, Switzerland

Bente Frølund
Faculty of Pharmaceutical Sciences
University of Copenhagen
Copenhagen, Denmark

D. Galaris
University of Ioannina
Ioannina, Greece

R.K. Guy
Department of Chemical Biology
  and Therapeutic
St. Jude Children's Research
  Hospital
Memphis, Tennessee

R.W. Hartmann
Saarland University
Saarbrücken, Germany

Eberhard Heller
Institute of Pharmacy and Food
  Chemistry
University of Würzburg
Würzburg, Germany

Ulrike Holzgrabe
Institute of Pharmacy and Food
  Chemistry
University of Würzburg
Würzburg, Germany

Costas Ioannides
School of Biomedical and Molecular
  Sciences
University of Surrey
Guildford, United Kingdom

Petra Kapkova
Institute of Pharmacy and Food
  Chemistry
University of Würzburg
Würzburg, Germany

A. Karlsson
Karolinska Institute
Stockholm, Sweden

Tetyana Khomenko
VA Medical Center
Long Beach, California
  and University of California, Irvine
Irvine, California

A.P. Kourounakis
Department of Pharmaceutical
  Chemistry
Faculty of Pharmacy
University of Athens
Athens, Greece

P.N. Kourounakis
Department of Pharmaceutical
  Chemistry
Faculty of Pharmacy
Aristotelian University of Thessaloniki
Thessaloniki, Greece

Uffe Kristiansen
Faculty of Pharmaceutical Sciences
University of Copenhagen
Copenhagen, Denmark

Povl Krogsgaard-Larsen
Faculty of Pharmaceutical Sciences
University of Copenhagen
Copenhagen, Denmark

Eva Kugelmann
Institute of Pharmacy and Food
  Chemistry
University of Würzburg
Würzburg, Germany

Didier M. Lambert
Université Catholique de Louvain
Brussels, Belgium

Rena Li
Sun Health Research Institute
Sun City, Arizona

Zhe Liang
Sun Health Research Institute
Sun City, Arizona

Tommy Liljefors
Faculty of Pharmaceutical Sciences
University of Copenhagen
Copenhagen, Denmark

Li-Fen Lue
Sun Health Research Institute
Sun City, Arizona

Giovanni Maga
Istituto di Genetica Molecolare
Pavia, Italy

A. Makriyiannis
Northeastern University
Boston, Massachusetts

U. Müller-Vieira
Saarland University
Saarbrücken, Germany

Lajos Nagy
VA Medical Center
Long Beach, California
  and University of California, Irvine
Irvine, California

K.J. Netter
Philipps University
Marburg, Germany

K.C. Nicolaou
Department of Chemistry and The
  Skaggs Institute for Chemical Biology
The Scripps Research Institute and
Department of Chemistry and
  Biochemistry
University of California, San Diego
La Jolla, California

Herbert Oelschläger
Friedrich-Schiller-Universität
Jena, Germany

O. Pelkonen
University of Oulu
Oulu, Finland

Jacques H. Poupaert
Université Catholique de Louvain
Brussels, Belgium

Marco Radi
Università degli Studi di Siena
Siena, Italy

H. Raunio
University of Oulu
Oulu, Finland

E.A. Rekka
Department of Pharmaceutical
  Chemistry
Faculty of Pharmacy
Aristotelian University of Thessaloniki
Thessaloniki, Greece

Joseph Rogers
Sun Health Research Institute
Sun City, Arizona

Alexander Roher
Sun Health Research Institute
Sun City, Arizona

Marwan Sabbagh
Sun Health Research Institute
Sun City, Arizona

Zsuzsanna Sandor
VA Medical Center
Long Beach, California
  and University of California, Irvine
Irvine, California

Walter Schunack
Freie Universität
Berlin, Germany

Andreas Seeling
Friedrich-Schiller-Universität
Jena, Germany

J.K. Seydel
Research Center Borstel
Leibniz Center for Medicine and
  Biosciences
Borstel, Germany

Yong Shen
Sun Health Research Institute
Sun City, Arizona

Holger Stark
Freie Universität
Berlin, Germany

Ronald Strohmeyer
Sun Health Research Institute
Sun City, Arizona

Sandor Szabo
VA Medical Center
Long Beach, California
  and University of California, Irvine
Irvine, California

P. Taavitsainen
University of Oulu
Oulu, Finland

Yvette Taché
Department of Medicine
University of California Los Angeles
and VA Greater Los Angeles Healthcare
  System
Los Angeles, California

M. Tenopoulou
University of Ioannina
Ioannina, Greece

Theoharis C. Theoharides
Tufts University School of Medicine
New England Medical Center
Boston, Massachusetts

Ganna Tolstanova
VA Medical Center
Long Beach, California
  and University of California, Irvine
Irvine, California

Alexandros Tselepis
University of Ioannina
Ioannina, Greece

M. Turpeinen
University of Oulu
Oulu, Finland

S. Ulmschneider
Saarland University
Saarbrücken, Germany

J. Uusitalo
University of Oulu
Oulu, Finland

K. Visser
Research Center Borstel
Leibniz Center for Medicine and
  Biosciences
Borstel, Germany

M. Voets
Saarland University
Saarbrücken, Germany

Douglas Walker
Sun Health Research Institute
Sun City, Arizona

Jesper Wengel
Department of Physics and Chemistry
University of Southern Denmark
Odense, Denmark

M. Wiese
Research Center Borstel
Leibniz Center for Medicine and
  Biosciences
Borstel, Germany

Ernst Wülfert
Hunter-Fleming Ltd.
Bristol, United Kingdom

Rational Drug Design Based Mainly on the Pathobiochemistry of the Disease

Examples of Atheromatosis and Alzheimer's Type Neurodegeneration

E. A. Rekka, A. P. Kourounakis, and P. N. Kourounakis

INTRODUCTION

The ever-increasing interest in better drugs, the high safety standards set by the health and drug authorities of all developed countries, and concern about the economic burden to society involved in the discovery of drugs and their therapeutic applications have changed dramatically the ways of drug discovery. The old processes, folk medicine (ethnopharmacy), serendipity, and massive pharmacological screening of organic compounds, although not completely abandoned, have been replaced by methods that lead to the desired molecules, resulting in drugs containing the expected properties, i.e., rational drug design has been applied.[1] In rational drug design there are three main steps. The *discovery of the lead compound* is the most difficult step. The *discovery of pharmacophore* follows, after a series of "surgical operations" on the lead molecule. *Optimization* involves further structural manipulations on the pharmacophore, which, in conjunction with physical organic measurements and biological testing, lead to the finding of the best compound for a particular purpose. Often, (quantitative) structure-activity relationships are also included. Of course, the discovery of a new medicine is still far from this stage. However, with the addition of studies on drug metabolism, and after the study of the molecular mechanism of action, the main work of a medicinal chemist in the discovery of new therapeutic agents is completed.

Among a number of approaches in rational drug design, a few of the most important study subjects are presented:

- The molecular mechanism of drug action.
- Drug-metabolizing enzyme action upon the structure of the drug molecule.
- Pathobiochemistry and pathophysiology of the target disease.
- A modern approach in drug design is the development of drugs starting from compounds stored in data banks. Structures, physicochemical properties, and biological activity are taken, a few are selected, and then the above-mentioned strategies are applied.

Herein, two examples of rational drug design will be presented, using basically the pathobiochemical approach. Other parameters, like drug metabolism and molecular mechanism of action, will also be covered.

DESIGN OF NOVEL ANTIDYSLIPIDEMIC AGENTS

Atheromatosis is the main contributor to cardiovascular problems, which are the most common cause of death in Western societies. Dyslipidemias are the prime cause of atheromatosis.[2] As low-density lipoprotein cholesterol (LDL) constitutes the main cholesterol carrier in the plasma, it influences cholesterol plasma levels and is closely related to the etiology of atheromatosis.[3]

In atheromatosis the following pathologic events take place: hypercholesterolemia, hypertriglyceridemia, oxidation of LDL to oxLDL, (a form not recognized by the LDL receptor), a free radical attack to the vascular endothelium, and the accumulation of macrophages that release superoxide anion radicals in atheromatic plaques.[4] Further injury is caused via nitric monoxide, peroxynitrite, hydrogen peroxide, and hydrolytic enzymes. Macrophages induce platelet-stimulating factors, generation of potentially toxic peroxidation products, which in turn aggravate the initial endothelial damage.[5] Oxidized LDL is a potent chemoattractive factor for human blood monocytes in atheromatic lesion and further induces eicosanoid production. Monocyte-derived macrophages are sources of growth factors derived of platelets, as well as smooth muscle cell growth factors.[6] Oxidized lipids induce production of interleukine-1. Lysophosphatidyl choline, a product of LDL oxidation, is chemoattractive for monocytes and T-lymphocytes, as well as intracellular adhesion molecules and various toxic products.[7] Furthermore, vascular endothelium is responsible for the biosynthesis of nitric monoxide, an important mediator of vascular tone, blood flow, and pressure, presenting antiatheromatic properties.[8] Thus, impairment of nitric monoxide production affects vasodilatation, control of platelet activation, interaction of endothelial cells and leukocytes, inhibition of the superoxide anion radical-mediated cell injury, and reduction of LDL oxidation.[9]

The authors, having extensive experience in regard to morpholine structures of type **1** (figure I.1), have shown that these agents were analgesics and anti-inflammatory agents.[10–12] They were also moderate antioxidants. We then increased their antioxidant activity by introducing the 2-diphenylyl substituent (structure **2**, figure I.1).[13]

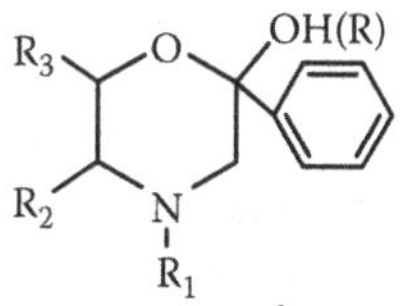

1

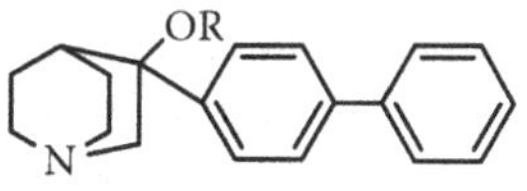

I

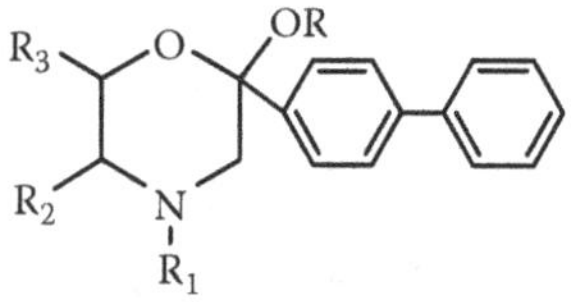

2

2a: $R_1 = CH_3$, $R_2R_3 = (CH_2)_4$, $R = H$

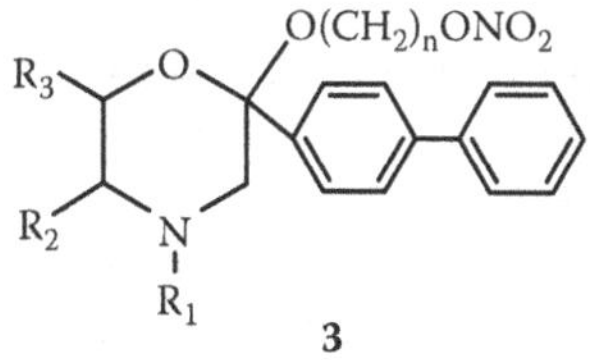

3

3a: $R_1 = CH_3$, R_2, $R_3 = H$
3b: $R_1 = CH_3$, $R_2R_3 = (CH_2)_4$
3c: $R_1R_2 = (CH_2)_4$, $R_3 = H$

FIGURE I.1 Investigated morpholine derivatives.

As compounds of type **2** have structural similarities with the known antidyslipidemic quinuclidines **I**,[14] we examined whether structures **2** could reduce dyslipidemia induced in animals by the administration of tyloxapol. Plasma total cholesterol (TC), LDL-cholesterol (LDL), and triglyceride (TG) levels were determined and compared with the untreated hyperlipidemic controls. It was found that the examined compounds were efficient antidyslipidemic agents, the best being compound **2a** (percent reduction of TC, LDL, and TG was 54, 49, and 51, respectively, at a dose of 26 μmol/kg, i.p., whereas for probucol, a known antidyslipidemic drug, the percent reduction of these values was 18, 11, and 18, respectively, under the same conditions). Compounds of the structure **2** significantly inhibited hepatic microsomal membrane lipid peroxidation (IC_{50} 250 μM, after 30 min of incubation, whereas probucol inhibited lipid peroxidation by 16% at 1 mM) and interacted with 2,2-diphenyl-1-picrylhydrazyl stable radical by about 80% at equimolar concentrations.[3]

We further modified structure **2** in order to add nitric monoxide-releasing activity (structure **3**). Most of the new compounds acquired NO-releasing properties (percent NO release 17.3 and 4.6, for **3a** and **3c**, respectively). In general, they presented an even better effect against lipid peroxidation (IC_{50} 120 and 70 μM for **3a** and **3b**, respectively) and antidyslipidemic activity (percent reduction of TC, LDL, and TG, 63, 85, 47, for **3a** and 20, 37, and 25 for **3c**, at the same dose).[15] The most active compound was **3a**. In an attempt to investigate the mechanism of these actions, we have found that they were potent squalene synthase inhibitors (IC_{50} 33 and 0.6 μM for **2a** and **3a**, respectively, using rabbit liver microsomes, and 63 or 1 μM, for **2a** and **3a**, respectively, using human liver microsomes).[16]

In addition, it has been found that **2a** and **3a** were able to reduce not only hyper-lipidemia, but also the development of atheromatic lesions.[17]

Thus, it can be concluded that, by the design of structures, which combine anti-inflammatory, antioxidant, antidyslipidemic and NO-liberating properties, new lead compounds have been developed against hyperlipidemia and atheromatosis, which may lead to better, safer, and more efficient therapeutic agents for cardiovascular diseases.

DESIGN OF STRUCTURES AGAINST NEURODEGENERATION OF ALZHEIMER'S TYPE

Neurodegenerative diseases like senile dementia of Alzheimer's type (SDAT) are among the greatest challenges in pharmacochemistry. It is estimated that there will be about 4×10^7 sufferers from this disorder by 2020.[18] SDAT, cardiovascular dis-orders, and cancer are the first three causes of morbidity worldwide. At present, there is no effective treatment for SDAT. The applied pharmacotherapy is based on the cholinergic hypothesis.[19] These drugs are of limited benefit, quite toxic, and fail to inhibit the progress of the disease. Furthermore, there is a challenge because the pathogenesis of SDAT is essentially not known. However, there is a series of patho-biochemical changes in the demented brain that could be used as a base for a rational design of drug molecules for SDAT. In addition to the cholinergic deficit, the patho-biochemical changes related to the present work are: inflammation, oxidative stress, hypercholesterolemia,[20] and reduced presence of the nerve growth factor (NGF).[21] Thus, there are several targets for a pharmacochemical approach against SDAT.

We have designed and synthesized structures **II** (figure I.2) that are constructed from a nonsteroidal anti-inflammatory drug (NSAID), which offers anti-inflammatory properties and suppresses the amyloid pathologic process; a substituted proline con-nected to the NSAID via an amide bond, offering nootropic properties;[22,23] and a cysteamine or L-cysteine ethyl ester moiety, which amidates the carboxylic group of proline, conferring antioxidant activity. Furthermore, the carboxylic group of pro-line was esterified with 4-methyl-2-methoxyphenol or 3-(3-pyridyl)propanol. These are two neuroprotective agents, the former acting via induction of NGF,[21] the latter by another, yet unknown mechanism.[24,25]

The novel compounds are expected to have anti-hyperlipidemic properties, since we have found that closely related compounds reduce blood cholesterol levels. Furthermore, they should have reduced gastrointestinal (GI) toxicity because the carboxylic group of the parent NSAID has been masked and also through their anti-oxidant properties, since the ulcerative GI tissue caused by NSAIDs suffers from oxidative stress.[26,27]

Almost all compounds possessed the expected properties:

- They demonstrated very satisfactory *in vivo* anti-inflammatory activity (car-rageenin or Freund's Complete Adjuvant [FCA]-induced paw edema model or FCA-induced arthritis). The best two compounds, **6** and **8**, presented 69% and 61% inhibition in the carrageenin test, respectively, whereas com-pound **4** inhibited completely (100%) the FCA-induced arthritis.[28,29]

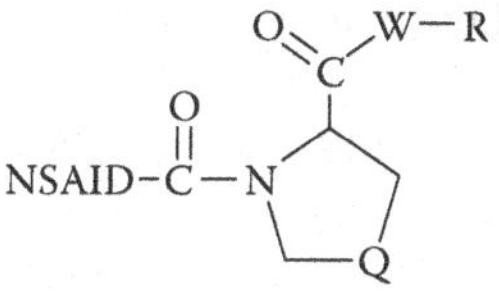

Comp.	NSAID	Q	W	R
1	Indomethacin	CH_2	NH	CH_2CH_2SH
2	Indomethacin	CH_2	NH	$CH(COOC_2H_5)CH_2SH$
3	Naproxen	CH_2	NH	CH_2CH_2SH
4	Naproxen	CH_2	NH	$CH(COOC_2H_5)CH_2SH$
5	Naproxen	CHOH	NH	CH_2CH_2SH
6	Naproxen	CHOH	NH	$CH(COOC_2H_5)CH_2SH$
7	Naproxen	$(CH_2)_2$	NH	$CH(COOC_2H_5)CH_2SH$
8	Ibuprofen	CH_2	NH	CH_2CH_2SH
9	Ibuprofen	CH_2	NH	$CH(COOC_2H_5)CH_2SH$
10	Ibuprofen	CH_2	O	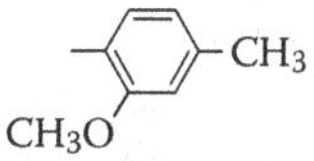
11	Ibuprofen	CH_2	O	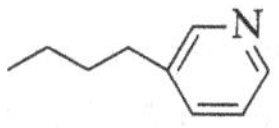

FIGURE I.2   Structures of the studied prospective nootropic molecules.

- They inhibited *in vitro* lipoxygenase activity, with IC_{50} values of 35 μM and 48 μM, for compounds **3** and **4**, respectively.[28]
- Compound **11** inhibited the production of cyclooxygenase-2 (COX-2) from spleenocytes from arthritic rats by 88%.[29]
- Some of the compounds inhibited lipid peroxidation very efficiently. Compounds **7**, **8**, and **9** demonstrated IC_{50} values of 122 μM, 19 μM, and 75 μM, respectively, after 45 min of incubation.[28, 29]
- They interacted with DPPH quite well. The best three compounds were **4**, **3**, and **7**, interacting with DPPH by 86%, 79%, and 77%, at equimolar concentration.[28, 29]
- Most of the compounds reduced tyloxapol-induced rat dyslipidemia. They decreased TC, LDL, and TG levels by about 70%, 55%, and 70%, respectively, at doses of 56 to 300 μmol/kg.[28, 29]
- Compounds **10** and **11** demonstrated moderate acetylcholinesterase inhibitory activity.[29]
- Compound **9** protected the brain from oxidative injury caused by ischemia/reperfusion. An increase in malondialdehyde (MDA) and a depletion of glutathione (GSH) indicated brain oxidative damage, whereas administration

of **9** reduced MDA toward normal values and increased GSH concentration completely back to normal values. [MDA: 0.80±0.13 (control); 1.73±0.31 (ischemia/reperfusion); 1.35±0.17 (ischemia/reperfusion plus **9**) nmol/ mg protein. GSH: 97.8±6.8 (control); 54.1±8.1 (ischemia/reperfusion); 90.3±10.0 (ischemia/reperfusion plus **9**) μmol/g brain].[29]

- We examined the effect of these compounds on the gastrointestinal tract, reporting mortality, gastrointestinal ulceration, and body weight loss and melena incidence. The administered dose was equimolar to that of the parent NSAID, which produced 50% mortality, given subcutaneously to rats for four days.[26,27] All tested compounds were almost free of gastrointestinal toxicity, a common, dangerous side effect of the NSAIDs. We attribute this mainly to two reasons: The carboxylic group of the parent drug has been masked in the designed compounds. Thus, the local irritation is prevented. Furthermore, the ulceration due to NSAIDs causes oxidative stress to the injured tissue. The antioxidant activity of the synthesized compounds protects the gastrointestinal tract from oxidative insult.[28,29]

In conclusion, we designed and synthesized compounds **1–11** aiming for agents that would acquire a series of biological properties able to prevent or restore a number of pathological changes implicated in SDAT and appearing in the demented brain. Most of them possess anti-inflammatory, antioxidant, and antidyslipidemic actions. Two of them moderately inhibit acetylcholinesterase. They were almost free of gastrointestinal toxicity, a property required for long-term application.[29] Thus, it is indicated that the synthesized compounds could be considered satisfactory lead structures for agents against pathological conditions involving inflammation, oxidative stress, and dyslipidemia, such as SDAT. The fact that compound **9** protected the brain from oxidative injury produced by ischemia/reperfusion justifies this molecular design. We believe that a multicausal disease such as SDAT can be better treated with multifunctional agents, acting at different causes and stages of the disease pathogenesis.

REFERENCES

1. Christofferson, R.E. and Marr, J.J., The management of drug discovery. In *Burger's Medicinal Chemistry and Drug Discovery. Vol 1: Principles and practice.* Wolf, A., Ed., Wiley Interscience, New York, 1995, pp. 9-36.
2. Chrysselis, M.C., Rekka, E.A., and Kourounakis, P.N., Antioxidant therapy and cardiovascular disorders, *Exp. Opin. Ther. Patents*, 11, 33, 2001.
3. Chrysselis, M.C., Rekka, E.A., and Kourounakis, P.N., Hypocholesterolemic and hypolipidemic activity of some novel morpholine derivatives with antioxidant activity, *J. Med. Chem.*, 43, 609, 2000.
4. Berliner, J.A. et al., Atherosclerosis: basic mechanisms. Oxidation, inflammation and genetics, *Circulation*, 91, 2488, 1995.
5. Pannala, A.S. et al., Interaction of peroxynitrite with carotenoids and tocopherols within low density lipoprotein, *FEBS Lett.*, 423, 297, 1998.
6. Ross, R., The pathogenesis of atherosclerosis: a perspective for the 1990s, *Nature*, 362, 801, 1993.
7. McMurray, H.F. et al., Oxidatively modified low density lipoprotein is a chemoattractant for human T lymphocytes, *J. Clin. Invest.*, 1004, 1993.

8. Ku, G. et al., Induction of interleukin 1 beta expression from human peripheral blood monocyte-derived macrophages by 9-hydroxyoctadecadienoic acid, *J. Biol. Chem.*, 267, 14183, 1992.

9. Eberhardt, R.T. and Loscalzo, J., Nitric oxide in atherosclerosis. In *Nitric Oxide and the Cardiovascular System*. Loscalzo, J. and Vita, J.A., Eds., Humana Press, New Jersey, 2000, pp. 273-295.

10. Rekka, E. and Kourounakis, P.N., Synthesis, physicochemical properties and biological studies of some substituted 2-alkoxy-4-methyl-morpholines, *Eur. J. Med. Chem.*, 24, 179, 1989.

11. Rekka, E. et al., Lipophilicity of some substituted morpholine derivatives synthesized as potential antinociceptive agents, *Arch. Pharm.*, 323, 53, 1989.

12. Hadjipetrou-Kourounakis, L. et al., Immunosuppression by a novel analgesic-opioid agonist, *Scand. J. Immunol.*, 29, 449, 1989.

13. Tani, E. et al., Modification of drug metabolic processes by compounds which intervene with the function of cytochrome P-450 as a mono-oxygenase and as an oxidase, *Arzneim.-Forsch./Drug Res.*, 39, 1399, 1989.

14. Brown, G.R. et al., Synthesis and activity of a novel series of 3-biarylquinuclidine squalene synthase inhibitors, *J. Med. Chem.*, 39, 2971 1996.

15. Chrysselis, M.C. et al., Nitric oxide releasing morpholine derivatives as hypolipidemic and antioxidant agents, *J. Med. Chem.*, 45, 5406, 2002.

16. Tavridou, A. et al., Pharmacological characterization *in vitro* of EP2306 and EP2302, potent inhibitors of squalene synthase and lipid biosynthesis, *Eur. J. Pharmacol.*, 535, 34, 2006.

17. Tavridou, A. et al., EP2306 [2-(4-biphenyl)-4-methyl-octahydro-1,4-benzoxazin-2-ol, hydrobromide], a novel squalene synthase inhibitor, reduces atherosclerosis in the cholesterol-fed rabbit, *J. Pharmacol. Exp. Ther.*, 323, 794, 2007.

18. Praticò, D., Oxidative imbalance and lipid peroxidation in Alzheimer's disease, *Drug Dev. Res.*, 56, 446, 2002.

19. Davis, K.L. and Samuels, S.C., Dementia and delirium. In *Pharmacological Management of Neurological and Psychiatric Disorders*. Enna, S.J. and Coyle, J.T., Eds., McGraw Hill, Columbus, OH, 1998, pp. 267-316.

20. Michaelis, M.L., Drugs targeting Alzheimer's disease: some things old and some things new, *J. Pharmacol. Exp. Ther.*, 304, 897, 2003.

21. Kourounakis, A. and Bodor, N., Quantitative structure activity relationships of catechol derivatives on nerve growth factor secretion in L-M cells, *Pharm. Res.*, 12, 1199, 1995.

22. Gudasheva, T.A. et al., Design of N-acylprolyltyrosine "tripeptoid" analogues of neurotensin as potential atypical antipsychotic agents, *Eur. J. Med. Chem.*, 31, 151, 1996.

23. Romanova, G.A. et al., Antiamnesic effect of acyl-prolyl-containing dipeptide (GVS-111) in compression-induced damage to frontal cortex, *Bull. Exp. Biol. Med.*, 130, 846, 2000.

24. Steiner, J.P. et al., Neurotrophic immunophilin ligands stimulate structural and functional recovery in neurodegenerative animal models, *Proc. Natl. Acad. Sci. U.S.A.*, 94, 2019, 1997.

25. Steiner, J.P. et al., Neurotrophic actions of nonimmunosuppressive analogues of immunosuppressive drugs FK506, rapamycin and cyclosporin A, *Nat. Med.*, 3, 421, 1997.

26. Kourounakis, P.N. et al., Reduction of gastrointestinal toxicity of NSAIDs via molecular modifications leading to antioxidant anti-inflammatory drugs, *Toxicology*, 144, 205, 2000.

27. Galanakis, D. et al., Synthesis and pharmacological evaluation of amide conjugates of NSAIDs with L-cysteine ethyl ester, combining potent antiinflammatory and antioxidant properties with significantly reduced gastrointestinal toxicity, *Bioorg. Med. Chem. Lett.*, 14, 3639, 2004.

28. Doulgkeris C.M. et al., Synthesis and pharmacochemical study of novel polyfunctional molecules combining anti-inflammatory, antioxidant, and hypocholesterolemic properties, *Bioorg. Med. Chem. Lett.*, 16, 825, 2006.
29. Siskou, I.C. et al., Design and study of some novel ibuprofen derivatives with potential nootropic and neuroprotective properties, *Bioorg. Med, Chem.*, 15, 951, 2007.

Part I

Chemical, Biochemical,
and Biological Aspects of
Pathophysiological Conditions

1 Inflammatory Mechanisms in Alzheimer's Disease and Other Neurodegenerative Disorders

Joseph Rogers, Thomas Beach, Rena Li,
Zhe Liang, Lih-Fen Lue, Alexander Roher,
Marwan Sabbagh, Yong Shen,
Ronald Strohmeyer, and Douglas Walker

CONTENTS

Twenty years ago it was recognized that inflammatory mechanisms play a patho-genic role in multiple sclerosis, but for other disorders the brain was generally con-sidered to be "immunologically privileged." This view began to change with the discovery that microglia in the Alzheimer's disease (AD) cortex were activated (e.g., immunoreactive for the major histocompatibility complex),[1,2] similar to peripheral macrophages at sites of inflammation. We now know that cytokines, complement, reactive oxygen and nitrogen species, and other markers of inflammation are highly upregulated in a wide range of neurodegenerative disorders, particularly AD.[3] This should not be surprising because, by definition, tissue damage occurs in neurode-generative disorders, and inflammation is a normal response to tissue damage wher-ever it occurs in the body. In addition, the chronic presence of highly inert, highly insoluble deposits is another classic inflammatory stimulus, and such deposits are present by the millions in the AD brain as amyloid β peptide (Aβ) plaques and neu-rofibrillary tangles.

Parkinson's disease (PD) also commonly features highly insoluble, highly inert deposits, the Lewy body,[4,5] and prion plaques have been reported to stimulate inflammation.[6–8] Given so many potential causes of brain inflammation in neurodegenerative disorders, the only real surprise is how long it took to recognize them!

In addition to the fact that the brain is not immunologically privileged, there are three other fundamental concepts that need to be understood before one can begin to make sense of the thousands of AD, PD, and other neuroscience papers that have been written about inflammation and neurodegenerative disorders. The first is that inflammatory mechanisms are highly interactive. The classical and alternative complement pathways include over 40 proteins and regulators, all of which depend on each other and can have a multitude of effects, including the induction of cytokines, chemokines, and inflammation-related growth factors. In turn, cytokines such as interleukin-1 (IL-1), IL-6, and tumor necrosis factor-α (TNF-α) induce each other, induce chemokines, induce growth factors, and induce complement. It is therefore important to examine carefully any paper that claims to have discovered the primary, all-important inflammatory mechanism in AD or some other neurological disease. Inflammation seldom, if ever, works that way.

Another fundamental concept is that inflammation includes both destructive and beneficial mechanisms. To fail to appreciate this fact can have serious consequences. For example, a recent clinical trial attempted to enhance inflammation in the AD brain by immunizing against Aβ. This did have the desired beneficial effect of promoting Aβ clearance, but it also injured several patients and killed another, presumably through additional inflammatory mechanisms that were simultaneously invoked.[9,10]

A third and final fundamental concept of inflammation in neurodegenerative disease is that inflammation is unlikely to be an etiology, a root initial cause, of any brain disorder. When we speak of the inflammatory *response*, we do so for a reason: inflammation almost always arises as a response to some other, more primary pathogen. This does not mean, however, that inflammation is unimportant. Especially in the brain, inflammation can often cause more damage than the agent that engendered it, as is typically the case in head trauma and encephalitis. Wounds and infections of the brain are etiologies, but the secondary inflammatory responses to them are often what kill the patient.

A mature view of inflammation in neurodegenerative disorders therefore requires that we recognize four simple but frequently unaddressed concerns: (1) that inflammatory mechanisms are likely to occur in any neurodegenerative disorder, (2) that these mechanisms are highly complex and interactive, (3) that they have both beneficial and destructive consequences, and (4) that they likely will require as much clinical attention as the etiology that gave rise to them. With this fundamental background, we can then begin to piece together the many basic research and clinical findings about inflammation in such diseases as AD and PD, and we can begin to develop rational therapies.

Most of the basic research on AD inflammation has centered on Aβ deposits and microglia. This follows from the initial discovery that microglia cluster within Aβ deposits and are activated there.[11–13] By going back and forth from *in situ* tissue sections of the postmortem AD brain to *in vitro* assays, much knowledge about these processes has been gained. For example, it is now known that microglia possess receptors that are responsive to Aβ, including the macrophage scavenger receptor,[14] the receptor for advanced glycation endproducts (RAGE)[15] and the formyl peptide receptor.[16–18] Aβ activation of the formyl peptide receptor, among other things, stimulates microglial secretion of IL-1,[17] and Aβ activation of RAGE, among other things, stimulates microglial secretion of M-CSF.[19] IL-1, and M-CSF, in turn, promotes the expression of many other inflammatory mediators through such transcription factors as NF-κB[20] and C/EBP.[21] C/EBP is, in fact, upregulated in the AD cortex.[22] There are many other Aβ-receptor interactions, as well as the secretory products and transcription factors they induce, that activate microglia and account for the colocalization of activated microglia with Aβ deposits.

Microglia and inflammation have also been implicated in PD pathogenesis.[23–27] Studies by Herrera and colleagues,[28,29] among others,[30–32] demonstrate that administration of the classic inflammatory stimulus lipopolysaccharide (LPS) causes a selective loss of dopamine in the substantia nigra, accompanied by massive activation of microglia, which cluster around deteriorating dopamine neurons.

These data have resonance with a number of other findings:

1. LPS is a common bacterial toxin, and bacterial exposure *in utero* may be a factor predisposing to PD.[33]
2. Many of the putative etiologies of PD, including head trauma,[34,35] influenza with encephalitis,[36,37] and environmental toxins such as pesticides and MPTP,[38–40] all have in common their ability to stimulate inflammation.
3. Substantia nigra neurons appear exquisitely sensitive to oxidative and nitric oxide insult,[41–43] and both are major weapons for inflammatory attack.
4. The substantia nigra is reported to have the highest density of microglia of any brain structure.[44,45]

Complement proteins are among the many inflammatory molecules produced by microglia.[46,47] In AD, both the classical[48–50] and the alternative[51] pathways are activated in an antibody-independent fashion by Aβ and by neurofibrillary tangles[52] through mechanisms similar to those for coat proteins of RNA tumor viruses. This produces complement anaphylatoxins that further fuel the inflammatory reaction, as well as complement opsonins that target Aβ for microglial phagocytosis.[49] The opsonins appear to play a major role in Aβ clearance, as inhibition of complement opsonin actions in transgenic mouse models of AD results in significantly increased Aβ deposition.[53]

Antibodies to Aβ can also opsonize Aβ for phagocytosis by microglia through microglial Fc receptors, as shown by passive and active Aβ immunization in human Aβ-overexpressing transgenic mice.[54,55] In addition to helping clear Aβ, C1q binding to Aβ potently aggregates it into the form seen in parenchymal senile plaques.[56–59] The terminal component of both classical and alternative complement pathway activation is the membrane attack complex (MAC), which forms on and lyses targeted and, sometimes, "innocent bystander" cells.[60] MAC fixation on neurite processes in the vicinity of Aβ deposits has been demonstrated in the AD brain.[50] Lastly, AD deficits in the critical complement regulator CD59 (membrane inhibitor of reactive lysis) may be permissive for complement damage.[61]

Cytokines are also abundantly increased in pathologically vulnerable regions of the AD brain.[3,62] They can be produced by microglia and astrocytes and, similar to complement, have multiple actions. Of particular importance, IL-1 appears to stimulate production of the Aβ precursor protein.[63] Moreover, the classic pro-inflammatory triad of IL-1, IL-6, and TNF-α, along with other cytokines and growth factors, may stimulate apoptosis and angiogenesis.[3]

Many of these processes can be modeled in neuron and glial cultures derived from rapid autopsies of AD and control patients.[2] As in the AD cortex, astrocytes in culture take up periplaque positions. Neurons in culture die when seeded on synthetic Aβ deposits, whereas neurons seeded outside the deposits extend neurites that either skirt the Aβ or are retracted. These behaviors are also seen in AD tissue sections. Microglia, as in the AD brain, cluster around and within Aβ deposits, where they increase their expression of a wide range of inflammatory mediators, including complement, cytokines, chemokines, growth factors, and nitric oxide.[2,15,19,46,64,65]

We have used *in vitro* culture models to investigate mechanisms of Aβ immunization and the possibility that certain nonsteroidal anti-inflammatory drugs (NSAIDs) might be a useful therapeutic adjunct to Aβ immunization. Indomethacin, in particular, is known to inhibit secretion of cytotoxic inflammatory factors that may have caused adverse reactions in Aβ-immunized patients. In addition to these potentially useful inflammation-inhibiting properties, our studies demonstrate that indomethacin does not materially impede microglial chemotaxis to or phagocytosis of Aβ.[66]

It is generally the case that chronic inflammation exacerbates disease. Nonetheless, the pathogenic relevance of inflammation in AD and PD continues to be an issue. The evidence for a pathogenic role of inflammation in these disorders essentially rests on the following grounds:

1. There is selective upregulation of inflammatory mechanisms in pathologically vulnerable but not pathologically spared regions of the AD and PD brain.[3,67,68]
2. The classical hallmarks of AD, Aβ deposits and neurofibrillary tangles, have in common a unique ability to directly interact with and stimulate inflammatory mechanisms.[3,48,49,63,67,69]
3. Inflammation is a common denominator of many putative etiologies of PD (e.g., head trauma, influenza with encephalitis, bacterial infections, environmental toxins such as pesticides and MPTP).[23,70]
4. Complement fixation and lysis of neurites occur in the AD cortex.[50]
5. Complement opsonization of deteriorating dopamine neurons occurs in the PD substantia nigra.[71]
6. Polymorphisms of inflammatory genes may increase susceptibility to AD and PD.[72–80]
7. Elevated C-reactive protein levels predict dementia onset as much as 20 years later.[81]
8. A rise of inflammation markers immediately precedes synapse loss and cognitive deterioration in AD.[82]
9. Basic research studies show protective effects of common anti-inflammatory drugs in animal models of PD.[83–86]

10. Nearly 20 epidemiologic studies suggest that common anti-inflammatory drugs may delay onset of AD.[87–89]

11. One small pilot trial has shown a beneficial effect of the NSAID indomethacin.[90]

Set against these findings, two large-scale clinical studies, one with the steroid anti-inflammatory drug prednisone,[91] and another with the nonsteroid anti-inflammatory drug hydroxyquinoline,[92] failed to find a therapeutic benefit in patients with AD. Two critical questions therefore remain: are we using the right anti-inflammatory drugs, and are we using them in the right patients?

The steroid prednisone has a well-known adverse reaction profile that includes confusion and agitation at the clinical level[93] and damage to hippocampal neurons at the basic science level.[94,95] These are obviously not desirable properties for the treatment of AD patients. Conversely, several NSAIDs that have not yet been tested in AD are reported to have such desirable actions as inhibition of COX I (without inhibition of COX II), inhibition of Aβ production, inhibition of lipoxygenase, and PPAR-γ facilitation. Indomethacin, which did show therapeutic effects in a limited AD trial,[90] is both a potent COX I[96] and COX II inhibitor,[97] an inhibitor of Aβ production,[98] a lipoxygenase inhibitor,[99] and a PPAR-γ agonist.[100] It is taken worldwide by millions of arthritis patients. Nonetheless, its gastrointestinal and renal adverse effects, which may be somewhat higher than other NSAIDs,[101] have apparently inhibited its further consideration in AD trials.

A second critical consideration in the application of anti-inflammatory drugs for AD is when and to whom they should be given. Anyone who has ever removed the brain of a patient who has suffered AD can attest that the already existing damage is unlikely to be reversed by any treatment now known or that will be known for the next 50 years. Moreover, AD inflammation, as previously noted, is a response to preexisting pathogens, such as Aβ and neurofibrillary tangles. Thus, our best hope for treatment is to remove the initiating stimuli or to quench the inflammation they produce before significant damage is done to the brain. Clinical trials using NSAIDs to prevent or delay AD in susceptible but still normal elderly patients are now underway, and should provide answers about their efficacy in the next few years.

ACKNOWLEDGMENTS

Supported by grants from the National Institute on Aging, the Alzheimer's Association, the Arizona Alzheimer's Disease Core Center, the Arizona Alzheimer's Research Center, and the Arizona Parkinson's Disease Research Center.

REFERENCES

1. Luber-Narod, J. and Rogers, J., Immune system associated antigens expressed by cells of the human central nervous system, *Neurosci. Lett.*, 94, 17, 1988.
2. Lue, L.F. et al., Characterization of glial cultures from rapid autopsies of Alzheimer's and control patients, *Neurobiol. Aging*, 17, 421, 1996.
3. Akiyama, H., Barger, S., and Barnum, S., Inflammation and Alzheimer's disease, *Neurobiol. Aging*, 21, 383, 2000.

4. Mackenzie, I.R., Activated microglia in dementia with Lewy bodies, *Neurology*, 55, 132, 2000.
5. Mackenzie, I.R., Cortical inflammation in dementia with Lewy bodies, *Arch. Neurol.*, 58, 519, 2001.
6. Betmouni, S., Perry, V.H., and Gordon, J.L., Evidence for an early inflammatory response in the central nervous system of mice with scrapie, *Neuroscience*, 74, 1, 1996.
7. Kim, J.I., Ju, W.K., and Choi, J.H., Expression of cytokine genes and increased nuclear factor-kappa B activity in the brains of scrapie-infected mice, *Brain Res. Mol. Brain Res.*, 73, 17, 1999.
8. Perry, V.H., Cunningham, C., and Boche, D., Atypical inflammation in the central nervous system in prion disease, *Curr. Opin. Neurol.*, 15, 349, 2002.
9. Schenk, D., Amyloid-beta immunotherapy for Alzheimer's disease: the end of the beginning, *Nat. Rev. Neurosci.*, 3, 824, 2002.
10. Selkoe, D.J. and Schenk, D., Alzheimer's disease: Molecular understanding predicts amyloid-based therapeutics, *Annu. Rev. Pharmacol. Toxicol.*, 43, 545–84, 2003.
11. Luber-Narod, J. and Rogers, J., Immune system associated antigens expressed by cells of the human central nervous system, *Neurosci. Lett.*, 94, 17, 1988.
12. McGeer, P.L., Itagaki, S., Tago, H., and McGeer, E.G., Reactive microglia in patients with senile dementia of the Alzheimer type are positive for the histocompatibility glycoprotein HLA-DR, *Neurosci. Lett.*, 79, 195, 1987.
13. Rogers, J., Luber-Narod, J., Styren, S.D., and Civin, W.H., Expression of immune system-associated antigens by cells of the human central nervous system relationship to the pathology of Alzheimer's disease, *Neurobiol. Aging*, 9, 339, 1988.
14. El Khoury, J. et al., Scavenger receptor-mediated adhesion of microglia to beta-amyloid fibrils, *Nature*, 382, 716, 1996.
15. Lue, L.F., Walker, D.G., Brachova, L., and Rogers, J., Involvement of microglial receptor for advanced glycation endproducts (RAGE) in Alzheimer's disease: identification of a cellular activation mechanism, *Exp. Neurol.*, 171, 29, 2001.
16. Lorton, D., beta-Amyloid-induced IL-1 beta release from an activated human monocyte cell line is calcium- and G-protein-dependent, *Mech. Ageing Dev.*, 94, 199, 1997.
17. Lorton. D. et al., beta-Amyloid induces increased release of interleukin-1 beta from lipopolysaccharide-activated human monocytes, *J. Neuroimmunol.*, 67, 21, 1996.
18. Lorton, D., Schaller, J., Lala, A., and De Nardin, E., Chemotactic-like receptors and Abeta peptide induced responses in Alzheimer's disease, *Neurobiol. Aging*, 21, 463, 2000.
19. Lue, L.F., Walker, D.G., and Rogers, J., Modeling microglial activation in Alzheimer's disease with human postmortem microglial cultures, *Neurobiol. Aging*, 22, 945, 2001.
20. Li, Q. and Verma, I.M., NF-kappaB regulation in the immune system, *Nat. Rev. Immunol.*, 2, 725, 2002.
21. Ramji, D.P. and Foka, P., CCAAT/enhancer-binding proteins: structure, function and regulation, *Biochem. J.*, 365, 561, 2002.
22. Li, R. et al., CCAAT/enhancer binding protein 5 expression and elevation in Alzheimer's disease, *Neurobiol. Aging*, 25, 991, 2004.
23. Czlonkowska, A. et al., Immune processes in the pathogenesis of Parkinson's disease— a potential role for microglia and nitric oxide, *Med. Sci. Monit.*, 8, 165, 2002.
24. Fiszer, U., Does Parkinson's disease have an immunological basis? The evidence and its therapeutic implications, *BioDrugs*, 15, 351, 2001.
25. McGeer, P.L. and McGeer, E.G., Glial cell reactions in neurodegenerative diseases: pathophysiology and therapeutic interventions, *Alzheimer Dis. Assoc. Disord.*, 12, 1, 1998.
26. Orr, C.F., Rowe, D.B., and Halliday, G.M., An inflammatory review of Parkinson's disease, *Prog. Neurobiol.*, 68, 325, 2002.
27. Vila, M., Jackson-Lewis, V., and Guegan, C., The role of glial cells in Parkinson's disease, *Curr. Opin. Neurol.*, 14, 483, 2001.

28. Castano, A., Herrera, A.J., Cano, J., and Machado, A., Lipopolysaccharide intranigral injection induces inflammatory reaction and damage in nigrostriatal dopaminergic system, *J. Neurochem.*, 70, 1584, 1998.
29. Herrera, A.J. et al., The single intranigral injection of LPS as a new model for studying the selective effects of inflammatory reactions on dopaminergic system, *Neurobiol. Dis.*, 7, 429, 2000.
30. Gao, H.M. et al., Microglial activation-mediated delayed and progressive degeneration of rat nigral dopaminergic neurons: relevance to Parkinson's disease, *J. Neurochem.*, 81, 1285, 2002.
31. Le, W. et al., Microglial activation and dopaminergic cell injury: an *in vitro* model relevant to Parkinson's disease, *J. Neurosci.*, 21, 8447, 2001.
32. Liu, Y., Qin, L., and Li, G., Dextromethorphan protects dopaminergic neurons against inflammation-mediated degeneration through inhibition of microglial activation, *J. Pharmacol. Exp. Ther.*, 305, 212, 2003.
33. Ling, Z., Gayle, D.A., and Ma, S.Y., In utero bacterial endotoxin exposure causes loss of tyrosine hydroxylase neurons in the postnatal rat midbrain, *Mov. Disord.*, 17, 116, 2002.
34. Lees, A.J., Trauma and Parkinson disease, *Rev. Neurol. (Paris)*, 153, 541, 1997.
35. Wiest, R.G., Burgunder, J.M., and Krauss, J.K., Chronic subdural haematomas and Parkinsonian syndromes, *Acta Neurochir. (Wien)*, 141, 753, 1999.
36. Casals, J., Elizan, T.S., and Yahr, M.D., Postencephalitic parkinsonism—a review, *J. Neural. Transm.*, 105, 645, 1998.
37. Moore, G., Influenza and Parkinson's disease, *Public Health Rep.*, 92, 79, 1977.
38. Di Monte, D.A., Lavasani, M., and Manning-Bog, A.B., Environmental factors in Parkinson's disease, *Neurotoxicology*, 23, 487, 2002.
39. Fukuda, T., Neurotoxicity of MPTP, *Neuropathology*, 21, 323, 2001.
40. Le Couteur, D.G. et al., Pesticides and Parkinson's disease, *Biomed. Pharmacother*, 53, 122, 1999.
41. Beal, M.F., Excitotoxicity and nitric oxide in Parkinson's disease pathogenesis, *Ann. Neurol.*, 44, S110, 1998.
42. Gerlach, M. et al., Nitric oxide in the pathogenesis of Parkinson's disease, *Adv. Neurol.*, 80, 239, 1999.
43. Greenfield, S.A., Cell death in Parkinson's disease, *Essays Biochem.*, 27, 103, 1992.
44. Kim, W.G. et al., Regional difference in susceptibility to lipopolysaccharide-induced neurotoxicity in the rat brain: role of microglia, *J. Neurosci.*, 20, 6309, 2000.
45. Lawson, L.J. et al., Heterogeneity in the distribution and morphology of microglia in the normal adult mouse brain, *Neuroscience*, 39, 151, 1990.
46. Lue, L.F. et al., Inflammatory repertoire of Alzheimer's disease and nondemented elderly microglia *in vitro*, *Glia*, 35, 72, 2001.
47. McGeer, P.L. and McGeer, E.G., Complement proteins and complement inhibitors in Alzheimer's disease, *Res. Immunol.*, 143, 621, 1992.
48. Jiang, H. et al., β-Amyloid activates complement by binding to a specific region of the collagen-like domain of the C1q A chain, *J. Immunol.*, 152, 5050, 1994.
49. Rogers, J., Cooper, N.R., and Webster, S., Complement activation by beta-amyloid in Alzheimer's disease, *Proc. Natl. Acad. Sci. U.S.A.*, 89, 10016, 1992.
50. Webster, S. et al., Molecular and cellular characterization of the membrane attack complex, C5b-9, in Alzheimer's disease, *Neurobiol. Aging*, 18, 415, 1997.
51. Strohmeyer, R., Shen, Y., and Rogers, J., Detection of complement alternative pathway mRNA and proteins in the Alzheimer's disease brain, *Brain Res. Mol. Brain Res.*, 81, 7, 2000.
52. Shen, Y. et al, Complement activation by neurofibrillary tangles in Alzheimer's disease, *Neurosci. Lett.*, 305, 165, 2001.

53. Wyss-Coray, T., Yan, F., and Lin, A.H., Prominent neurodegeneration and increased plaque formation in complement-inhibited Alzheimer's mice, *Proc. Natl. Acad. Sci. U.S.A.*, 99, 10837, 2002.

54. Bard, F., Cannon, C., and Barbour, R., Peripherally administered antibodies against amyloid beta-peptide enter the central nervous system and reduce pathology in a mouse model of Alzheimer disease, *Nat. Med.*, 6, 916, 2000.

55. Schenk, D., Barbour, R., and Dunn, W., Immunization with amyloid-beta attenuates Alzheimer-disease-like pathology in the PDAPP mouse, *Nature*, 400, 173, 1999.

56. Webster, S., Bonnell, B., and Rogers, J., Charge-based binding of complement component C1q to the Alzheimer amyloid beta-peptide, *Am. J. Pathol.*, 150, 1531, 1997.

57. Webster, S., Glabe, C., and Rogers, J., Multivalent binding of complement protein C1Q to the amyloid beta-peptide (A beta) promotes the nucleation phase of A beta aggregation, *Biochem. Biophys. Res. Commun.*, 217, 869, 1995.

58. Webster, S., O'Barr, S., and Rogers, J., Enhanced aggregation and beta structure of amyloid beta peptide after coincubation with C1q, *J. Neurosci. Res.*, 39, 448, 1994.

59. Webster, S. and Rogers, J., Relative efficacies of amyloid beta peptide (Aβ) binding proteins in A beta aggregation, *J. Neurosci. Res.*, 46, 58, 1996.

60. Cooper, N.R. and Nemerow, G.R., Complement effector mechanisms in health and disease, *J. Invest. Dermatol.*, 85, 39, 1985.

61. Shen, Y., Yang, L.-B., and Rogers, J., Deficiency of complement defense protein CD59 may contribute to neurodegeneration in Alzheimer's disease, *J. Neurosci.*, 20, 7505, 2000.

62. Griffin, W.S., Sheng, J.G., and Royston, M.C., Glial-neuronal interactions in Alzheimer's disease: the potential role of a "cytokine cycle" in disease progression, *Brain Pathol.*, 8, 65, 1998.

63. Goldgaber, D., Harris, H.W., and Hla, T., Interleukin 1 regulates synthesis of amyloid beta-protein precursor mRNA in human endothelial cells, *Proc. Natl. Acad. Sci. U.S.A.*, 86, 7606, 1989.

64. Rogers, J. and Lue, L.F., Microglial chemotaxis, activation, and phagocytosis of amyloid beta-peptide as linked phenomena in Alzheimer's disease, *Neurochem. Int.*, 39, 333, 2001.

65. Rogers, J. et al., Elucidating molecular mechanisms of Alzheimer's disease in microglial cultures, *Ernst Schering Res. Found. Workshop*, 39, 25, 2002.

66. Strohmeyer, R. et al., Microglial responses to amyloid beta peptide opsonization and indomethacin treatment, *J. Neuroinflammation*, 2, 1, 2005.

67. Griffin, W.S. et al., Interleukin-1 expression in different plaque types in Alzheimer's disease: significance in plaque evolution, *J. Neuropathol. Exp. Neurol.*, 54, 276, 1995.

68. Lue, L.F. and Rogers, J., Full complement activation fails in diffuse plaques of the Alzheimer's disease cerebellum, *Demeintia*, 3, 308, 1992.

69. Aksenov, M.Y. et al., Alpha 1-antichymotrypsin interaction with A beta (1-42) does not inhibit fibril formation but attenuates the peptide toxicity, *Neurosci. Lett.*, 217, 117, 1996.

70. McGeer, P.L., Yasojima, K., and McGeer, E.G., Inflammation in Parkinson's disease, *Adv Neurol.*, 86, 83, 2001.

71. Yamada, T., McGeer, P.L., and McGeer, E.G., Lewy bodies in Parkinson's disease are recognized by antibodies to complement proteins, *Acta Neuropathol. (Berl.)*, 84, 100, 1992.

72. Du, Y., Dodel, R.C., and Eastwood, B.J., Association of an interleukin 1 alpha polymorphism with Alzheimer's disease, *Neurology*, 55, 480, 2000.

73. Ehl, C., Kolsch, H., and Ptok, U., Association of an interleukin-1 beta gene polymorphism at position -511 with Alzheimer's disease, *Int. J. Mol. Med.*, 11, 235, 2003.

74. Grimaldi, L.M., Casadei, V.M., and Ferri, C., Association of early-onset Alzheimer's disease with an interleukin-1 alpha gene polymorphism, *Ann. Neurol.*, 47, 361, 2000.

75. Kolsch, H., Ptok, U., and Bagli, M., Gene polymorphisms of interleukin-1 alpha influence the course of Alzheimer's disease, *Ann. Neurol.*, 49, 818, 2001.

76. Licastro, F., Pedrini, S., and Ferri, C., Gene polymorphism affecting alpha 1-antichy-motrypsin and interleukin-1 plasma levels increases Alzheimer's disease risk, *Ann. Neurol.*, 48, 388, 2000.
77. McGeer, P.L., Yasojima, K., and McGeer, E.G., Association of interleukin-1 beta poly-morphisms with idiopathic Parkinson's disease, *Neurosci. Lett.*, 326, 67, 2002.
78. Nicoll, J.A., Mrak, R.E., and Graham, D.I., Association of interleukin-1 gene polymor-phisms with Alzheimer's disease, *Ann. Neurol.*, 47, 365, 2000.
79. Rebeck, G.W., Confirmation of the genetic association of interleukin-1A with early onset sporadic Alzheimer's disease, *Neurosci. Lett.*, 293, 75, 2000.
80. Rogers, J., An IL-1 alpha susceptibility polymorphism in Alzheimer's disease—new fuel for the inflammation hypothesis, *Neurology*, 55, 464, 2000.
81. Schmidt, R. et al., Early inflammation and dementia: a 25-year follow-up of the Hono-lulu-Asia Aging Study, *Ann. Neurol.*, 52, 168, 2002.
82. Lue, L.F. et al., Soluble amyloid beta peptide concentration as a predictor of synaptic change in Alzheimer's disease, *Am. J. Pathol.*, 155, 853, 1999.
83. Aubin, N. et al., Aspirin and salicylate protect against MPTP-induced dopamine deple-tion in mice, *J. Neurochem.*, 71, 1635, 1998.
84. Ferger, B. et al., Salicylate protects against MPTP-induced impairments in dopaminer-gic neurotransmission at the striatal and nigral level in mice, *Naunyn Schmiedebergs Arch. Pharmacol.*, 360, 256, 1999.
85. Kurkowska-Jastrzebska, I. et al., Indomethacin protects against neurodegeneration caused by MPTP intoxication in mice, *Int. Immunopharmacol.*, 2, 1213, 2002.
86. Teismann, P. and Ferger, B., Inhibition of the cyclooxygenase isoenzymes COX-1 and COX-2 provide neuroprotection in the MPTP-mouse model of Parkinson's disease, *Synapse*, 39, 167, 2001.
87. McGeer, P.L., Schulzer, M., and McGeer, E.G., Arthritis and anti-inflammatory agents as possible protective factors for Alzheimer's disease: a review of 17 epidemiologic studies, *Neurology*, 47, 425, 1996.
88. Zandi, P.P. et al., Reduced incidence of AD with NSAID but not H_2 receptor antago-nists: the Cache County Study, *Neurology*, 59, 880, 2002.
89. Zandi, P.P. and Breitner, J.C., Do NSAIDs prevent Alzheimer's disease? And, if so, why? The epidemiological evidence, *Neurobiol. Aging*, 22, 811, 2001.
90. Rogers, J., Kirby, L.C., and Hempelman, S.R., Clinical trial of indomethacin in Alzheimer's disease, *Neurology*, 43, 1609, 1993.
91. Aisen, P.S., Davis, K.L., and Berg, J.D., A randomized controlled trial of prednisone in Alzheimer's disease. Alzheimer's Disease Cooperative Study, *Neurology*, 54, 588, 2000.
92. Van Gool, W.A. et al., Effect of hydroxychloroquine on progression of dementia in early Alzheimer's disease: an 18-month randomised, double-blind, placebo-controlled study, *Lancet*, 358, 455, 2001.
93. Lewis, D.A. and Smith, R.E., Steroid-induced psychiatric syndromes. A report of 14 cases and a review of the literature, *J. Affect. Disord.*, 5, 319, 1983.
94. Sapolsky, R.M., The possibility of neurotoxicity in the hippocampus in major depres-sion: a primer on neuron death, *Biol. Psychiatry*, 48, 755, 2000.
95. Uno, H., Eisele, S., and Sakai. A., Neurotoxicity of glucocorticoids in the primate brain, *Horm. Behav.*, 28, 336, 1994.
96. Rothermich, N.O., An extended study of indomethacin, *J. Clin. Pharmacol. JAMA*, 195, 531, 1966.
97. Yamamoto, T. and Nozaki-Taguch, N., Analysis of the effects of cyclooxygenase (COX)-1 and COX-2 in spinal nociceptive transmission using indomethacin, a non-selective COX inhibitor, and NS-398, a COX-2 selective inhibitor, *Brain Res.*, 739, 104, 1996.
98. Weggen, S., Eriksen, J.L., and Das, P., A subset of NSAIDs lower amyloidogenic Abeta42 independently of cyclooxygenase activity, *Nature*, 414, 212, 2001.

99. Randall, R.W. et al., Inhibition of arachidonic acid cyclo-oxygenase and lipoxygenase activities of leukocytes by indomethacin and compound BW755C 1980, *Agents Actions*, 43, 176, 1994.

100. Lehmann, J.M. et al., Peroxisome proliferator-activated receptors alpha and gamma are activated by indomethacin and other non-steroidal anti-inflammatory drugs, *J. Biol. Chem.*, 272, 3406, 1997.

101. Smyth, C.J., Indomethacin—its rightful place in treatment, *Ann. Intern. Med.*, 72, 430, 1970.

2 Treatment Development Strategies for Alzheimer's Disease

Ernst Wülfert

CONTENTS

KEYWORDS

Alzheimer's Disease, Acetylcholine, Glutamate, Neurotrophins, Glucocorticoids, β-amyloid, Free Radicals, Apolipoprotein E, Neurotoxicity, Neuroprotection

INTRODUCTION

Alzheimer's disease (AD) is a progressive neurodegenerative disease afflicting a significant proportion of the elderly population. Estimated incidence rates increase from 1% at 65 to between 25 and 40% at age 85 and older. The number of new cases appearing among previously unaffected individuals may be as high as 3% annually in people over age 85.[1] The disease is characterized clinically by a progressive loss of memory function and mental impairment (dementia), and histopathologically by the presence, within widespread regions of the brain, of degenerative lesions known as senile plaques (SP) and neurofibrillary tangles (NFT).[2] Studies in postmortem brain material have revealed multiple disturbances of neurotransmitters, their metabolites, receptors, and enzymes involved in their metabolism. Deficits have been observed for classical neurotransmitter systems such as the cholinergic, serotonergic, dopaminergic, as well as excitatory amino acids and peptides.[3]

THE CHOLINERGIC TREATMENT HYPOTHESIS

Loss of basal forebrain cholinergic neurons is a significant feature of the disease and appears to be correlated with the memory deficits. Experimental lesions of cholinergic systems also produce memory deficits in animals. Treatments for the memory loss in AD have therefore focused on cholinergic replacement approaches (the cholinergic hypothesis).

Much research has been devoted to the design and testing of acetylcholinesterase inhibitors (AChE-I) as a means of preserving acetylcholine (ACh) in synapses where large numbers of cholinergic nerve terminals have been lost. Tacrine is the first drug to be approved for symptomatic treatment of memory impairment in Alzheimer patients. Although the drug produces small but significant benefit in approximately 40% of patients,[4] its marginal level of efficacy and high level of liver toxicity is likely to restrict its use to a small group of patients. Several chemical analogues of tacrine have been reported to be in advanced stages of clinical trials (an exception is velnacrine, which has been dropped on the basis of its unfavorable risk/benefit ratio), but data confirming their therapeutic potential are still lacking (Figure 2.1).

Physostigmine is another extensively studied AChE-I. Early studies with this substance were encouraging,[5] but its rapidly evanescent effect prohibited further development. Heptylphysostigmine was designed as a prodrug of the naturally occurring substance to increase its half-life and therefore produce more long-lasting effects. However, no clinical data have been published confirming the clinical efficacy of this drug in Alzheimer patients.

Compound E-2020 has been reported to be a selective and long-lasting inhibitor of AChE and is presently undergoing clinical evaluation.[6] Huperzine and galanthamine are more recent AChE-I that have demonstrated potent activity in animals, but unequivocal clinical data are still lacking. Clinical data should confirm ultimately whether a constantly maintained inhibition of AChE is critical to expressing therapeutic effects. Increasing ACh levels in the synaptic cleft could be expected to limit release of the neurotransmitter through activation of presynaptic muscarinic M_2 autoreceptors. One could therefore hypothesize that AChE-I endowed with M_2 antagonistic properties should offer some advantage over the classical AChE-I.

Activation of postsynaptic receptors with muscarinic M_1 agonists represents a more direct approach to compensate for the low levels of ACh released due to loss of cholinergic nerve terminals. Administration of arecoline has been shown to improve memory in a small number of Alzheimer patients treated for two weeks.[7] Although these findings are encouraging, their clinical significance must be confirmed in larger patient populations before any conclusions can be drawn. Other cholinergic agonists with selective M_1 properties are presently in clinical trials, but data confirming their therapeutic efficacy in Alzheimer patients are not yet available (Figure 2.2).

Whereas muscarinic agonists may cause downregulation of muscarinic receptors, chronic nicotine administration, on the other hand, induces upregulation of nicotinic receptors. Because activation of nicotinic presynaptic receptors increases the release of ACh, it could be hypothesized that nicotinic agonists should improve memory deficits. Subcutaneous injection of nicotine has been reported to cause significant dose-related improvement in attention in Alzheimer patients,[8] but larger

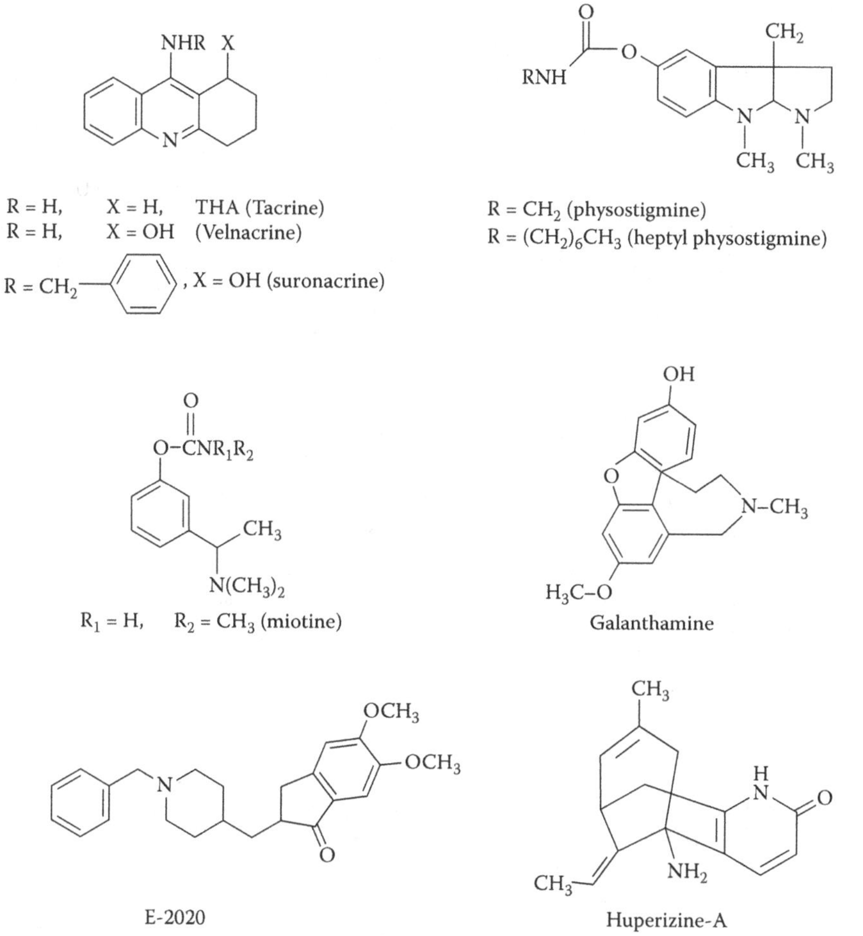

FIGURE 2.1 Acetylcholinesterase inhibitors (ACh-I) evaluated clinically for their potential in Alzheimer's disease.

treatment studies have not confirmed that nicotinic agonists improve cognitive deficits in patients.

Further to being controlled by autoreceptors, hippocampal ACh release is also enhanced by activation of serotonin 5-HT$_3$ receptors.[9] The finding that ondansetron, a 5-HT$_3$ antagonist, may improve rather than impair cognition in the elderly[10] does, however, caution against extrapolating from neurochemical data obtained with drugs in animals to treatment effects in AD.

Numerous compounds have been reported to enhance the release of ACh in animals, but their mode of action is unclear (Figure 2.3). Tacrine has been shown both to inhibit AChE and to release ACh via activation of nicotinic receptors in brain tissue from AD patients. The marginal level of efficacy observed with the drug does, however, suggest that drugs aimed at direct or indirect activation of cholinergic receptors may only have limited therapeutic potential in AD.

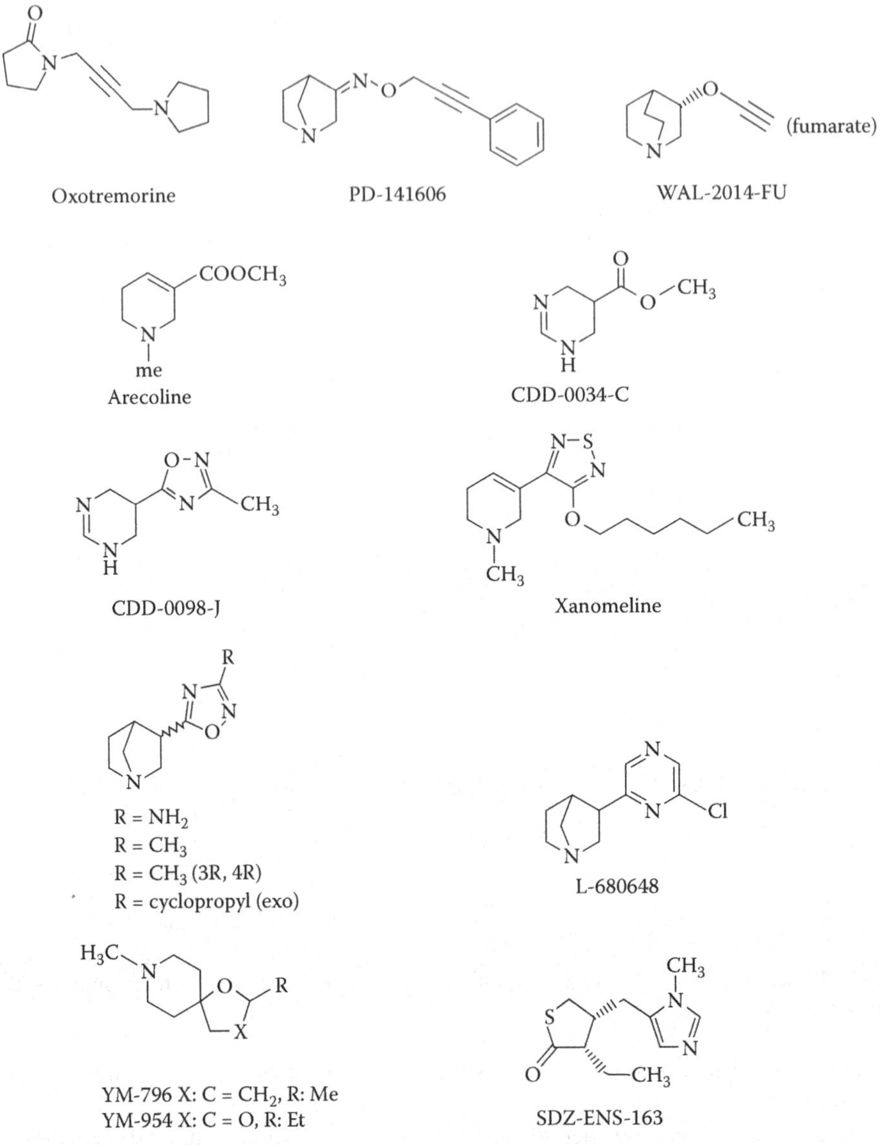

FIGURE 2.2 Recently reported M_1-selective agonists.

Difficulties in confirming the clinical potential of drugs activating cholinergic functions in animals may, at least in part, be due to the experimental models used. The behavioral deficits studied in animals to select "cholinergic" drugs for clinical trials are not caused by neurodegenerative processes similar to those encountered in AD (models lack "construct validity"). It is also doubtful whether the deficits observed in animal models represent the symptoms to be treated (models lack "face validity"). Obviously, the failure in confirming efficacy in clinical trials with drugs affecting animal behavior indicates that the animal models being used have poor predictive validity.[11] Furthermore, AD is a complex disease involving numerous neuro-chemical deficits. Successful treatment with cholinergic drugs might therefore require subgroups of patients characterized by cholinergic hypofunction only.

Finally, a more basic consideration is that most cholinergic agents tested in AD patients (except THA) lack functional specificity, i.e., they produce a variety of effects at the cellular and molecular level through the activation of different effector systems. Whether all these effects, or a combination of some, or whether

R = H
R = NH_2 Du Pont Merck

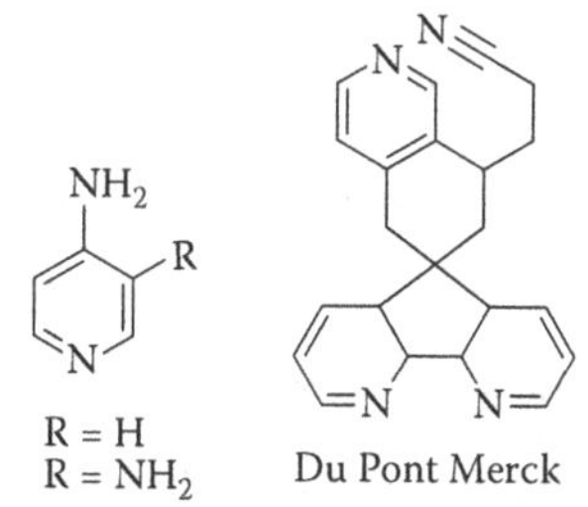

FIGURE 2.3 Compounds facilitating release of ACh.

the activation of a unique signal translation system is required for successful treatment, is not established. The muscarinic receptors belong to the superfamily of G-protein-coupled receptors. Several structurally different human muscarinic receptors (mAChR) subtypes have been cloned and expressed. The m_1, m_2, m_3, and m_4 AChRs correspond to the M_1, M_2, M_3, and M_4 AChRs defined pharmacologically. These different subtypes are closely related in sequence and include seven hydrophobic transmembrane helices. Because they are G-protein-coupled receptors, they contain two binding domains, a ligand-binding pocket and a G-protein-binding intracellular domain. Unlike antagonists, agonists stimulate the receptor via coupling with G-proteins. Thus, selectivity of an agonist is not intrinsic to the agonist–receptor interaction alone but also derives from the properties of the agonist–receptor–G-protein(s) complexes.

Activation of the transfected m_1AChR subtype has been linked to *several* signal transaction pathways, such as phosphoinositide (PI) hydrolysis, arachidonic acid (AA) release, and c-AMP accumulation. Both in Chinese hamster ovary cells and in rat pheochromocytoma (PC 12) cells stably infected with m_1AChR, carbachol, ACh and other nonselective agonists elicited all three responses (PI hydrolysis, AA-release, and c-AMP accumulation) in a dose- and time-dependent manner.[12] The muscarinic m_1 subtype of the AChR must therefore have been coupled to different G-proteins in these cells. Compounds AF 102B and AF 150(S) have been reported to be partial agonists, whereas pilocarpine, AF 150, AF 151, and AF 151(S) were full agonists in stimulating PI hydrolysis and arachidonic acid release, but they all failed to elevate c-AMP levels[12] (Figure 2.4). These findings suggest (a) that m_1AChRs may be coupled to more than one signal transduction system, (b) that muscarinic agonists could be designed to activate a distinct set of G-proteins, and (c) that such functional selectivity could result from a *unique* complex between agonist, receptor, G-protein, and a "selected" effector system. It could be speculated that the simultaneous activation of PKA by c-AMP would attenuate IP_3-mediated Ca^{2+} increases by m_1AChR activation. Thus, m_1 agonists which do not increase c-AMP levels should be capable of producing more substantial elevations in intracellular Ca^{2+} and hence facilitate more powerfully some of the neurotrophic-like effects associated with cholinergic activation.[13]

In the case of a flexible agonist, mutual conformational changes can occur in both receptor and ligand, leading to multiple complexes allowing various ligand–receptor–G-protein interactions and hence activation of different signaling pathways.

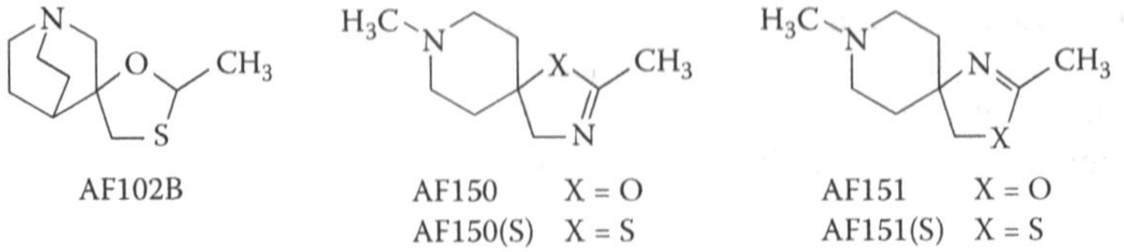

AF102B

AF150   X = O
AF150(S) X = S

AF151   X = O
AF151(S) X = S

FIGURE 2.4 m_1-Selective functional agonists.

Flexible agonists may therefore allow the receptor to exhibit "promiscuity." On the contrary, rigid muscarinic agonists would not allow multiple conformational slates of the ligand–receptor–G-protein complexes. A new class of "functionally specific" cholinergic drugs could be anticipated if this hypothesis is correct. Whether such drugs will be effective in the treatment of AD remains to be shown.

AD AND THE GLUTAMATERGIC SYSTEM

Some investigators have implicated glutamate in the pathogenesis of AD. The link between AD and excitatory amino acid (EAA) abnormalities is based on a substantial amount of indirect data, much of which was acquired from studies of brain specimens from AD patients. The predominant view submits that excitotoxicity may be a causative factor in neuronal pathology associated with AD and that treatment efforts should be aimed at preventing EAA neurotoxicity. This approach would suggest the use of antagonists of the glycine receptor—site of the NMDA receptor, partial AMPA antagonists, or partial NMDA antagonists (Figure 2.5). Others believe that the data suggest that the cognitive deficits of AD are related to loss of glutamate function and recommend that treatment efforts for AD patients be directed at increasing glutamatergic activity[14] (Figure 2.6). Protecting against EAA neurotoxicity by glutamate receptor antagonists may, however, further impair cognitive function caused by hypoactivity of

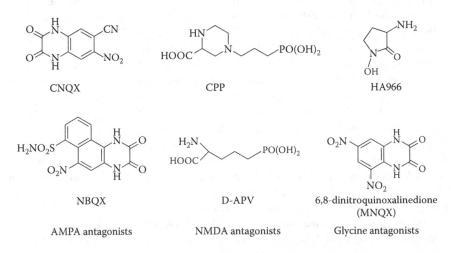

CNQX

CPP

HA966

NBQX

D-APV

6,8-dinitroquinoxalinedione
(MNQX)

AMPA antagonists

NMDA antagonists

Glycine antagonists

FIGURE 2.5 Classical antagonists of AMPA, NMDA, and glycine-NMDA receptors.

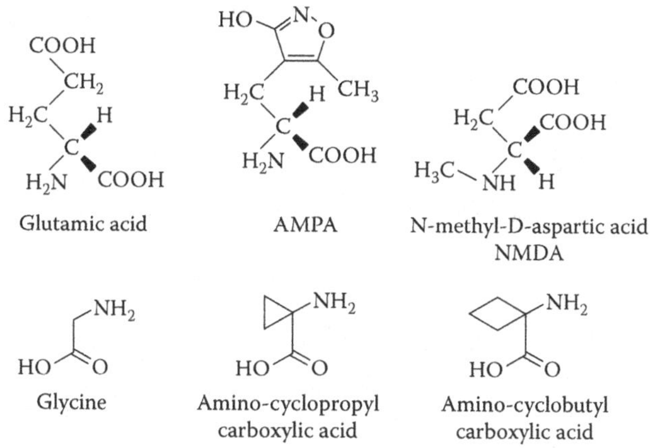

FIGURE 2.6   Classical agonists of glutamate receptors (AMPA, NMDA, and glycine-NMDA).

glutamatergic pyramidal neurons, whereas activating glutamate receptors might facilitate neurodegenerative processes in already-compromised neurons.

Despite the interest in a possible role for glutamate in the pathogenesis of AD, the data thus far have not defined such a role. Although AD usually begins with disturbance of memory, it progresses to a state of diffuse cognitive and intellectual dysfunction and widespread pathology far beyond brain structures associated with memory function. Thus, the AD patient does not appear to be an appropriate model for studying the role that glutamate may play in brain pathology that causes loss of cognitive function. However, recent human and animal investigations have more directly implicated EAA neurotoxicity in the pathogenesis of other brain disorders such as stroke, hypoglycemia, trauma, and epilepsy in which, unlike AD, learning and memory are more selectively impaired. Enhanced glutamatergic transmission has, however, been proposed to explain some of the effects of a class of drugs widely used to treat memory disturbances in the elderly. These drugs, called "nootropes," appear to prolong the time of opening of ion channels activated by a subset of glutamate receptors (AMPA-receptors)[15] (Figure 2.7).

It has been proposed that release of neurotransmitters is modulated by a wide range of inhibitory mechanisms (e.g., feedforward, feedback, and γ-aminobutyrate-mediated inhibition). Because inhibitory (hippocampal) interneurons are negatively modulated by serotonergic "inhibitory" receptors, release of neurotransmitters could, at least in theory, be enhanced by antagonists at the inhibitory 5-HT$_{1A}$ receptor. This approach might offer the potential advantage of "activating" neuronal circuits indirectly, independently of specific neurotransmitters involved (e.g., cholinergic, glutamatergic, etc.).[16]

NEUROTROPHIN-ENHANCING DRUGS

Several nerve growth factors (neurotrophic factors) known collectively as "neurotrophins" have been identified and their growth-promoting functions (also required

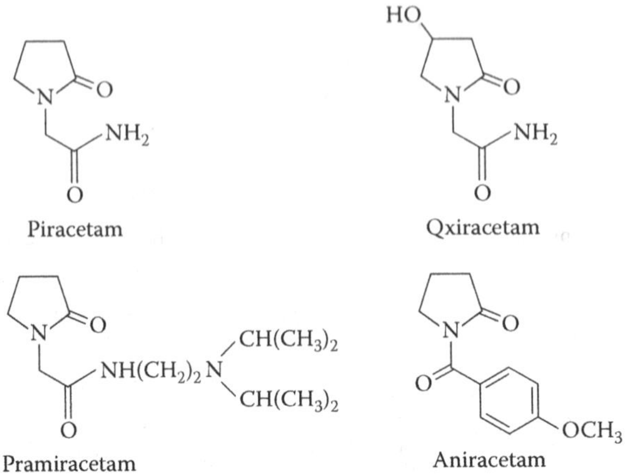

FIGURE 2.7 Chemical structures of classical nootropics.

for neuronal survival) have been described.[17] A variety of *in vitro* and *in vivo* studies have suggested that certain neurotrophins may be useful for the treatment of various neurological disorders such as AD. In particular, exogenously supplied nerve growth factor (NGF) has shown efficacy in animal models of degeneration of basal forebrain cholinergic neurons.[18] Because no alterations of NGF mRNA levels have been detected in AD brain tissue, there is no evidence to suggest that the cholinergic neuronal degeneration observed is a result of decreased NGF production. In contrast to results with NGF, two studies have clearly demonstrated decreased levels of brain-derived neurotrophic factor (BDNF) mRNA in AD hippocampal tissue relative to nondiseased tissue.[19] Although these findings do not prove a direct causal role for BDNF in AD, decreased BDNF expression during the course of the disease may be predicted to accelerate neurodegenerative changes. Because these neuropeptides do not cross the blood–brain barrier, search for compounds that could functionally mimic the effect of neurotrophins, facilitate their effects, or enhance their biosynthesis has become a major focus of interest.

The potential for upregulating endogenous neurotrophin expression was first suggested by studies demonstrating increased NGF and BDNF mRNA levels following recurrent seizures[20,21] The increases in NGF and BDNF mRNA are dramatic and have been shown to be followed by increased NGF and BDNF protein levels.[22,23] The increased NGF and BDNF synthesis following seizures is thought to be mediated by activation of glutamate receptors and subsequently increased intracellular calcium levels.[24] Activation of both NMDA and non-NMDA ionotropic receptors has been shown to induce NGF and BDNF mRNA expression in the hippocampus.[24,25] Increased BDNF synthesis following treatment of cultured cortical neurons with the metabotropic receptor agonist (*1S,3R*)-ACPD may also be responsible for the attenuating effect of this drug on excitatory amino acid toxicity[26] (Figure 2.8).

The muscarinic acetylcholine receptor is another potential pharmacological target. Pilocarpine, a potent muscarinic agonist, produces increased expression of NGF and BDNF mRNA but only at doses that produce behavioral seizures.[27] Similar to (*1S,3R*)-ACPD, which increases NGF and BDNF mRNA levels at doses that do not

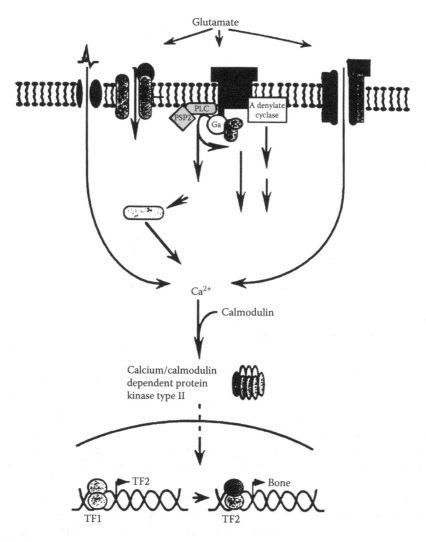

FIGURE 2.8 Potential receptors and signaling pathways involved in regulating transcription of the BDNF gene.

produce seizures, drugs may conceivably be developed that are capable of selectively activating certain muscarinic receptors without inducing seizures and, at the same time, upregulating NGF and BDNF.

The intracellular signaling mechanisms leading to altered neurotrophin expression are unknown. Calcium-calmodulin-dependent protein kinase type II, however, is likely to be functionally important in regulating BDNF mRNA expression following increases in intracellular calcium concentration. This has been suggested by studies with cultured rat hippocampal neurons in which calmodulin antagonists prevented increased BDNF mRNA levels following kainic acid treatment.[24] Activation of Cam II kinase by increased intracellular calcium following glutamate

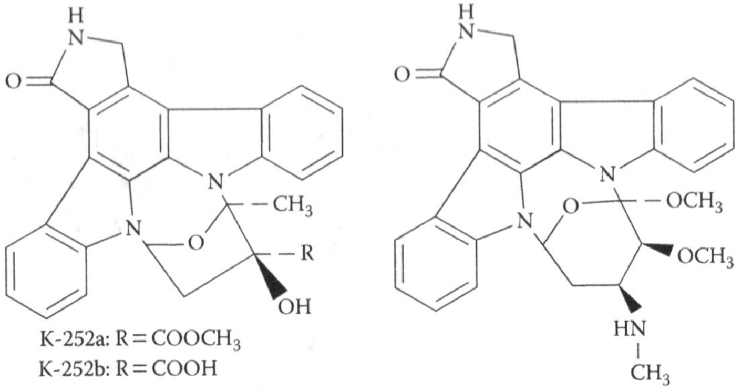

K-252a: R = COOCH$_3$
K-252b: R = COOH

Staurosporine

FIGURE 2.9 Comparison of the chemical structures of K-252a, b, c, and d and staurosporine. The aglycon structure identical to K252-c is shared by all compounds.

receptor activation may result in the phosphorylation of transcription factors, which could directly, or indirectly, enhance transcription of the BDNF gene. Activation of muscarinic receptors may produce similar effects via activation of the PLC pathway, which also increases intracellular calcium.

A totally different class of "indirect" neurotrophic agents demonstrating neuroprotective or neuroregenerative activity not mediated via the activation of either glutamatergic or cholinergic receptors has been reported.[28] A family of indolocarbazole compounds (aglycones) was among the first discovered and studied for their neuronal-growth-promoting effects. A major focus of this approach has been on the molecule K-252a, a microbial alkaloid, and more recently on a cell-impermeant analog K-252b (Figure 2.9). Although these compounds, at concentrations greater than 200 nM, inhibit trophic actions of all the neurotrophins, K-252b at concentrations between 100 pM and 30 nM, enhances the effect of the neurotrophin NT-3-induced survival *in vitro*. How these drugs produce their neurotrophic effects is unclear. It is interesting to note, however, that these compounds bear close similarity to staurosporine, a well-known inhibitor of protein kinase C. Confirmation of the therapeutic value of this approach must, however, await clinical trials.

STRESS, GLUCOCORTICOIDS, AND AD

Stress induces the release of corticotrophin-releasing hormone (CRH) by the hypothalamus, which triggers the release of adrenocorticotrophic hormone (ACTH) by the pituitary, which in turn stimulates adrenocortical secretion of glucocorticoids. Circulating titers of corticosteroids finally suppress subsequent release of CRH and ACTH through steroid-sensitive receptor sites in the brain and in the pituitary, thus preventing prolonged exposure to elevated levels of cortisol following stress. The adrenal cortical axis therefore acts as a closed-loop feedback system where circulating titers of glucocorticoids act to maintain secretion of corticosteroids in-between narrow limits. However, approximately 50% of individuals with AD present higher cortisol titers than patients without dementia, and these levels increase proportionately as the dementia increases.[29] Furthermore, AD patients also demonstrate resistance to inhibition by dexamethasone of ACTH release caused by CRH, suggesting a higher incidence of stress-adaptation failure in AD. Studies have shown that glucocorticoids induce a generalized state of metabolic vulnerability in hippocampal neurons and that cumulative exposure to glucocorticoids such as occurs in aged rats, or following chronic stress, is a cause of loss of hippocampal neurons. These effects appear to be mediated through brain glucocorticoid receptors (GR), probably via inhibition of glucose transport in hippocampal neurons.[30] Taken together, these findings would suggest that the higher incidence of chronic stress-adaptation in AD could be of specific etiopathological significance. Search for specific agents enhancing feedback inhibition of CRH and ACTH secretion, i.e., restoring stress-adaptation of the adrenocortical axis or inhibiting the effect of corticoids (cortisol) on glucose transport may hold promise for future understanding of neurodegenerative mechanisms and drug discovery.

ANTI-INFLAMMATORY THERAPY IN AD

Most data available suggest that AD is a chronic disease, characterized by a long preclinical period during which several initiation factors (genetic, environmental, or both) cause neuronal death. The aging process probably acts as a promoting factor for the disease.[31] Immunohistochemical evidence points to a chronic inflammatory state of the brain in AD. The senile plaques are associated with reactive microglial cells that contribute to the inflammatory response, including the production of the cytokine interleukin-I (IL-I) (both IL-Iα and IL-Iβ). There is an approximately 30-fold increase in the level of IL-I in the brain and cerebrospinal fluid of AD and Down syndrome patients compared to aged matched controls. IL-Iβ is a potent and rapid inducer of glial phospholipase A_2 (PLA_2) and cyclooxygenase (COX) leading to the production of both prostaglandins and thromboxanes.[32,33] Interestingly, a significant increase in the formation of prostaglandin D_2 in the frontal cortex of AD patients has been detected. Anti-inflammatory drugs such as indomethacin are powerful inhibitors of cytokine (IL-Iβ and TNF-α) dependent increases in prostaglandin production (Figure 2.10). Long-term anti-inflammatory therapy might therefore retard the development of AD. This notion has been tested by comparing the prevalence of AD in the general population with that in patients with rheumatoid arthritis (RA) because such patients receive anti-inflammatory drugs and often contract arthritis before the age of risk for AD. The data published showed an unexpectedly low prevalence of AD in patients with RA.[34] Several reports have now demonstrated that anti-inflammatory drugs may retard or even prevent progress of AD symptoms.[35] Although further studies with anti-inflammatory drugs are necessary before definite conclusions can be drawn about the possible preventive effects of such treatments in AD, research has opened a new approach to the design of drugs for treating neurodegenerative diseases. Inhibitors of IL-I transcription, such as E-5090,[36] and of posttranslational processing of IL-I, such as pentamidine,[37] are potential candidates for the study of brain inflammatory processes (Figure 2.11). Compounds of unknown mechanism, but screened in cellular models to decrease cytokine production and/or

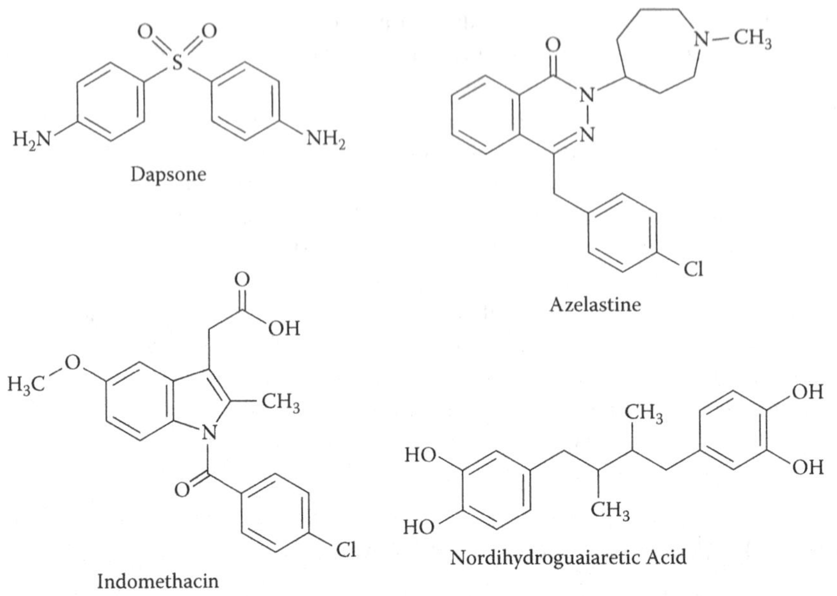

FIGURE 2.10 Anti-inflammatory agents under experimental and clinical evaluation for their potential to inhibit the signal transduction cascade initiated by microglial cytokines.

secretion, have been described, e.g., danazol, which decreases both IL-Iβ and TNF-α synthesis. Zinc protoporphyrin sodium is a purported IL-I receptor antagonist that reduces ischemia brain lesion in the rat.[38] The anti-inflammatory agent sulfasalazine is a TNF-α receptor antagonist, but its effect in brain inflammatory processes has not been reported. An alternative approach involves inhibition of phosphodiesterase with pentoxifylline, which increase c-AMP levels and suppresses TNF-α (see also zardaverine[39]) and also IL-Iα and IL-Iβ production.[40] Propentofylline (HWA 285) (Figure 2.12), a structural analog of pentoxifylline, has been reported to improve symptoms in chronically treated (6 to 12 months) AD patients.[41]

Whether the clinical improvement observed was caused by suppression of microglial hyperactivity, and hence the inflammatory disease process, has yet to be established. Activated microglial cells also produce increased amounts of superoxide anions, which are potentially neurotoxic via oxidation of cellular proteins and cell membrane. Free radical scavengers may, therefore, offer some interest in retarding the progression of AD.

NEUROFIBRILLARY TANGLES, AMYLOID DEPOSITION, FREE RADICALS, AND NEUROTOXICITY

Replacement strategies are aimed at alleviating clinical symptoms (memory loss), not at preventing the development of the disease. Discovering drugs that can inhibit the progression of AD will therefore largely require an understanding of the molecular causes and mechanisms underlying the neurodegenerative processes responsible

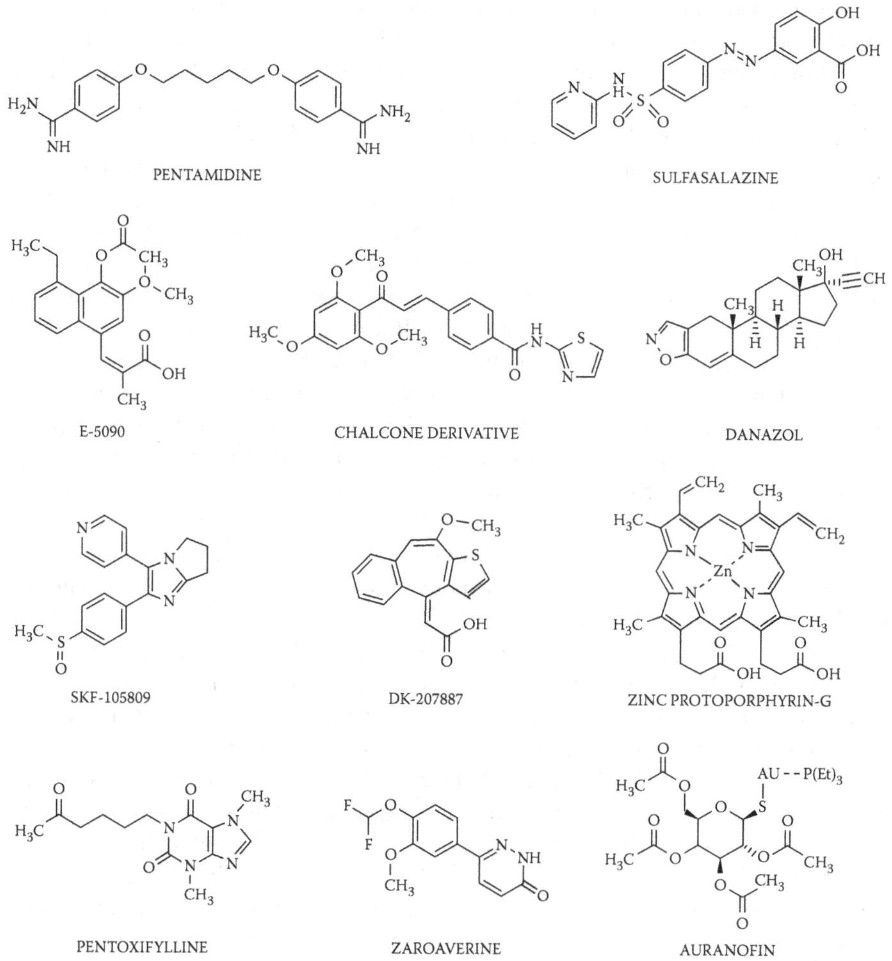

FIGURE 2.11 Candidate compounds for inhibition of cytokine synthesis and antagonism of cytokine receptors.

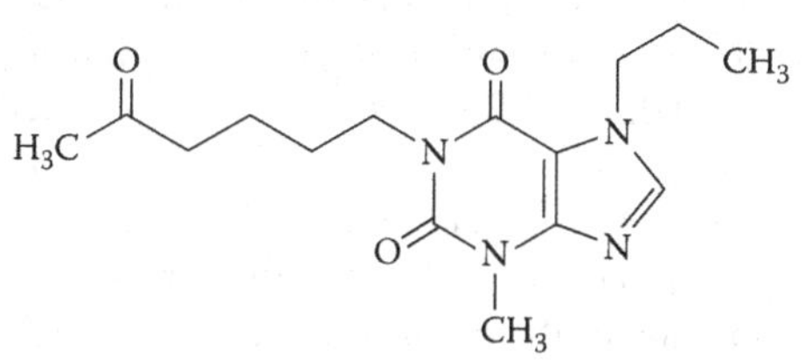

FIGURE 2.12 Propentofyllin.

for the neurochemical deficits and neuronal losses observed. Both the regional distribution and the amount of neurofibrillary tangles (which are composed of paired helical filaments, PHFs) correlate with the degree of dementia.[42] It has been shown that PHFs are composed of protein tau, a protein normally associated with the microtubular network of the cell. Tau isolated form PHFs (AD) appear to be in a hyperphosphorylated state when compared to normal soluble tau. The region of tau that normally binds to the microtubule contains a serine residue (Ser-262), which is of particular interest because phosphorylation of this residue has only been observed in tau isolated from AD brain.[43] Furthermore, because phosphorylation at this site alone dramatically reduces the affinity for microtubules *in vitro,*[44] tau is unable to bind, and therefore self-assembles into PHFs. Because tau binding stabilizes the otherwise dynamically unstable microtubules, decreased binding of tau leads to microtubule instability, disruption of the neuronal microtubule network, impaired axonal transport, and eventually nerve cell degeneration and death. Both a decrease in phosphatase activity and an increase in kinase activity may account for hyperphosphorylated tau in AD.[45] A protein kinase that regulates tau–microtubule interactions and dynamic instability of microtubules by phosphorylation at the AD-specific serine 262 has been described.[46] Design of specific inhibitors of this enzyme might offer a new approach to prevention of PHF formation, microtubule disruption, and neuronal death.

The role of senile plaques in the development of AD is not clearly understood. A 4.2-kD peptide (called βAP) is the principal constituent of the cerebral amyloid deposit or senile plaque (SP). Both βAP and SP may cause the brain damage associated with AD, although most available evidence indicates that βAP-induced neurotoxicity is probably caused by aggregated forms of the peptide. Some studies suggest that βAP (or its aggregated form) increases intracellular calcium, allowing neurons to become more vulnerable to the effects of excitotoxins such as glutamate that form free radicals and can kill healthy cells.[47] The neurotoxic activity of βAP resides within amino acids 25–35. The 4.2-kD peptide is derived from a much larger molecule, the beta amyloid precursor protein (βAPP), a transmembrane glycoprotein located on chromosome 21. Two alternative pathways of metabolism of this protein have been identified.[48] One pathway involves an enzymatic cleavage at the cell surface (within the βAPP sequence) that results in liberation of secreted forms of βAPP (APPs) into the extracellular milieu, thereby precluding the generation of βAP. The alternative pathway apparently involves internalization of cell surface βAPP via coated pit-mediated endocytosis and processing in an acidic (possibly lysosomal) compartment. This pathway can result in liberation of intact βAP from cells and precipitation of the amyloid deposit. Because βAP is the product of a "normal" metabolic pathway but accumulates excessively only during aging and more dramatically in AD, it seems likely that in AD there is either an increased generation of βAP from βAPP or decreased degradation of βAP. The finding that one of the pathological correlates of AD is inflammation of the brain, and that IL-I alters the mRNA encoding for APP and enhances the processing of APP in cell culture,[49,50] does suggest that the formation of neuritic plaques might be driven by glial cytokines.[51] However, investigators have still not determined whether amyloid deposition is the initial event triggering the development of AD or whether the amyloid deposit is an "inflammatory response" and therefore occurs as a result of other biological processes causing neuronal death. Nevertheless, understanding the mechanisms that control the secretion and metabolism of βAPP into βAP and APP^s, the precipitation of βAP into senile plaques, and the role of these proteins in the neurodegenerative process becomes critically important for rational design of drugs to treat AD.

Characterization of proteases involved in lysosomal βAPP processing will be an important component of research aimed at design of anti-AD drugs. Inhibitors of lysosomal protease aimed at one or more of these enzymes or agents affecting H$^+$/Na$^+$ transport in lysosomes might retard amyloid deposition and slow the disease process. A number of studies have determined that βAP spontaneously fragments into free radicals in solution and that oxygen does seem to be a requirement for radical generation. The reactive oxygen species are generated by a specific 11-amino-acid sequence (25–35) of the β-amyloid during cell free incubation. This is also the sequence responsible for neurotoxicity of βAP. These radicals, which may directly damage cell membranes, also combine with each other into aggregates that form the SP.[52] Because both ascorbic acid and trolox (Vit. E analog) inhibit aggregation of monomeric βAP, antioxidants may have therapeutic potential in AD. The finding that excitotoxicity caused by overstimulation of glutamatergic receptors (AMPA plus NMDA) is also associated with significant production of oxygen radicals further supports a neuroprotective role for antioxidants.

Heparan sulfate proteoglycan (HSPG) is a component of the senile plaque and also coprecipitates βAP *in vitro*.[53] To examine the hypothesis that sulfated (or sulfonated) compounds could interfere with amyloid formation *in vivo*, Kisilevski et al.[54] prepared a series of related, simple, small-molecule, anionic sulfonates or sulfates and assessed their effects in acute and chronic murine models of rapid inflammation-associated (AA) amyloidogenesis. When administered orally, these compounds substantially reduced AA amyloid progression. They also interfered with heparan-sulfate-stimulated β-peptide fibril aggregation *in vitro*. Congo red (Figure 2.13) (an aromatic disulfonate), which has long been used as a histochemical stain for amyloid, has also been reported to inhibit βAP neurotoxicity in primary rat hippocampal culture by inhibiting fibril formation. Congo red has also been reported to inhibit the pancreatic islet cell toxicity of diabetes-associated amylin, another type of amyloid fibril, suggesting a potential therapeutic approach to AD.[55] Whether sulfonates or sulfates will prevent amyloid formation in the brain remains to be established.

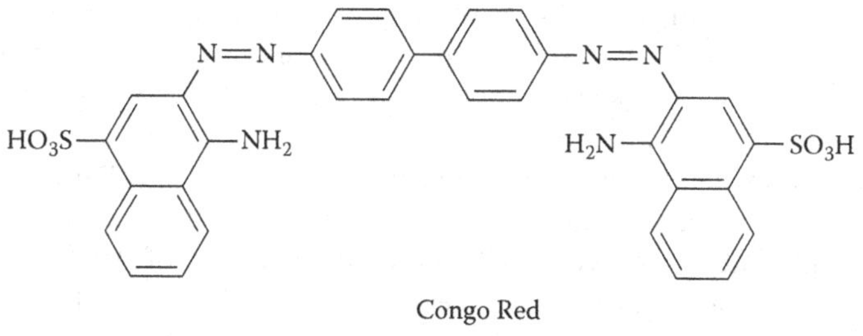

Congo Red

FIGURE 2.13 Congo red.

Resolution of the tridimensional solution structure of residues 1–28 of βAP has demonstrated that the side chains of histidine-13 and lysine-16 are close, residing on the same face of the α-helical structure of the peptide.[56] Their proximity may constitute a binding motif with HSPG and eventually a "molecular pattern" for the design of drugs that could specifically interfere with βAP–HSPG binding and precipitation of the amyloid.

AMYLOID DEPOSITION, APOLIPOPROTEIN E, AND FREE RADICALS

Apolipoprotein E is an important cholesterol-transporting protein and a major constituent of very low density lipoproteins (VLDL). There are three common alleles (e2, e3, and e4) and two less common alleles (ε1 and ε5). The common alleles provide six possible genotypes. Apo ε3/ε3 is the most common genotype, occurring in about 65% of the world population. The apo ε4/ε4 genotype occurs in 2 to 3% of the population. It has been shown that there is a strong and apparently robust association of late-onset and sporadic AD with the e4 allele of apolipoprotein E.[57] The frequency of the ε4 allele is around 0.5 in affected individuals from families where several other members also have late-onset AD. It is around 0.40 in patients unselected for the presence of a family history, and around 0.12 in control populations. Corder et al.[58] reported that the risk was increased as a function of the inherited dose of the gene (ApoE-4), and that the mean age of onset was lowered with each ApoE-4 (ε4) allele. However, some 40% of people with AD do not possess an ε4 allele, so it is neither necessary nor sufficient to cause the disease. Therefore, the ApoE gene probably modifies the risk and timing of expression of the disease. The presence of apolipoprotein E in both senile plaques and neurofibrillary tangles does, however, point to it playing a direct part in the pathogenesis of AD.[59] Nicoll et al.[60] have shown that the ε4 allele is also associated with deposition of amyloid β-protein following head injury. This finding provides further evidence linking apolipoprotein E ε4 allele with βAP deposition *in vivo* and suggests that environmental and genetic factors for AD may act additively.

ApoE-4 differs from ApoE-3 in having an arginine in position 112 instead of cysteine. As a result of this, glutamic acid 109 forms a salt bridge with arginine 112, and the arginine 61 side chain is displaced to a new position when compared to the structure of ApoE-3.[61] Whether these minor structural differences will explain the role of Apo E in AD remains to be established.

Studies have demonstrated that ApoE-4 interacts with and precipitates βAP *in vitro* more readily than ApoE-3.[62] Binding of βAP to ApoE-4 was observed in minutes, whereas binding to ApoE-3 required hours. Oxygen-mediated complex formation was implicated because binding was increased in oxygenated buffer and prevented by reduction with dithiothreitol or 2-mercaptoethanol. This finding again suggests that antioxidants may have therapeutic potential in AD (Figure 2.14).

A specific amino-acid sequence (244–272) in ApoE-4 appears to be critically required for the interaction with βAP and the precipitation of ApoE-3–βAP complexes. Drugs designed to bind to this specific site might inhibit the formation of SP and hence offer a new treatment approach in AD. Studies also suggest that there is a negative association, or protective effect, of the ε2 allele of apolipoprotein E in AD.[63]

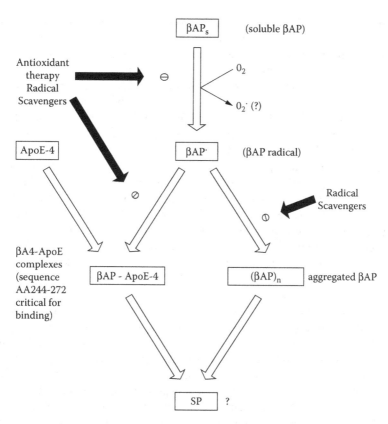

FIGURE 2.14   Fragmentation of βAP, interaction between βAP and ApoE-4 (244–271), and possible formation of senile plaques (SP).

Because AD is an important cause of morbidity in the elderly, this protective effect of apoE ε2 on risk for AD may explain the reported positive association of apoE ε2 with longevity.[64]

ACETYLCHOLINE, APP PROCESSING, AND NEUROTROPHIC AND NEUROPROTECTIVE EFFECTS

Activation of protein kinase C (PKC) by phorbol esters has been shown to increase secretion of APP^s by cells in culture and concomitantly reduce the production of βAP.[65] This would imply that the amyloidogenic and the nonamyloidogenic pathways for processing APP are interrelated. Activation of muscarinic (m_1 and m_3) receptors that increases phosphatidyl inositol turnover (via phospholipase C, PLC) and PKC activity has been shown to mimic the effects of phorbol esters on APP processing.[66] Agents or conditions that lead to increased phosphorylation would therefore be expected to accelerate normal processing and secretion of APP. Studies of a series of rigid cholinergic m_1 agonists [AF 102B and congeners of AF 150(S)] have demonstrated that activation of m_1-receptors stimulated both APP^s secretion and tau dephosphorylation and potentiated the neurotrophic effects of NGF in PC12(MI) cells.[67] These findings suggest that cholinergic replacement therapy, or any agents favoring phosphorylation via depolarization or second-messenger systems, might beneficially affect the metabolism of APP and eventually slow the course of the disease. Testing this hypothesis would obviously require long-term treatment of Alzheimer patients in an early stage of the disease with well-tolerated cholinergic drugs. It remains to be established whether cholinergic hypoactivity caused by the loss of cholinergic nerve terminals contributes to plaque formation via reduced activation of cholinergic m_1 receptors and PKC.

The processed form of βAPP appears to play a critical role in new synapse formation (neuronal plasticity) and also in protecting against excessive increases in cellular calcium caused by activation of glutamate receptors.[47] These neurotrophic and neuroprotective effects are mediated via different amino acid sequences in the APP molecule. Using synthetic peptides and deletion constructs of the protein (βAPP$_{695}$), it has been demonstrated[68] that the active site responsible for growth and differentiation effects is entirely included within a 17 amino-acid (17-mer) domain from Ala-319 to Met-335. A pentameric peptide RERMS (5-mer) spanning the region Arg-328 to Ser-332 was also found to be active, but the concentration needed was tenfold higher than that of the 17-mer. Displacement studies indicate that the neurotrophic effects of APP are receptor mediated. Because the 17-mer sequence has been conserved during evolution (100% homology in rat, mouse, monkey, and human), it probably serves a basic biological function. The neuroprotective effects of AP can be mimicked by synthetic peptides spanning area 444–592. These peptides do not express neurotrophic effects, indicating that neuroprotective and growth-promoting sequences are localized in different domains of the protein. Design of drugs mimicking the effects of these neurotrophic and neuroprotective domains may open a new area to drug discovery. Ultimate success in designing drugs for prevention of the development of AD will, however, critically depend on the availability of experimental models reflecting behavioral changes caused by the etiopathogenic mechanisms characterizing the disease to be treated.

REFERENCES

1. Kokmen, E. et al., Is the incidence of dementing illness changing? A 25-year time trend study in Rochester, Minnesota (1960–1984), *Neurology*, 43, 1887, 1993.
2. Weksler, M.E. et al., The immune system, amyloid-beta peptide, and Alzheimer's disease, *Immunol. Rev.*, 205, 244, 2005
3. Katzman, J.B. et al., In *Neuropeptides in Neurologic and Psychiatric Disease,* Martin, J.B. and Barchas, J.D., Eds., Raven Press, New York, 1986, pp. 279–286.
4. Kaufer, D.I., Cummings, J.L., and Christine, D., Effect of tacrine in Alzheimer's disease: or non-effect? *Med. J. Aust.*, 162, 52, 1995.
5. Coelho, F. and Birks, J., Physostigmine for Alzheimer's disease, *Cochrane Database Syst. Rev.*, 2, CD001499, 2001.
6. Ohnishi, A. et al., Comparison of the pharmacokinetics of E2020, a new compound for Alzheimer's disease, in healthy young and elderly subjects, *J. Clin. Pharmacol.*, 33, 1086, 1993.
7. Soncrant, T.T. et al., Memory improvement without toxicity during chronic, low dose intravenous arecoline in Alzheimer's disease, *Psychopharmacology*, 112, 421, 1993.

8. White, H.K. and Levin, E.D., Four-week nicotine skin patch treatment effects on cognitive performance in Alzheimer's disease, *Psychopharmacology*, 143, 158, 1999.

9. Consolo, S. et al., Serotonergic facilitation of acetylcholine release *in vivo* from rat dorsal hippocampus via serotonin 5-HT3 receptors, *J. Neurochem.*, 62, 2254, 1994.

10. Dysken, M., Kuskowski, D.M., and Love, S., Ondansetron in the treatment of cognitive decline in Alzheimer dementia, *Am. J. Geriatr. Psychiatry*, 10, 212, 2002.

11. Sarter, M., Taking stock of cognition enhancers, *Trends Pharmacol. Sci.*, 12, 456, 1991.

12. Gurtwitz, D. et al., Discrete activation of transduction pathways associated with acetylcholine m1 receptor by several muscarinic ligands, *Eur. J. Pharmacol.*, 267, 21, 1994.

13. Fisher, A. et al., Selective signaling via unique M1 muscarinic agonists, *Ann. N.Y. Acad. Sci.*, 695, 300, 1993.

14. Bowen, D.M. et al., Treatment of Alzheimer's disease, *J. Neurol. Neurosurg. Psychiatry*, 55, 328, 1992.

15. Isaacson, J.S. and Nicoll, R.A., Aniracetam reduces glutamate receptor desensitization and slows the decay of fast excitatory synaptic currents in the hippocampus, *Proc. Natl. Acad. Sci. U.S.A.*, 88, 10936, 1991.

16. Bowen, D.M. et al., "Traditional" pharmacotherapy may succeed in Alzheimer's disease, *Trends Neurosci.*, 15, 84, 1992.

17. Ebendal, T., Function and evolution in the NGF family and its receptors, *J. Neurosci. Res.*, 32, 461, 1992.

18. Carswell, S., The potential for treating neurodegenerative disorders with NGF-inducing compounds, *Exp. Neurol.*, 124, 36–42, 1993.

19. Murray, K.D. et al., Differential regulation of brain-derived neurotrophic factor and type II calcium/calmodulin-dependent protein kinase messenger RNA expression in Alzheimer's disease, *Neuroscience*, 60, 37, 1994.

20. Isackson, P.J. et al., BDNF mRNA expression is increased in adult rat forebrain after limbic seizures: temporal patterns of induction distinct from *NGF*, *Neuron*, 6, 937, 1991.

21. Ernfors, P. et al., Increased levels of messenger RNAs for neurotrophic factors in the brain during kindling epileptogenesis, *Neuron*, 7, 165, 1991.

22. Wetmore, C., Olson, L., and Bean, A.J., Regulation of brain-derived neurotrophic factor (BDNF) expression and release from hippocampal neurons is mediated by non-NMDA type glutamate receptors, *J. Neurosci.*, 14, 1688, 1994.

23. Bengzon, J. et al., Widespread increase of nerve growth factor protein in the rat forebrain after kindling-induced seizures, *Brain Res.*, 587, 338, 1992.

24. Zafra, F. et al., Regulation of brain-derived neurotrophic factor and nerve growth factor mRNA in primary cultures of hippocampal neurons and astrocytes, *J. Neurosci.*, 12, 4793, 1992.

25. Gwag, B.J. and Springer, J.E., Activation of NMDA receptors increases brain-derived neurotrophic factor (BDNF) mRNA expression in the hippocampal formation, *Neuroreport*, 5, 125, 1992.

26. Pizzi, M. et al., Attenuation of excitatory amino acid toxicity by metabotropic glutamate receptor agonists and aniracetam in primary cultures of cerebellar granule cells, *J. Neurochem.*, 61, 683, 1993.

27. da Penha Berzaghi, M. et al., Cholinergic regulation of brain-derived neurotrophic factor (BDNF) and nerve growth factor (NGF) but not neurotrophin-3 (NT-3) mRNA levels in the developing rat hippocampus, *J. Neurosci.*, 13, 3818, 1993.

28. Knusel, B. and Hefti, F., K-252 compounds: modulators of neurotrophin signal transduction, *J. Neurochem.*, 59, 1987, 1992.

29. Deshmukh, V. and Desmukh, S.V., Stress-adaptation failure hypothesis of Alzheimer's disease, *Med. Hypotheses*, 32, 293, 1990.

30. Virgin, C.E. et al., Glucocorticoids inhibit glucose transport and glutamate uptake in hippocampal astrocytes: implications for glucocorticoid neurotoxicity, *J. Neurochem.*, 57, 1422, 1991.

31. Katzman, R. and Kawas, C., In *AD: The Epidemiology of Dementia and AD,* Terry, R.D., Katzman, R., and Tsick, K.I., Eds., Raven Press, New York, 1994, pp. 105–119.

32. Giulian, D. and Corpuz, M., Microglial secretion products and their impact on the nervous system, *Adv. Neurol.*, 59, 315, 1993.

33. Carman-Krzan, M. and Wise, B.C., Arachidonic acid lipoxygenation may mediate interleukin-1 stimulation of nerve growth factor secretion in astroglial cultures, *J. Neurosci. Res.*, 34, 225, 1993.

34. McGeer, P.L. et al., Anti-inflammatory drugs and Alzheimer's disease, *Lancet*, 335, 1037, 1990.

35. Breitner, J.C.S. et al., Inverse association of anti-inflammatory treatments and Alzheimer's disease: initial results of a co-twin control study, *Neurology*, 44, 227, 1994.

36. Tanaka, M. et al., 3-(5-Alkyl-4-hydroxy-3-methoxy-1-naphthalenyl)-2-methyl-2-propenoic acids as orally active inhibitors of IL-1 generation, *J. Med. Chem.*, 34, 2647, 1991.

37. Rosenthal, G.J. et al., Pentamidine: an inhibitor of interleukin-1 that acts via a posttranslational event, *Toxicol. Appl. Pharmacol.*, 107, 555, 1991.

38. Yamasaki, Y. et al., Possible involvement of interleukin-1 in ischemic brain edema formation, *Neurosci. Lett.*, 142, 45, 1992.

39. Schade, F.U. and Schudt, C., The specific type III and IV phosphodiesterase inhibitor zardaverine suppresses formation of tumor necrosis factor by macrophages, *Eur. J. Pharmacol.*, 230, 9, 1993.

40. Chao, C.C. et al., Cytokine release from microglia: differential inhibition by pentoxifylline and dexamethasone, *J. Infect. Dis.*, 166, 847, 1992.

41. Rother, M. et al., HWA 285 (propentofylline)—a new compound for the treatment of both vascular dementia and dementia of the Alzheimer type, *Ann. N.Y. Acad. Sci.*, 777, 404, 1996.

42. Hyman, B.T., Studying the Alzheimer's disease brain: insights, puzzles, and opportunities, *Neurobiol. Aging*, 15 (suppl. 2) S79, 1994.

43. Hasegawa, M. et al., Protein sequence and mass spectrometric analyses of tau in the Alzheimer's disease brain, *J. Biol. Chem.*, 267, 17047, 1992.

44. Biernat, J. et al., Phosphorylation of Ser262 strongly reduces binding of tau to microtubules: distinction between PHF-like immunoreactivity and microtubule binding, *Neuron*, 11, 153, 1993.

45. Roush, W., Protein studies try to puzzle out Alzheimer's tangles, *Science*, 267, 793, 1995.

46. Drewes, G. et al., Microtubule-associated protein/microtubule affinity-regulating kinase (p110mark). A novel protein kinase that regulates tau-microtubule interactions and dynamic instability by phosphorylation at the Alzheimer-specific site serine 262, *J. Biol. Chem.* 270, 7679, 1995.

47. Mattson, M.P. et al., beta-Amyloid precursor protein metabolites and loss of neuronal Ca2+ homeostasis in Alzheimer's disease, *Trends Neurosci.*, 16, 409, 1993.

48. Selkoe, D.J., Physiological production of the beta-amyloid protein and the mechanism of Alzheimer's disease, Trends *Neurosci.*, 16, 403, 1993.

49. Yang, F. et al., IL-1 beta decreases expression of amyloid precursor protein gene in human glioma cells, *Biochem. Biophys. Res. Commun.*, 191, 1014, 1993.

50. Buxbaum, J.D. et al., Cholinergic agonists and interleukin 1 regulate processing and secretion of the Alzheimer beta/A4 amyloid protein precursor, *Proc. Natl. Acad. Sci. U.S.A.*, 89, 10075, 1992.

51. Mrak, R.E., Sheng, J.G., and Griffin, W.S., Glial cytokines in Alzheimer's disease: review and pathogenic implications, *Hum. Pathol.*, 26, 816, 1995.

52. Hensley, K. et al., A model for beta-amyloid aggregation and neurotoxicity based on free radical generation by the peptide: relevance to Alzheimer's disease, *Proc. Natl. Acad. Sci. U.S.A.*, 91, 3270, 1994.

53. Snow, A.D. et al., An important role of heparan sulfate proteoglycan (Perlecan) in a model system for the deposition and persistence of fibrillar A beta-amyloid in rat brain, *Neuron*, 12, 219, 1994.

54. Kisilevski, R. et al., Arresting amyloidosis *in vivo* using small-molecule anionic sulphonates or sulphates: implications for Alzheimer's disease, *Nat. Med.*, 1, 143, 1995.

55. Lorenzo, A. and Yankner, B.E., Beta-amyloid neurotoxicity requires fibril formation and is inhibited by congo red, *Proc. Natl. Acad. Sci. U.S.A.*, 91, 12243, 1994.

56. Talafous, J. et al., Solution structure of residues 1-28 of the amyloid beta-peptide, *Biochemistry*, 33, 7788, 1994.

57. Corder, E.H. et al., Gene dose of apolipoprotein E type 4 allele and the risk of Alzheimer's disease in late onset families, *Science*, 261, 921, 1993.

58. Corder, E.H. et al., Protective effect of apolipoprotein E type 2 allele for late onset Alzheimer disease, *Nat. Genet.*, 7, 180, 1994.

59. Wisniewsky, T. and Frangione, B., Apolipoprotein E: a pathological chaperone protein in patients with cerebral and systemic amyloid, *Neurosci. Lett.*, 135, 235, 1992.

60. Nicoll, J.A.R., Roberts, G.W., and Graham, D.I., Apolipoprotein E epsilon 4 allele is associated with deposition of amyloid beta-protein following head injury, *Nat. Med.*, 1, 135, 1995.

61. Dong, L.M. et al., Human apolipoprotein E. Role of arginine 61 in mediating the lipoprotein preferences of the E3 and E4 isoforms, *J. Biol. Chem.*, 269, 22358, 1994.

62. Strittmatter, W.J. et al., Binding of human apolipoprotein E to synthetic amyloid beta peptide: isoform-specific effects and implications for late-onset Alzheimer's disease, *Proc. Natl. Acad. Sci. U.S.A.*, 90, 8098, 1993.

63. Talbot, C. et al., Protection against Alzheimer's disease with apoE epsilon 2, *Lancet*, 343, 1432, 1994.

64. Schachter, F. et al., Genetic associations with human longevity at the APOE and ACE loci, *Nat. Genet.*, 6, 29, 1994.

65. Fukushima, D. et al., Activation of the secretory pathway leads to a decrease in the intracellular amyloidogenic fragments generated from the amyloid protein precursor, *Biochem. Biophys. Res. Comm.*, 194, 202, 1993.

66. Nitsch, R.M. et al., Release of Alzheimer amyloid precursor derivatives stimulated by activation of muscarinic acetylcholine receptors, *Science*, 258, 304, 1992.

67. Fisher, A. et al., M1 agonists for the treatment of Alzheimer's disease. Novel properties and clinical update, *Ann. N.Y. Acad. Sci.*, 777, 189, 1996.

68. Roch, J.M. et al., Biologically active domain of the secreted form of the amyloid beta/A4 protein precursor, *Ann. N.Y. Acad. Sci.*, 695, 149, 1993.

3 Lipoprotein-Associated Phospholipase A$_2$ as A New Prognostic Factor for Coronary Artery Disease

Alexandros D. Tselepis

CONTENTS

INTRODUCTION

Disturbed lipoprotein metabolism, and mainly elevated plasma low-density lipoprotein cholesterol (LDL-cholesterol) concentration, is a major component of the guidelines for the prevention of coronary artery disease (CAD). However, plasma LDL-cholesterol levels are insufficient to identify individuals with incident CAD events because approximately 50% of all CAD events occur in subjects with normal or even low LDL-cholesterol levels.[1] This has led to the hypothesis that other factors may be involved in the pathogenesis and progress of atherosclerosis and CAD. Recent advances in cardiovascular research point to a critical role of inflammatory processes in the etiology of CAD. Discovery of novel inflammatory biomarkers followed, which may be useful as additional screening tools for the identification of individuals at increased risk of CAD. One such novel inflammatory biomarker is platelet-activating factor acetylhydrolase (PAF-acetylhydrolase), also referred to as lipoprotein-associated phospholipase A$_2$ (Lp-PLA$_2$). This chapter is focused on the main structural and catalytic features of plasma Lp-PLA$_2$, on the association of the enzyme with distinct lipoprotein particle subspecies, on its cellular sources, and finally on the potential significance of Lp-PLA$_2$ in atherosclerosis and cardiovascular disease.

BIOCHEMICAL CHARACTERISTICS OF Lp-PLA$_2$

The presence of an enzyme in human plasma that catalyzes the hydrolysis of the sn-2 ester bond of the potent proinflammatory phospholipid, platelet-activating factor (PAF) (and which thus attenuates its bioactivity) was first demonstrated by Farr et al. in 1980.[2] This enzyme was named PAF-acetylhydrolase (PAF-AH) (EC 3.1.1.47). Plasma PAF-acetylhydrolase is categorized in the subfamily VIIA of phospholipases A$_2$, and its activity does not require calcium. Unlike other PLA$_2$, PAF-acetylhydrolase is specific for short acyl groups (Cn < 6) at the sn-2 position of the phospholipid substrate. With the exception of PAF, PAF-acetylhydrolase can also effectively hydrolyze oxidized phospholipids produced by peroxidation of phosphatidylcholines (PC) containing an sn-2 polyunsaturated fatty acyl residue (OX-PC) (Figure 3.1). Such oxidized phospholipids are formed during the oxidative modification of LDL and play key roles in several aspects of atherogenesis.[3] Early studies had shown that PAF-acetylhydrolase is an interfacial enzyme; however, more recent work has revealed that PAF-acetylhydrolase accesses its substrates only from the aqueous phase, and thus this enzyme may hydrolyze other lipid esters that are partially soluble in the aqueous phase.[4] Indeed, with the exception of its PLA$_2$ activity, PAF-acetylhydrolase can equally hydrolyze short-chain diacylglycerols, triacylglycerols, and acetylated alkanols, and also displays a PLA$_1$ activity. Consequently, PAF-acetylhydrolase possesses broad substrate specificity toward lipid esters containing short acyl chains.[4]

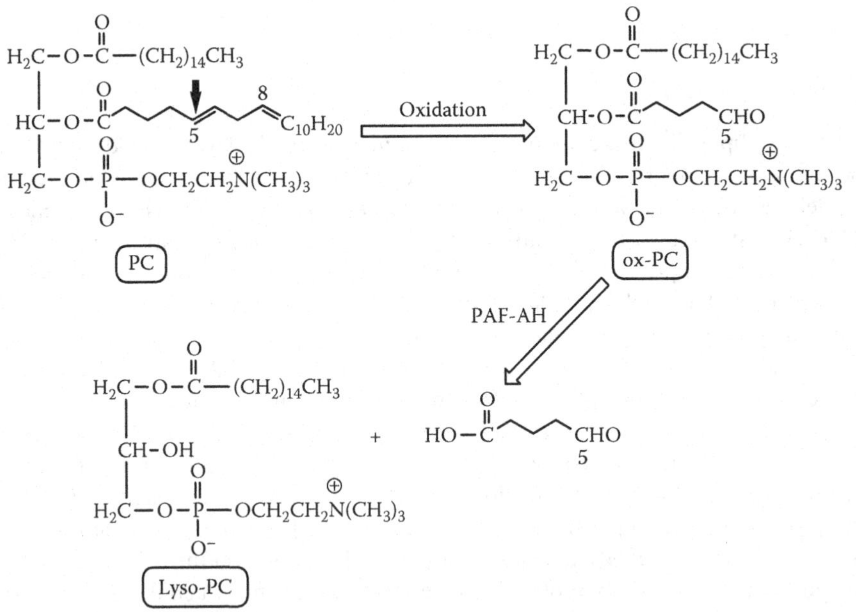

FIGURE 3.1   Hydrolysis of oxidized phospholipids by Lp-PLA$_2$ (PAF-acetylhydrolase [PAF-AH]).

The cloning of a cDNA encoding the human plasma PAF-acetylhydrolase was reported by Tjoelker et al in 1995.[5] The cDNA contains an open reading frame predicted to encode a 441-amino-acid protein that is cleaved between Lys-41 and Ile-42 to generate a mature enzyme with a calculated molecular mass of 45.4 kDa. The primary structure of PAF-acetylhydrolase is unique, with the exception of the Gly-His-Ser-Phe-Gly consensus sequence in the catalytic site of the enzyme, which is similar to that found in serine esterases and lipases. With the use of site-directed mutagenesis, it was shown that Ser-273 of the Gly-His-Ser-Phe-Gly motif as well as Asp-296, and His-351 are essential for catalytic activity.[6] The linear orientation and spacing of Ser-273, Asp-296, and His-351 are consistent with the α/β hydrolase conformation. These features indicate that PAF-acetylhydrolase is structurally and mechanistically different from PLA$_2$, consistent with the recently proposed broad substrate specificity of the latter enzyme. Plasma PAF-acetylhydrolase is N-glycosylated and contains about 9 kDa of a heterogeneous asparagines-conjugated sugar chain involving sialic acid.[7,8]

PAF-acetylhydrolase in human plasma is complexed to lipoproteins; thus, it was recently named Lp-PLA$_2$. Lp-PLA$_2$ is mainly associated with the apolipoprotein B (apoB)-containing lipoproteins and primarily with LDL, whereas a small proportion (<20% of total enzyme activity) is associated with high-density lipoprotein (HDL). Within these lipoprotein pools, it appears that the enzyme preferentially associates with small dense LDL (sdLDL) and with the very high-density lipoprotein-1 subfraction.[9]

Using site-directed mutagenesis, Stafforini et al.[10] demonstrated that Lp-PLA$_2$ interacts directly with apo B-100, and that the carboxyl terminus of this apolipoprotein is required for such interaction. On the Lp-PLA$_2$ side, residues Try-115, Leu-116, and Tyr-205 are critical to association of the enzyme with LDL.[10]

In contrast to the well-studied association between LDL and Lp-PLA$_2$, there is a paucity of data concerning the association of this enzyme with HDL. We have provided evidence that removal of the carbohydrate content of the macrophage-derived Lp-PLA$_2$ (which represents one of the major sources of the enzyme pool present in plasma) enhances enzyme association with HDL subfractions. These results provide evidence that a factor contributing to the preferential association of Lp-PLA$_2$ with LDL versus HDL could be the degree of enzyme glycosylation.[8] The major features of the structure, catalytic properties, and plasma transport of Lp-PLA$_2$ are summarized in Table 3.1.

CELLULAR SOURCES OF HUMAN PLASMA Lp-PLA$_2$

Early studies demonstrated that peripheral blood-monocyte-derived macrophages[11] and the human hepatocarcinoma cell line HepG2[12] secrete the plasma form of Lp-PLA$_2$. Other cell types that secrete this enzyme are neutrophils, differentiated HL-60 cells, activated bone-marrow-derived mast cells, and activated platelets.[13-15] The Lp-PLA$_2$ mRNA has been detected in differentiated macrophages, as well as in thymus, tonsils, and human placenta, but not in liver cells and monocytes. Such a cellular location of Lp-PLA$_2$ mRNA is consistent with a macrophage origin of the enzyme.[5] Indeed, more recent data suggested that hepatocytes do not appear to significantly

TABLE 3.1

Biochemical Characteristics of Lp-PLA$_2$ (PAF-Acetylhydrolase)

1. 45.4-kDa monomeric protein
2. Contains N-linked heterogeneous sugar chain(s), 9 kDa, involving sialic acid
3. The cDNA encodes a 441-amino acid protein containing a secretion signal sequence (Met-1–Ala-17)
4. The catalytic site contains the Gly-His-Ser-Phe-Gly consensus sequence characteristic of lipases and esterases
5. Ser-273, Asp-296, and His-351 are essential for catalytic activity consistent with an α/β hydrolase conformation
6. Expresses Ca2+-independent PLA2 activity toward PAF and oxidized phospholipids
7. Expresses lipase and PLA1 activities
8. Enzyme sources: cells of hematopoietic origin (monocytes/macrophages, hepatic kupffer cells, mast cells, platelets)
9. Enzyme expression is primarily regulated by the differentiation state of the cell and by proinflammatory mediators
10. Plasma transport; 80–85% LDL (primarily sdLDL); 15–20% HDL
11. PAF-AH binds directly to -COOH terminal of LDL-Apo B100 (PAF-AH residues: Tyr 205, Tryp 115, Leu 116)

contribute to the plasma pool of Lp-PLA$_2$. Further studies have revealed that cells of hematopoietic origin constitute the primary source of circulating Lp-PLA$_2$. Human peripheral blood monocytes do not express Lp-PLA$_2$; however, upon differentiation into macrophages *in vitro*, Lp-PLA$_2$ mRNA becomes detectable, and enzyme activity is secreted into the culture medium.[8,11] Consistent with this finding, macrophages in human atherosclerotic lesions express plasma PAF-AH.[16] Furthermore, the mRNA of plasma Lp-PLA$_2$ has been detected in lipid-laden macrophages in human atherosclerotic lesions.[16]

Overall, the above studies strongly suggest that the main source of the plasma Lp-PLA$_2$ is macrophages. Our finding that the glycosylation of Lp-PLA$_2$ secreted from human blood-monocyte-derived macrophages resembles that of the plasma enzyme is consistent with this possibility.[8] The promoter of the gene coding for the plasma form of Lp-PLA$_2$ contains seven MS2 and eleven STAT-binding consensus sequences.[17] The multiple MS2 binding sites indicate that expression of the Lp-PLA$_2$ gene is under tight differentiation control, whereas the STAT consensus sequences may mediate the effects of inflammatory mediators on Lp-PLA$_2$ gene expression. The cellular expression of plasma Lp-PLA$_2$ is regulated by various factors, including the differentiation state of the cell and the degree of activation by proinflammatory mediators. Future studies of the regulation of Lp-PLA$_2$ gene expression will clarify the effects of proinflammatory mediators on Lp-PLA$_2$ expression.

ROLE OF Lp-PLA$_2$ IN ATHEROSCLEROSIS

Lp-PLA$_2$ may play a significant role in atherogenesis and cardiovascular disease due to its role in the metabolism of bioactive lipids such as PAF and oxidized phospholipids. PAF activates leukocytes and platelets and enhances leukocyte adhesion to the vessel wall. Furthermore, PAF is a vasoactive mediator that may be synthesized locally at the site of endothelial injury during thrombosis. Equally, PAF accumulates in the atherosclerotic plaques of subjects with advanced CAD, suggesting that this phospholipid mediator actively participates in the pathophysiology of atherosclerosis.[18] In addition, PAF itself, as well as proinflammatory and vasoactive oxidized phospholipids, are formed in LDL during oxidation and are believed to play central roles in the formation of atherosclerotic plaques.[19,20] Consequently, by degrading such phospholipids, Lp-PLA$_2$ could act as a potent antiatherogenic enzyme. Indeed, various studies *in vitro* support such a role. Thus, it has been shown that Lp-PLA$_2$, by degrading oxidized phospholipids, inhibits the oxidative modification of the apoB moiety of LDL, suggesting that intact but not hydrolyzed phospholipids are required for apoB modification. Furthermore, Lp-PLA$_2$ inhibits several proatherogenic activities of oxidized LDL *in vitro* in which oxidized phospholipids are involved.[21]

In contrast to the foregoing hypothesis involving an anti-oxidant, anti-inflammatory, and anti-atherogenic role for LDL-associated Lp-PLA$_2$, other studies have suggested that it can be a proinflammatory and proatherogenic enzyme. This is supported by the observation that, during the hydrolysis of oxidized phospholipids, Lp-PLA$_2$ generates lysophosphatidylcholine (lyso-PC),[19] a phospholipid that participates in several aspects of plaque formation.[22] In addition to lyso-PC, oxidized fatty acids exhibiting proatherogenic activities are liberated during hydrolysis of oxidized phospholipids by Lp-PLA$_2$.[22] Moreover, the enrichment of sdLDL with Lp-PLA$_2$ has, as a consequence, the enhanced production of lyso-PC during oxidation of this particle compared to large-light ones (the large, light LDL subspecies), in both normolipidemic and hypercholesterolemic patients. Thus, under the action of Lp-PLA$_2$, the sdLDL particles are enriched with proatherogenic lyso-PC.[23,24]

Consistent with a proatherogenic role for LDL-associated Lp-PLA$_2$ are recent *in vitro* data showing that inhibition of endogenous Lp-PLA$_2$ prior to LDL oxidation by serine esterase inhibitors or by the specific Lp-PLA$_2$ inhibitor SB222657 inhibited the rise in lyso-PC levels in oxidized LDL and diminished the capacity of oxidized LDL to induce apoptosis and toxicity in human macrophages.[25]

Overall, from the foregoing controversial results, it is not evident as to whether the equilibrium between the degradation and generation of proatherogenic phospholipids observed in studies *in vitro* of isolated LDL may favor the antiatherogenic or proatherogenic role of LDL-associated Lp-PLA$_2$. Thus, whether Lp-PLA$_2$ exhibits a pro- or antiatherogenic role *in vivo* needs further investigation.

Contrasting findings have been made *in vivo* concerning plasma Lp-PLA$_2$ activity, which mainly reflects LDL-associated Lp-PLA$_2$. Lp-PLA$_2$ activity increases gradually with age and exhibits an almost fivefold variation in healthy adult populations. About 60% of this variation can be accounted for by genetic factors.[26] Plasma LDL concentration exerts a major influence on Lp-PLA$_2$ activity. Indeed, several studies have demonstrated a strong correlation between enzyme activity and plasma LDL-cholesterol or apoB levels.[26,27] We have shown that plasma Lp-PLA$_2$ activity, and that specifically associated with LDL particles in patients with primary hypercholesterolemia without clinical evidence of CAD preferentially increases relative to that associated with HDL, in parallel with increase in the severity of hypercholesterolemia. In fact, the highest levels are seen in homozygous familial hypercholesterolemia, in which the lowest rate of LDL clearance is observed among all forms of primary hypercholesterolemia.[27] The dependence of plasma Lp-PLA$_2$ activity on LDL clearance rate is further supported by the effect of lipid-lowering therapy with HMG-CoA reductase inhibitors (statins) in patients with hyperlipidemia. We have showed that atorvastatin therapy in patients with primary hypercholesterolemia, as well as in those with combined hyperlipidemia, significantly reduces total plasma- and LDL-associated Lp-PLA$_2$ activity; this effect occurred in parallel with reduction in plasma LDL-cholesterol levels.[28] This observation is also supported by the positive correlation observed between the reduction of plasma LDL cholesterol levels and that of plasma Lp-PLA$_2$ activity. It is well documented that patients with primary hypercholesterolemia exhibit premature atherosclerosis, mainly as a result of high plasma LDL levels. The elevated plasma Lp-PLA$_2$ activity in these patients is indicative of the high plasma LDL levels, and thus, plasma Lp-PLA$_2$ may be considered as a marker of atherogenesis and cardiovascular risk.[29] Indeed, recent data from clinical trials have consistently provided evidence that plasma Lp-PLA$_2$ mass or activity could represent a potentially new marker for cardiovascular risk.

Lp-PLA$_2$ AS A NEW MARKER FOR CARDIOVASCULAR RISK

The association of Lp-PLA$_2$ with risk for CAD has been investigated in a number of different human populations. Data from large Caucasian population studies consistently report a positive association of plasma Lp-PLA$_2$ mass or activity and risk for CAD. The West of Scotland Coronary Prevention Study (WOSCOPS)[30] reported that, in men with increased plasma LDL-cholesterol levels (174–232 mg/dL), plasma levels of Lp-PLA$_2$ mass were significantly associated with development of CAD events. This association was independent of other traditional risk factors or markers of inflammation including C-reactive protein (CRP). More recently, in a prospective case-cohort study, the Atherosclerosis Risk in Communities study (ARIC), authors reported that circulating levels of Lp-PLA$_2$ mass were associated with incident CAD in apparently healthy middle-aged men and women after adjustment for age, sex, and race.[31] Similarly, in the monitoring of trends and determinants in cardiovascular disease study (MONICA),[32] increased plasma levels of Lp-PLA$_2$ mass were found to be predictive of future coronary events in apparently healthy middle-aged men with moderately increased total cholesterol levels, independently of CRP. The Rotterdam study, which measured Lp-PLA$_2$ activity instead of enzyme mass, reported an increased risk (hazard ratio 1.76) for incident CAD in individuals with high levels of Lp-PLA$_2$ activity.[33] These studies clearly demonstrate a positive association of Lp-PLA$_2$ with risk for CAD. However, whether Lp-PLA$_2$ plays a causal role in atherosclerosis or if it is simply a marker of risk needs to be investigated further.

REFERENCES

1. Kannel, W.B., Range of serum cholesterol values in the population developing coronary artery disease, *Am. J. Cardiol.*, 76, 69C, 1995.

2. Far, R.S. et al., Preliminary studies of an acid-labile factor (ALF) in human sera that inactivates platelet activating factor (PAF), *Clin. Immunol. Immunopathol.*, 15, 318, 1980.
3. Tselepis, A.D. and Chapman, M.J., Inflammation, bioactive lipids and atherosclerosis: potential roles of a lipoprotein-associated phospholipase A2, platelet activating factor-acetylhydrolase, *Atheroscler. Suppl.*, 3, 57, 2002.
4. Min, J.H. et al., Membrane-bound plasma platelet activating factor acetylhydrolase acts on substrate in the aqueous phase, *Biochemistry*, 38, 12935, 1999.
5. Tjoelker, L.W. et al., Anti-inflammatory properties of a platelet activating factor acetylhydrolase, *Nature*, 374, 549, 1995.
6. Tjoelker, L.W. et al., Plasma platelet-activating factor acetylhydrolase is a secreted phospholipase A2 with a catalytic triad, *J. Biol. Chem.*, 270, 25481, 1995.
7. Tew, D.C. et al., Purification, properties, sequencing, and cloning of a lipoprotein-associated, serine-dependent phospholipase involved in the oxidative modification of low-density lipoproteins, *Arterioscler. Thromb. Vasc. Biol.*, 16, 591, 1996.
8. Tselepis, A.D. et al., N-linked glycosylation of macrophage-derived PAF-AH is a major determinant of enzyme association with plasma HDL, *J. Lipid Res.*, 42, 1645, 2001.
9. Tselepis, A.D. et al., PAF-degrading acetylhydrolase is preferentially associated with dense LDL and VHDL-1 in human plasma. Catalytic characteristics and relation to the monocyte-derived enzyme, *Arterioscler. Thromb. Vasc. Biol.*, 15, 1764, 1995.
10. Stafforini, D.M. et al., Molecular basis of the interaction between plasma platelet-activating factor acetylhydrolase and low density lipoprotein, *J. Biol. Chem.*, 274, 7018, 1999.
11. Stafforini, D.M. et al., Human macrophages secrete platelet-activating factor acetylhydrolase, *J. Biol. Chem.*, 265, 9682, 1990.
12. Satoh, K. et al., Platelet-activating factor (PAF) stimulates the production of PAF-acetylhydrolase by the human hepatoma cell line, HepG2, *J. Clin. Invest.*, 87, 476, 1991.
13. Narahara, H., Frenkel, R.A., and Johnston, J.M., Secretion of platelet-activating factor acetylhydrolase following phorbol ester-stimulated differentiation of HL-60 cells, *Arch. Biochem. Biophys.*, 301, 275, 1993.
14. Nakajima, K.I. et al., Activated mast cells release extracellular type platelet-activating factor acetylhydrolase that contributes to autocrine inactivation of platelet-activating factor, *J. Biol. Chem.*, 272, 19708, 1997.
15. Goudevenos, J. et al., Platelet-associated and secreted PAF-acetylhydrolase activity in patients with stable angina. Sequential changes of the enzyme activity after angioplasty, *Eur. J. Clin. Invest.*, 31, 15, 2000.
16. Hakkinen, T. et al., Lipoprotein-associated phospholipase A$_2$, platelet-activating factor acetylhydrolase, is expressed by macrophages in human and rabbit atherosclerotic lesions, *Arterioscler. Thromb. Vasc. Biol.*, 19, 290, 1999.
17. Cao, Y. et al., Expression of plasma platelet-activating factor acetylhydrolase is transcriptionally regulated by mediators of inflammation, *J. Biol. Chem.*, 273, 4012, 1998.
18. Evangelou, A.M. Platelet-activating factor (PAF): implications for coronary heart and vascular diseases, *Prost. Leukot. Essent. Fatty Acids.*, 50, 1, 1994.
19. Liapikos, Th.A. et al., Platelet-activating factor formation during oxidative modification of low-density lipoprotein when PAF-acetylhydrolase has been inactivated, *Biochim. Biophys. Acta.*, 1212, 353, 1994.
20. Heery, J.M. et al., Oxidatively modified LDL contains phospholipids with platelet activating factor-like activity and stimulates the growth of smooth muscle cells, *J. Clin. Invest.*, 96, 2322, 1995.
21. Watson, A.D. et al., Effect of platelet activating factor acetylhydrolase on the formation and action of minimally oxidized low density lipoprotein, *J. Clin. Invest.*, 95, 774, 1995.
22. Macphee, C.H. et al., The lipoprotein-associated phospholipase A2 generates two bioactive products during the oxidation of low density lipoprotein: studies using a novel inhibitor, *Biochem. J.*, 338, 479, 1999.

23. Karabina, S.-A.P. et al., Distribution of PAF-acetylhydrolase activity in human plasma low-density lipoprotein subfractions, *Biochim. Biophys. Acta.*, 1213, 34, 1994.
24. Karabina, S.-A.P. et al., Increased activity of platelet-activating factor acetylhydrolase in low-density lipoprotein subfractions induces enhanced lysopbosphatidylcholine production during oxidation in patients with heterozygous familial hypercholesterolaemia, *Eur. J. Clin. Invest.*, 27, 595, 1997.
25. Carpenter, K.L.H. et al., Inhibition of lipoprotein-associated phospholipase A$_2$ diminishes the death-inducing effects of oxidized LDL on human monocyte-macrophages, *FEBS Lett.*, 505, 357, 2001.
26. Guerra, R. et al., Determinants of plasma platelet-activating factor acetylhydrolase: heritability and relationship to plasma lipoproteins, *J. Lipid Res.*, 38, 2281, 1997.
27. Tsimihodimos, V. et al., Altered distribution of PAF-acetylhydrolase activity between LDL and HDL as a function of the severity of hypercholesterolemia, *J. Lipid Res.*, 43, 256, 2002.
28. Tsimihodimos, V. et al., Atorvastatin preferentially reduces LDL-associated platelet activating factor acetylhydrolase activity in dyslipidemias of type IIA and IIB, *Arterioscler. Thromb. Vasc. Biol.*, 22, 306, 2002.
29. Elisaf, M. and Tselepis, A.D., Effect of hypolipidemic drugs on lipoprotein-associated platelet activating factor acetylhydrolase. Implication for atherosclerosis, *Biochem. Pharmacol.*, 66, 2069, 2003.
30. Packard, C.J. et al., Lipoprotein-associated phospholipase A2 as an independent predictor of coronary heart disease, *N. Engl. J. Med.*, 343, 1148, 2000.
31. Ballantyne, C.M. et al., Lipoprotein-associated phospholipase A2, high-sensitivity C-reactive protein, and risk for incident coronary heart disease in middle-aged men and women in the Atherosclerosis Risk in Communities (ARIC) Study, *Circulation*, 109, 837, 2004.
32. Koenig, W. et al., Lipoprotein-associated phospholipase A2 adds to risk prediction of incident coronary events by C-reactive protein in apparently healthy middle-aged men from the general population. Results from the 14-year follow-up of a large cohort from Southern Germany, *Circulation*, 110, 1903, 2004.
33. Oei, H.S. et al., Lipoprotein-associated phospholipase A$_2$ activity is associated with risk of coronary heart disease and ischemic stroke. The Rotterdam study, *Circulation*, 111, 570, 2005.

4 New Molecular Targets for the Prevention and Treatment of Gastrointestinal Ulcers and Inflammation

Sandor Szabo, Ganna Tolstanova, Lajos Nagy, Longchuan Chen, Tetyana Khomenko, Xiaoming Deng, and Zsuzsanna Sandor

CONTENTS

SUMMARY

Advances in basic sciences, especially in cell and molecular biology, and insights into
the pathogenesis, using animal models, provide new medicinal chemistry and pharmacologic implications for complex and multifactorial disorders such as ulcers and
inflammatory diseases of upper and lower gastrointestinal (GI) tract. Until recently,
the pathogenesis of GI ulceration and inflammation has been investigated mostly
from the point of view of aggressive factors, and therapeutic interventions affected
the healing process only indirectly. In this review we summarize mostly our data
on the prevention and healing of GI ulcers and inflammation. SH compounds, cysteine proteases, and ETs have been identified as new modulators of gastroduodenal
ulceration with pharmacologic implications. The new gastroprotective drugs derived
from animal experiments have probably even more clinical relevance, e.g., the discovery of potent gastroprotective effect of SH drugs and other antioxidants, of cysteine protease inhibitors, and especially that of pyrazole and thiazole derivatives that
exert a very long-lasting gastroprotection (e.g., 24–48 h after a single dose). We also
describe peptide or gene therapy related to angiogenic growth factors such as bFGF,
PDGF, and VEGF for direct ulcer treatment, which is now possible without affecting HCl and pepsin secretion. Studies performed in animal models and humans also
demonstrate a key role of these endogenous angiogenic peptides in ulcer healing. We
thus conclude that stimulation of cell proliferation is the most consistent mechanism
of ulcer healing by growth factors either with peptides or gene transfer. Furthermore, enhancement by VEGF of angiogenesis and granulation tissue production is
sufficient for ulcer healing. Hence, growth factors are potent, endogenously derived
anti-ulcer agents that directly stimulate ulcer healing in which angiogenesis seems
to be the most important process. In comparison with peptide growth factors, gene
therapy with single or double doses is more efficient for ulcer healing. Thus, VEGF
and PDGF gene therapies seem to be the new option to achieve a rapid ulcer healing
in the upper and lower GI tract. A bigger clinical impact may be derived from our
discovery of angiogenic steroids that accelerate the healing of experimental chronic
duodenal ulcers and ulcerative colitis. Because some of the angiosteroids have limited glucocorticoid potency, and hence anti-inflammatory action, these new steroids
might be ideal anti-IBD drugs that may be designed to exert a major angiogenic
and limited anti-inflammatory action or vice versa, or any combination of the two
effects, depending on the stages and severity of IBD.

INTRODUCTION

Recent conceptual and methodologic developments in molecular and cellular biology led to breakthroughs in the understanding of the etiology, pathogenesis, prevention, and treatment of diseases. Gene expression and gene therapy studies are
especially promising, e.g., the recent discovery of "cancer genes,"[1] and hopefully

will be followed by similar advancements in "ulcer genes" and ulcer-related transcription factors.[2] Once all the genes are identified and their bases are sequenced, it will be possible to induce virtually any human protein—valuable natural pharmaceuticals, e.g., by gene therapy, as well as new molecules designed specifically to block disease-producing proteins, e.g., by siRNA.[3,4] Based on discovering targeting genes, scientists can create new genetically determined animal models of diseases, which is the ideal approach to investigating the natural course of disease development, and then test new cures in these animal models. Only our understanding of precise molecular defects as the underlying causes of diseases will allow us to develop highly effective etiologic treatment. For instance, infectious diseases may be treated today by specific antibiotics that kill causative agents such as bacteria. These possibilities will open a new chapter in medicinal chemistry and pharmacology.

Another new emerging research development is the role of stem cells, potentially both in ulcer and cancer research. Recent investigations suggest that gastric carcinoma in *H. pylori*-infected mice may develop not from epithelial cells but from circulating bone-derived pluripotent stem cells.[5] Because both embryonic and organ-specific stem cells may be differentiated into various cell types and tissues, it is not difficult to predict that local stimulation or inhibition of stem cell production may be a novel pathway to treat ulcer or malignant tumors, respectively.

The GI tissue, containing a complex interplay of epithelial, vascular, neural, connective tissue, and smooth-muscle elements, is one of the most complicated structures of the mammalian organism. The complex structural organization is matched with a multitude of functions, e.g., digestion, secretion, absorption, and defense against a wide range of exogenous and endogenous aggressive agents, e.g., HCl, bile acids, digestive enzymes, and microbes. It is thus not surprising that the mechanisms of prevention and treatment of GI ulcers and inflammation are complex and multifactorial or pluricausal. This is one of the reasons for the virtual lack of etiologic treatment of GI diseases and for the high rate of recurrence of ulcers in both the upper and lower GI tract.

Ulcer disease, i.e., gastric and duodenal ulcers and inflammatory bowel disease (IBD), e.g., Crohn's disease and ulcerative colitis, are very prevalent disorders, especially because some of the diseases are related to stress, environmental exposure, and excessive alcohol and drug abuse, and belong to the five most costly GI diseases burdening our health-care system.[6] Despite advancements in the etiology and pathogenesis of these disorders, especially after the discovery of the role of *H. pylori* in gastroduodenal ulceration, and mutation NOD2 and IL-23 genes in the pathogenesis of IBD, the molecular mechanisms of these diseases are still poorly understood. Hence, virtually no etiologic and effective therapy is available. Furthermore, we are also acknowledging the importance of reflux esophagitis and chronic gastritis, but because of our lack of expertise in these emerging areas of pharmacologic targets, we cannot cover this topic in our focused overview.

The subject of this chapter, nevertheless, is to provide an overview of the advances in anti-ulcer and anti-inflammatory prevention and therapy of GI diseases, which have been investigated approximately during the last three decades of ulcer research, and mostly published from our laboratory—initially, from the very influential three years at the Institute of Experimental Medicine and Surgery of Dr. Hans Selye at the University of Montreal, and subsequently more than two decades at the Brigham and Women's Hospital and Harvard Medical School in Boston, MA, and from the last 12 years at the VA Medical Center, Long Beach and University of California, Irvine, CA.

In this overview, we use the two traditional concepts of preventive medicine, i.e., prevention of lesions occurrence and treating existing diseases, i.e., therapy to relieve symptoms and signs of disorders, with an expectation of accelerated healing.

PREVENTION OF GI ULCERS AND INFLAMMATION

The GI tract is one of the most susceptible of organs to environmental factors. The GI mucosa is almost permanently open and often exposed to attack by alcohol, drugs, microbial components, and other pathogens. Hans Selye, the father of biologic stress, was the first to describe the "triad of stress,"[7] and it is not surprising that severe physical and psychological stress may cause gastritis and ulceration in the upper GI tract, and according to some studies, even in the small or large intestines as well. In this context, search for new approaches to the prevention of GI lesions continues to attract more attention even after more than two decades following the introduction of the concept of gastric cytoprotection,[8] or gastroprotection in ulcer prevention, focused on the prevention of acute hemorrhagic mucosal lesions without suppressing normal gastric functions, such as acid secretion. Based on this concept, new compounds have been developed to provide protection of the GI mucosa. After the description of prostaglandin-induced gastric cytoprotection in rats by Andre Robert (former Ph.D. student of Hans Selye),[8] Paul Guth soon established that gastric cytoprotection is not unique to prostaglandins because it is also exerted by pentagastrin and cimetidine.[9] We then hypothesized that, if chemicals so different in structure and action demonstrate cytoprotection, there must be some endogenous mediators (e.g., antioxidants, sulfhydryls [SH], or certain hormones such as glucocorticoids) of this novel action of prostaglandins (Figure 4.1). This prediction proved to be correct because our laboratory soon described the gastroprotective effect of SH compounds[10] and glucocorticoids.[11]

SH COMPOUNDS

SH-containing chemicals are a group of the most widely distributed intracellular and extracellular protective agents in the organism and have been implicated in the protection against chemically induced lesions in virtually all the major organs and tissues.[10,12] The beneficial effect of SH is usually attributed to glutathione, which is the largest nonprotein SH fraction, although protein SH groups are also crucial to maintenance of cell membrane integrity and permeability as well as to release and activity of certain enzymes and regulatory peptides.[10] Glutathione and protein SH groups may directly act as antioxidants, scavengers of free radicals, and regulators of membrane integrity, and secretory and enzyme activities. In contrast, prostaglandins, which prevent hemorrhagic mucosal lesions in the stomach and may attenuate certain damage in the intestine, liver, and pancreas, cannot directly enter into protective chemical reactions, and may only indirectly initiate gastroprotective reactions

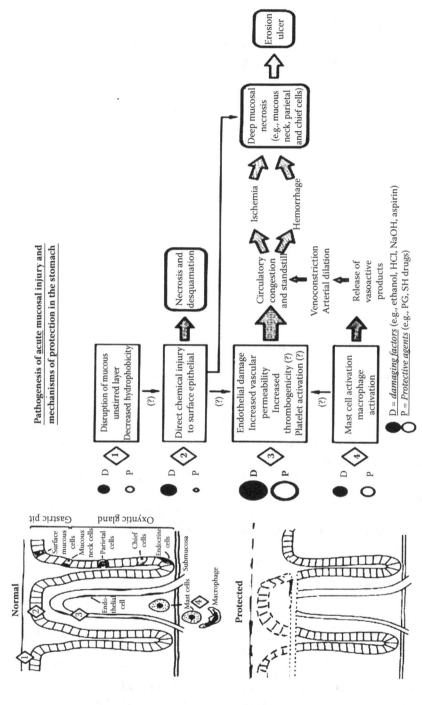

FIGURE 4.1 Schematic presentation of the pathogenesis of acute mucosal injury and mechanisms of protection in the stomach. The size of black and white dots implies significance of anatomic components in damage and protection, respectively. (Reproduced with permission from *Szabo, S. and Mozsik, G., Eds., New Pharmacology of Ulcer Disease*, Elsevier Science, 1987.)

TABLE 4.1

List of SH Compounds that Decrease Various Forms of Acute Gastric Erosions and Ulcers

Compounds with Reduced SH or SH in Various States of Oxidation	Cause of Gastric Lesions
1. N-Acetylcysteine	1. Aspirin
2. BAL or dimercaprol	2. Diclofenac
3. Cystamine	3. Ethanol
4. Cysteamine	4. Indomethacin
5. Cysteine	5. Phenylbutazone
6. Cystine	6. Piroxicam
7. Dimercaptosuccinic acid	7. Restraint and cold
8. Glutathione	
9. Methionine	
10. Penicillamine	
11. Sucralfate	
12. Thioctic (lipoic) acid	
13. Thimerosal	
14. Thiodipropionic acid	
15. Thiosulfate sodium	

Source: Szabo, S. and Mozsik, G., Eds., *New Pharmacology of Ulcer Disease*, Elsevier Science, 1987. With permission.

such as mucus and bicarbonate secretion, prevention of early vascular endothelial damage,[13] and maintenance of optimal blood flow[14,15] and smooth muscle tone.[16,17] A number of our investigations on different animal models of GI ulcers and inflammation showed the beneficial effect of both naturally occurring SH-containing amino acids (e.g., L-cysteine, L-cystine, methionine) and synthetic SH drugs that are clinically available (e.g., N-acetyl-L-cysteine/Mucomyst, dimercaprol, penicillamine) for gastroprotection (Table 4.1). This was confirmed with Mucomyst and aspirin in human volunteers. On the other hand, subcutaneously (s.c.) administered SH blockers such as iodoacetamide or N-ethylmaleimide 10 min after administration of protective agents such as prostaglandin $E_{2\beta}$ abolished gastric cytoprotection.[18]

In most of the pharmacologic studies, SH compounds were given by intragastric (i.g.) gavage or parenterally in experimental animals 30 min before administration of damaging agents as described in standard protocols for cytoprotection studies. In an attempt to evaluate efficacy of some SH compounds for preventing aspirin-induced hemorrhagic gastric erosions, SH were administered parallel with aspirin. Four SH compounds (i.e., N-acetylcysteine, methionine, sodium thiosulfate, and thiodipropionic acid) demonstrated a dose-dependent effect even in the coadministration regimen in rats.[13] These studies open new therapeutic possibilities where the protective SH compounds may be given together with aspirin-related drugs.

PYRAZOLE AND THIAZOLE DERIVATIVES

During structure–activity studies on the gastroprotective action of antioxidants, we discovered an unusually long-lasting gastroprotective effect of pyrazole and thiazole derivatives.[19,20] Pyrazole is a known inhibitor of enzyme alcohol dehydrogenase (ADH) in the GI tract and antagonist of H_2-histamine receptors. We used different animal models of gastric lesions to confirm this new gastroprotective effect of pyrazole and thiazole derivatives. The 4-methylpyrazole, 4-iodopyrazole (inhibitors of ADH), and 3,5-dimethylpyrazole (noninhibitor of ADH) were administered 30 min before 1 mL 100% ethanol by gavage. Animals were sacrificed 1 h after ethanol. The 4-methylpyrazole produced highly significant ($p < 0.05$) dose-related decreases in ethanol-induced gastric erosions, whereas 3,5-dimethylpyrazole and 4-iodopyrazole (50 mg/100 g) completely abolished the hemorrhagic lesions. It was concluded that both ADH inhibitor and noninhibitor derivatives of pyrazole exert potent gastroprotective effects against ethanol-induced hemorrhagic gastric erosions.[20]

In order to demonstrate that gastroprotection by pyrazole derivatives is not limited to alcohol-induced injury, additional agents capable of inducing gastric lesions, such as HCl and aspirin, were tested. Here, 50 mg/100 g 3-methylpyrazole was administered i.g. 0.5, 24, or 48 h prior to i.g. gavage of either 0.6 N HCl or 10 mg/100 g aspirin. Rats were sacrificed 1 h after this treatment, and gastric damage evaluated. The 3-methylpyrazole offered virtually complete protection against acid-induced damage—3-methylpyrazole given at 0.5 h and 24 h earlier reduced the damage to 0% and 1.2%, respectively, versus 2.1% induced by aspirin. It was concluded that pyrazole derivatives are effective against gastric lesions induced by acid and aspirin. Thus, the protective effect of a single dose of pyrazole derivative against gastric erosions and ulcers or alcohol lasted 24–48 h, whereas the previous gastroprotective agents (e.g., prostaglandins, SH) exerted effect only for 3–6 h.

GLUCOCORTICOIDS

Despite the extensive literature about the ulcerogenic effect of *large* doses of glucocorticoids, we postulated in the late 1970s that the effects of these steroids might be biphasic, like that of other critical natural substances (e.g., oxygen, pH, calcium). To investigate the possible role of hormones in gastric mucosal protection, the effect of prostaglandin $F_{2\beta}$ and SH-containing dimercaprol or cysteamine on ethanol-induced gastric erosions, and of cimetidine on gastric erosions caused by aspirin was studied in intact, adrenalectomized, medullectomized, ovariectomized, or thyroidectomized rats.[11] Cimetidine was administered at a low dose that did not inhibit gastric acid secretion. Adrenalectomized animals failed to exhibit the usual mucosal protective effect of prostaglandin $F_{2\beta}$, SH, or cimetidine. Ovariectomy or thyroidectomy did not influence gastroprotection by these agents. The inhibition by total adrenalectomy of mucosal protection was not reversed by large i.g. doses or by parenteral administration of prostaglandin $F_{2\beta}$. Adrenal medullectomy alone significantly diminished (by approximately one-third) ethanol-induced gastric mucosal injury; prostaglandin $F_{2\beta}$ or SH drugs produced significant additional protection. Replacement therapy with glucocorticoids (i.e., triamcinolone, corticosterone) but not with mineralocorticoids (i.e., deoxycorticosterone, 9α-fluorocortisol) restored the gastroprotective effect of prostaglandin $F_{2\beta}$ and SH in adrenalectomized rats. The generation of prostaglandin E_2-like and prostaglandin I_2-like activity in the gastric mucosa was unaltered by adrenalectomy.

These studies suggest a permissive role for glucocorticoids in gastric mucosal protection induced by prostaglandins, SH, and cimetidine, and indicate that physiologic levels of corticosterone exert gastroprotection. These implications were further investigated recently by Filaretova et al., who demonstrated that surgically or chemically induced corticosterone deficiency aggravated the ulcerogenic effect of stress and indomethacin.[21,22]

ENDOPEPTIDASES AND CYTOPROTECTION

Endopeptidases/proteolytic enzymes play a crucial role in the physiologic and pathologic events of cell and tissue metabolism. Their activity is exerted through an interaction with extracellular matrix proteins, growth factor receptors, cell adhesion molecules, and cytokines. Furthermore, aspartic-, metallo-, serine-, and cysteine (thiol) proteases may play a key role in the development of numerous diseases such as emphysema, lung inflammation, arthritis, and hematologic illnesses.[23,24] We tested the hypothesis that an imbalance between endopeptidases and their endogenous inhibitors underlies acute gastric injury.[18,25,26] These results revealed a rapid activation and release of cysteine proteases cathepsin B, L, and H in the first 5 min in luminal perfusate after 75% ethanol or ammonia (the main product of *H. pylori*, which is a recognized mediator of gastric mucosal injury). We extracted and partially isolated acid and thermostable inhibitors of cathepsin B in the gastric mucosa, and found rapid inactivation of tissue protease inhibitors and activation of cathepsin B in the early phase of ethanol- or ammonia-induced hemorrhagic mucosal lesions. Negative correlations between cysteine proteases and protease inhibitors were also found (Figure 4.2). Pretreatment with cysteine protease inhibitors (i.e., SH alkylators iodoacetate, *N*-ethylmaleimide, or butyrophenone) dose- and time-dependently decreased the ethanol-induced mucosal injury and inhibited activity of cathepsin B. Our fluorescent histochemical studies with the glandular stomach revealed cathepsin B activity in the epithelial cells in the upper and lower part of the mucosa and muscle wall. In contrast, frozen sections preincubated in iodoacetate revealed no fluorescence at all. We concluded that one of the possible directly or indirectly acting endogenous mediators of gastric mucosal injury is the cysteine proteases–endogenous protease inhibitors system. Furthermore, inhibition of cysteine proteases might be a new pharmacologic target to achieve gastroprotection.

Recently, we showed an increased proteolytic activity of another class of endopeptidases: matrix-metalloproteinase (MMP2 and MMP9) in duodenum and colonic mucosa before manifestation of GI lesions induced by cysteamine or iodoacetamide administration, respectively (unpublished data). Others detected enhanced MMPs activity during indomethacine-induced gastric ulcers.[27] Hence, MMPs might be also potentially new targets for prevention of GI lesions, especially because variations in MMPs gene were associated with the development of gastric ulcer after *H. pylori* infection.[28]

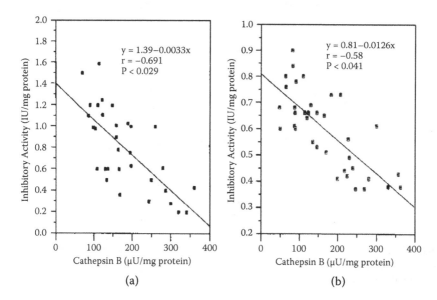

FIGURE 4.2 Correlations between cathepsin B and cysteine protease inhibitory activities in the rat gastric mucosa after 75% ethanol (a) or 1% ammonia–water (b) administration. (From Nagy, L., Kusstatcher, S., Hauschka, P.V., and Szabo, S., *J. Clin. Inv.*, 98, 1047, 1996. With permission.)

VASCULAR-RELATED COMPOUNDS

Early vascular damage is an important event in the process leading to hemorrhagic mucosal injury, and vascular lesions have been shown to be early pathogenic factors in gastric hemorrhagic erosions. Our investigations were the first to demonstrate that microvascular injury, leading to increased vascular permeability and capillary stasis, precedes the development of chemically induced hemorrhagic mucosal lesions in the stomach.[13] Vascular damage is more amenable to protection by prostaglandins and SH than the diffuse surface mucosal cell injury. Vascular and mucosal lesions may be the result of direct toxicity of damaging agents (e.g., ethanol, HCl, NaOH) and the release of vasoactive amines, leukotrienes, and endothelins (ET). Therefore, the prevention of chemically induced vascular damage in the gastric mucosa may be a new cellular and molecular target in gastroprotection.

ENDOTHELINS

One of the very early molecular or biochemical changes in experimental duodenal ulceration is the rapid local release and approximately tenfold increase in duodenal mucosal concentration of ET-1 in 15 min after cysteamine administration, much preceding the development of duodenal ulcer.[29] This was followed (in 30 min to 1 h) by rapid increase of mRNA and protein levels of immediate early genes such as egr-1. Plasma and gastric mucosal ET-1 concentrations were also elevated during ethanol-, indomethacin-, and stress-induced gastric ulceration.[30–32] Local gastric

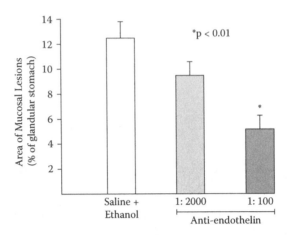

FIGURE 4.3 Effect of ET-3 antiserum on the hemorrhagic mucosal lesions induced by 75% ethanol in the rat. (From Morales, R.E., Johnson, B.R., and Szabo, S., *FASEB J.*, 6, 2354, 1992. With permission.)

ischemia-reperfusion resulted in gastric mucosal damage accompanied by increased tissue ET-1 concentration.[30] ET-1 levels were also elevated in the gastric mucosa of patients with peptic ulcer.[33] Thus, the release of ET-1 seems to be a common mechanism of gastric and duodenal ulceration.

In our investigations, intravascular infusion of ETs caused endothelial damage in gastroduodenal mucosal capillaries and venules, and sensitized the mucosa to chemically induced hemorrhagic erosions. Pretreatment of rats with the neutralizing anti-ET antibodies dose-dependently decreased the severity and incidence of cysteamine-induced duodenal ulcers as well as ethanol-induced gastric mucosal lesions[29,34] (Figure 4.3). Furthermore, pretreatment with the ET-A, -B receptors antagonist bosentan also improved signs of cysteamine-induced duodenal ulcers. In addition to inducing vasoconstriction and endothelial injury, ET-1 has a dose-dependent mitogenic effect, and acts, for the most part, as a paracrine regulator.

ETs elicit their biologic effects via the complex signaling pathways, e.g., ET-1 stimulates 1,4,5-inositoltriphosphate formation and intracellular calcium mobilization,[35,36] and activates protein kinase C and a nonreceptor tyrosine kinase.[36,37] Because neutralization of ET-1 prevented the chemically induced acute gastric and duodenal ulcers in rats, anti-ET chemicals may represent a new class of gastroprotective and antiulcer agents, operating on the basis of new pathogenic elements, i.e., affecting vascular factors and not the epithelial elements of ulceration.

LEUKOTRIENES

Leukotrienes are less potent vasoconstrictor than ET and have been implicated in the pathogenesis of gastric ulceration by our[14] and other studies.[38,39] Our studies performed in rats indicated that intra-arterial infusion of LTC4 or LTD4 in the stomach caused vascular injury as revealed by monastral blue deposition in the damaged blood vessels. Infusion of leukotrienes alone caused no hemorrhagic mucosal lesions but aggravated the damage caused by 25, 50, or 100% ethanol and 0.2 N HCl given i.g. The ethanol-induced mucosal lesions were slightly diminished by the lipoxygenase inhibitor L-651,392 and markedly decreased by eicosapentaenoic acid, which competes with arachidonic acid as a substrate for 5-lipoxygenase.[14]

These data and results from other laboratories demonstrating increased levels of leukotrienes in the gastric mucosa after administration of ethanol and decreased release after pretreatment with prostaglandins or SH-related agents suggest a mediatory role for leukotrienes in the pathogenesis of vascular injury and mucosal lesions in the stomach.

GROWTH FACTORS

Growth factors (bFGF, PDGF, VEGF) stimulate important cellular elements of ulcer healing such as angiogenesis, granulation tissues formation, and reepithelialization, but their specificity and potency vary. Here we summarize some of our original experiments demonstrating the activity of growth factors in the prevention of GI lesions and gastroprotection. Actually, our gastroprotective studies were performed *after* our discovery of the potent ulcer healing effect of angiogenic growth factors.[40–42]

PDGF was tested first for the prevention of acute ethanol-induced gastric erosions and subsequently for the acceleration of healing of indomethacin-induced gastric ulcers.[43] In our studies, groups of fasted rats were given PDGF at doses of 500 ng/100 g, 1 or 2.5 µg/100 g s.c. or by i.g. gavage, 30 min prior to the per os administration of 1 mL of 75% ethanol. As a positive control, an additional group of rats received the SH-containing taurine (50 mg/100 g). All animals were killed 1 h after receiving ethanol and the area of hemorrhagic mucosal lesions in the glandular stomach was measured by computerized stereomicroscopic planimetry. The results indicated that only 2.5 µg/100 g of PDGF administered i.g. reduced the area of acute mucosal lesions ($p < 0.05$), whereas pretreatment with taurine resulted in about 50% reduction of gastric damage ($p < 0.05$).[44,45]

bFGF administration by i.g. gavage 30 min prior to three ulcerogenic doses of cysteamine on the third day did not markedly reduce the size of acute duodenal ulcers detected in rats. More extensive dose–response studies revealed that the ulcer preventive effect of bFGF remained very modest at best.[42,44] Similar negative results were obtained with the prevention of ethanol-induced gastric mucosal lesions. Human recombinant (hr) bFGF had no effect on the development of ethanol-induced acute gastric erosions in rats when given prior to ethanol administration.[46] These results suggest that a mucosal protective effect may not be involved in the healing effect of bFGF. However, the previously mentioned hr-bFGF prevented the indomethacin-induced relapse of acetic acid gastric ulcer model in rats when given before or with indomethacin.[47,48]

VEGF/VPF, in contrast to bFGF and PDGF, is highly specific for endothelial cells, and it is the only growth factor that increases vascular permeability (hence the additional name vascular permeability factor or VPF). Our earlier experiments demonstrated a slightly increased vascular permeability in the gastric mucosa by structurally diverse gastroprotective (cytoprotective) agents (Figure 4.4).[49,50] These

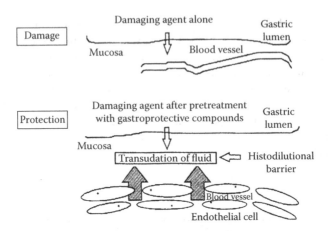

FIGURE 4.4 The histodilutional barrier as a mechanism of gastric mucosal protection. (From Szabo, S., Vincze, A., Sandor, Z., Jadus, M., Gombos, Z., Pedram, A., Levin, E., Hagar, J., and Iaquinto, G., *Dig. Dis. Sci.*, 43(Suppl), 40S, 1998. With permission.)

protective compounds produced a perivascular edema around subepithelial capillaries in the gastric mucosa and delayed the absorption of radiolabeled toxic chemicals, and hence minimized the toxic exposure of endothelial cells and markedly reduced or eliminated vascular injury, resulting in maintenance of mucosal blood flow and gastroprotection. Because of the known vascular effect of VEGF/VPF in increasing vascular permeability, we tested its role in gastroprotection. In an acute gastroprotection study, groups of fasted rats were given VEGF/VPF by gavage at 0.1 or 1 μg/100 g, and 30 min later, 1 mL of 75% ethanol. Pretreatment with VEGF/VPF significantly decreased gastric mucosal lesions. Histologically, this gastroprotection was accompanied by marked reduction or complete absence of hemorrhagic erosions and virtually complete reepithelialization in rats given 1 μg of VEGF/VPF.[29]

TREATMENT OF GASTROINTESTINAL ULCERS AND INFLAMMATION

GASTRODUODENAL ULCER

Growth Factors

Although growth factors are also new modulators of ulcer disease, these peptides are relatively large molecules (18–45 kDa) and structurally different from the small molecular mediators and modulators. Furthermore, whereas the small molecules usually exert multiple effects, the growth factors often have a single mechanism of actions, i.e., stimulating the proliferation of a single or several types of cells. Another advantage of growth-factor therapy is that they are active in nanogram quantities, and their molar potency is 2–7 million times superior to cimetidine-like antisecretory drugs. Thus, by molar potency, growth factors are among the most potent drugs in modern pharmacology, and because of these unique characteristics, their influence on GI ulcers is reviewed in this section.

bFGF

bFGF is an 18-kDa polypeptide that was first isolated under this name from the brain as a fibroblast stimulator,[51] but it was later found to be identical to the most potent heparin-binding angiogenic stimulator.[52] Indeed, bFGF is a direct mitogen for vascular endothelial cells, fibroblasts, smooth-muscle cells, certain epithelial cells, and neural cells. It has diverse roles in wound healing, tissue regeneration, and embryonic development, and probably in carcinogenesis as well. The high affinity of bFGF for the components of the extracellular matrix, such as heparin and heparan sulfate proteoglycans, might sustain a microenvironment with high availability of bFGF to the target cells.

Because biologic activity of bFGF is rapidly degraded at pH 2.0 or below, we used an acid-stable form of bFGF, produced by site-specific mutagenesis in which the second and third of the four cysteines within the peptide are replaced by serine residues (bFGF-CS23 mutein). Our laboratory was the first to demonstrate that per os treatment with either naturally occurring bFGF or its acid-resistant mutein bFGF-CS23 accelerated the healing of cysteamine-induced chronic duodenal ulcers.[53]

To induce chronic duodenal ulcers, nonfasted Sprague–Dawley female rats (160–200 g) were given cysteamine-HCl (25 mg/100 g) three times at 4-h intervals by gavage. To randomize rats with equally severe penetrating or perforated duodenal ulcers, the animals were anesthetized and laparotomized on the third day, when treatment started with saline vehicle, bFGF-wild, or bFGF-CS23, 100 ng/100 g twice daily by gavage for 21 days. At autopsy, the size of duodenal ulcers was measured and evaluated by stereomicroscopic planimetry, and histological sections were taken. Only the acid-resistant mutein bFGF-CS23 promoted a ninefold increase of angiogenesis and significantly decreased both size ($p < 0.01$ versus control) and prevalence ($p < 0.05$ versus control) of the remaining chronic ulcers. The results revealed that the ulcer-healing effect of bFGF derivatives was more potent than that of orally administered cimetidine (10 mg/100 g, twice a day for 3 weeks) (Figure 4.5). The potency of bFGF-CS23 was more than one million times greater than cimetidine on a weight-for-weight basis. Gastric acid and pepsin levels were unaffected by bFGF-CS23 when the growth factor was administered as a single dose, but when it was administered for three weeks, the volume of gastric juice and acid concentration was significantly increased.

Because of these structural similarities of heparin and the antiulcer drug sucralfate (which resembles the repeating disaccharide units of heparin), we hypothesized that bFGF may play a role in the mechanisms of action of this drug, i.e., it accelerates gastric and duodenal ulcer healing *without* reducing gastric acid secretion. We found that sucralfate bound bFGF *in vivo* and *in vitro* with higher affinity than heparin did, and that it could protect bFGF from degradation by acid and so increased bioavailability of endogenous bFGF for ulcer healing.[40,54–56] We also showed that the combined treatment of rats with bFGF and sucralftate or cimetidine at a dose of 10 mg/100 g and with low doses of bFGF-CS23 (10 and 50 ng/100 g) synergistically accelerated chronic duodenal ulcer healing.[57] Acid-stable bFGF-CS23 (25 ng/100 g) was also effective in accelerating the healing of chronic chemically induced gastritis. Surprisingly, sucraftale did not have any effect at low doses (5 mg/100 g) on

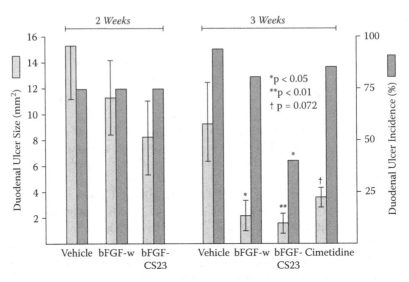

FIGURE 4.5   Effect of treatment with bFGF-w, bFGF-CS23, or cimetidine on the healing of chronic duodenal ulcers induced by cysteamine in rats ($n = 10$–14). Results are expressed as means ± SEM. Statistical comparison is with appropriate vehicle. (From Szabo, S., Folkman, J., Vattay, P., Morales, R.E., Pinkus, G.S., and Kato, K., *Gastroenterology*, 106, 1106, 1994. With permission.)

chronic gastritis when it was administered alone. However, sucraftale in combination with bFGF-CS23 repaired mucosal injury significantly more effectively than either agent alone.[45]

One of the few human studies of growth factors in GI diseases examined the use of bFGF-CS23 for healing of NSAID-associated gastric ulcers that were resistant to, or that relapsed after, conventional treatment. Five patients with nine ulcers were orally treated. After 4 weeks, four ulcers had healed, and there was significant reduction in the area of others.[58]

Novel derivatives of hr-bFGF were also tested in the treatment of experimental chronic duodenal ulcers in rats.[59] These included Ser-78,96-hr-bFGF (which is bioequivalent to hr-bFGF-CS23), CMC-hr-bFGF (a carboxymethyl cysteine derivative of hr-bFGF), and PEG-hr-bFGF (a polyethylene glycol derivative of hr-bFGF). Oral administration of these novel derivatives for 21 d accelerated the healing of cysteamine-induced chronic duodenal ulcer, and PEG-hr-bFGF was more active than the other analogues.[59]

In a series of other experiments, i.g. administration of a hr-bFGF twice a day for 2 weeks significantly accelerated the healing of acetic-acid-induced gastric ulcers in rats: the size of the ulcers decreased while regeneration of the mucosa was enhanced.[46] We also tested the hypothesis that the quality of ulcers healed after stimulation of angiogenesis, and treatment with bFGF was superior and resistant to recurrence in collaborative studies with Dr. H. Satoh,[48] i.e., both the acetic-acid-induced gastric and the nonperforated cysteamine-induced duodenal ulcers healed relatively rapidly (e.g., in 2–4 weeks), and the indomethacin-induced ulcer recurrence was decreased by about 60% in bFGF-treated rats with healed gastric or duodenal ulcers. Our data also support a physiologic role for endogenous bFGF in ulcer healing. Endogenous bFGF was visualized by immunostaining techniques in the normal rat duodenum after administration of the duodenal ulcerogen cysteamine, and bFGF immunoreactivity was time-dependently reduced in the mucosa, submucosa, and muscularis propria.[56,60] The localization of bFGF immunoreactivity was also studied in acetic-acid-induced experimental gastric ulcers in rats. bFGF binding sites, observed in the basal portion of the gastric mucosa in the control rats, were markedly increased in the regenerated tissues around the acetic-acid-induced ulcer.[61]

Western blot analysis of rat duodenal mucosa revealed a rapid depletion of mucosal cytoplasmic 18-kDa bFGF and increased synthesis of nuclear 21–25 kDa bFGF in cysteamine-induced ulceration in rats. At 12 h, we found a two- to threefold increase in nuclear forms of bFGF, with a 50% decrease in the cytoplasmic form, whereas 48 h after administration of cysteamine, the samples showed a trend toward the pattern seen in control rats. We used ELISA methods to actually measure the changes in duodenal mucosal concentration of bFGF during duodenal ulcer development in rats.[59,62] Endogenous levels of duodenal mucosal bFGF started to increase at 12 h; they reached maximal values at 48 h, when they were significantly higher than the basal value. Despite i.g. administration of the duodenal ulcerogen cysteamine, bFGF concentrations in the gastric mucosa did not change.

Studies with neutralizing anti-bFGF antibodies in cysteamine-induced chronic duodenal ulcer models also support a role for endogenous bFGF in spontaneous ulcer healing, i.e., daily injection of anti-bFGF antibodies markedly delayed the healing of experimental duodenal ulcers.[62] These results suggest a local pathogenetic role of endogenous bFGF in the natural history of duodenal ulcers and reinforce the potential therapeutic role of angiogenic growth factors to accelerate and better heal chronic GI ulcers.

Because the mechanism of delayed ulcer healing in the presence of *H. pylori* in humans is only partially understood, we postulated that decreased local bioavailability of growth factors (such as that arising after proteolytic degradation peptides or receptors) might be one of the mechanisms of poor healing of *H. pylori*-positive ulcers. Indeed, *H. pylori*, unlike other Gram-negative bacterial lysates and supernatants, decreased the bioavailability of bFGF and PDGF.[63–65] Pretreatment with nontoxic concentrations of the supernatants abolished the proliferative response of NIH 3T3 fibroblasts to bFGF and PDGF.[66] Because fibroblasts are key elements in ulcer healing, the *H. pylori*-induced impairment of proliferation and a decreased bioactivity of growth factors in the presence of the bacteria may be an important mechanism of interference of bacterial infection with ulcer healing. Results from other laboratories demonstrated that PDGF (bFGF only marginally) is susceptible to proteolytic degradation by *H. pylori* proteases.[67]

PDGF

PDGF was originally described as a product of platelets, but it is also synthesized and secreted by activated macrophages, endothelial cells, fibroblasts, and keratinocytes.[68] It consists of two disulfide-linked polypeptides: chain A (14 kDa) and chain B (17 kDa) sharing 60% similarity.[69, 70] Thus, the PDGF dimer has three isoforms:

PDGF-AA, -AB, and –BB. PDGF is a potent mitogen for a variety of cell types such as fibroblasts, osteoblasts, arterial smooth-muscle cells, and glial cells.[71] PDGF also attracts pericytes. It exerts chemotactic activity for endothelial cells *in vitro* and is angiogenic *in vivo*.[72] PDGF probably binds to heparan sulfate proteoglycans, associated with the cell surface and extracellular matrix.[73]

PDGF also stimulates protein synthesis and amino acid transport,[74] and enhances the number of low-density lipoprotein (LDL) receptors.[75] Furthermore, it stimulates the production of collagen by fibroblasts in culture, which is essential for tissue repair.[76]

Soon after recognizing the potent ulcer-healing effect of bFGF, our laboratory tested whether PDGF also accelerated the healing of experimental duodenal and gastric ulcers. Duodenal ulcer was induced by cysteamine as in the previous experiments with bFGF. That is, from the third day after ulcer induction, rats were treated by gavage with PDGF-BB at 100 or 500 ng/100 g twice daily for 3 weeks. At autopsy, ulcer size was evaluated macroscopically and with stereomicroscopic planimetry. Macroscopic and microscopic evaluations revealed that PDGF accelerated the healing of chronic cysteamine-induced ulcers. Specifically, ulcer sizes were 2.5 ± 1.1 mm^2 ($p < 0.05$) and 2.0 ± 1.4 mm^2 ($p < 0.048$), respectively, in the 100 and 500 ng/100 g groups, versus 16.9 ± 6.8 mm^2 in the control group. It is important to stress that gastric acid secretion was *not* influenced by any dose of PDGF.[77] Furthermore, the natural healing of cysteamine-induced duodenal ulcers was significantly delayed when the animals were treated daily with neutralizing monoclonal anti-PDGF antibodies.[78] Oral treatment with PDGF also dose-dependently decreased the severity of iodoacetamide-induced gastritis and accelerated the repair of gastric mucosa after indomethacin-induced damage.[43,44]

As with bFGF, ELISA, and Western blot, studies were performed to investigate the changes of PDGF concentration in duodenal mucosa during duodenal ulcer development in a rat model. The duodenal mucosal PDGF-AB concentration was elevated at 12 h and reached its peak 24 h after the administration of duodenal ulcerogen cysteamine. It decreased to the control value at 48 h after the first dose of cysteamine and decreased further during the healing phase of experimental duodenal ulcer. The PDGF elevation in the early phase of duodenal ulcer formation, however, seemed to be mainly the release of the growth factor from presynthesized pools because we could not detect PDGF mRNA expression in the duodenum.[60,78]

VEGF

VEGF is a heparin-binding glycoprotein that occurs in several molecular forms, consisting of 121, 145, 165, 183, 189, and 206 amino acids derived from the same gene by alternative mRNA splicing.[79] The VEGF family of secreted growth factors acts specifically on vascular endothelial cells to increase vascular permeability and to stimulate endothelial cell proliferation, migration, and tube formation (angiogenesis).[80–82] Its unique combination of properties is largely due to its specific receptors on endothelial cells. VEGF also increases blood flow and apparently prevents endothelial cell apoptosis. Its expression is highly regulated by hypoxia, providing a physiologic feedback mechanism to accommodate insufficient tissue oxygenation by promoting blood vessel formation.

Our previous studies demonstrated that bFGF and PDGF, which stimulate angiogenesis and granulation tissue production, accelerated experimental duodenal ulcer healing, and in this respect they are more than two million times as potent as cimetidine on a molar basis.[40,42,63] We also described in this chapter that VEGF/VPF exerts gastroprotection based on its ability to increase vascular permeability. Because VEGF/VPF exerts both acute (e.g., vascular permeability) and chronic actions (e.g., endothelial cells proliferation, migration), we have recently tested the hypothesis that stimulation of angiogenesis alone is sufficient for chronic ulcer healing. VEGF/VPF was used in cysteamine-induced chronic duodenal ulcer model, i.e., 1 μg/100 g of the peptide or vehicle was given once daily by gavage for 21 d. The results revealed that oral treatment of rats with VEGF/VPF accelerated the healing of cysteamine-induced chronic duodenal ulcers, and that this healing was accompanied by complete mucosal restoration and by stimulation of angiogenesis and granulation tissue production. Because VEGF/VPF is highly specific for endothelial cells, i.e., its receptors present only in this cell type,[83] it is unlikely to have an effect on gastric secretion.

Thus, conceptually and scientifically, one of the most important implications seems to be that the modulation of vascular factors by highly specific molecules such as VEGF/VPF appears to be sufficient for both acute gastroprotection and chronic duodenal ulcer healing. In acute gastroprotection this reinforces the vascular components such as the prevention of severe vascular injury and maintenance of mucosal blood flow, whereas in chronic ulcer healing, the angiogenesis-dependent granulation tissues that replace the necrotic debris provides sufficient scaffolding for epithelial cell migration and proliferation to complete the healing process.

GENE THERAPY AND GENE MODULATION: PDGF, VEGF

The data reviewed in the previous sections showed a potent ulcer healing and gastroprotective effect of orally administered growth factors such as bFGF, PDGF, or VEGF.[40,42,53,84,85] The healing effect of these peptides is 2–7 million times more potent, on molar basis, than that of antisecretory drugs such as histamine-2 receptor antagonists and proton-pump inhibitors.[44,86] The mechanism of action of the growth factors did not involve inhibition of gastric acid or stimulation of mucus-bicarbonate secretion. This might be related to the fact that these growth factors stimulate with varying potency virtually all the cellular elements needed for ulcer healing, e.g., epithelial cell proliferation and migration (bFGF>PDGF), fibroblast proliferation (bFGF>PDGF), and angiogenesis (VEGF>bFGF>>PDGF). Nevertheless, i.g. administration of peptide growth factors is limited by acid-proteolytic degradation in the stomach. Furthermore, large-scale production of human recombinant proteins is still an expensive process. Some of these problems may be overcome by gene transfer of the cDNA of angiogenic growth factors into the lesion directly. Gene therapy, i.e., direct injection of naked DNA (ND) and using viral vectors for gene transfer with growth factors (e.g., VEGF), has been investigated in the treatment of ischemic diseases in both animal models and clinical settings such as limb ischemia[87,88] and cardiovascular diseases.[89]

The main interest of our research work has been related to the study of vascular factors in mucosal injury and repair,[28,33,41,44,49] especially because the ensuing healing of chronic ulcers depends upon the formation of vascularized granulation tissue. Because our pharmacologic experiments demonstrated a potent ulcer healing effect of bFGF, VEGF, and PDGF peptides, we wanted to examine the effects of gene therapy with adenoviral vector (AV) and ND of VEGF and PDGF on the healing of experimental chronic duodenal ulcers in rats. That is, we hypothesized that a single dose of gene therapy related to angiogenic growth factors may be enough to accelerate the healing of duodenal ulcers through enhancement of synthesis of endogenous angiogenic growth factors. Thus, we compared the effects of intraduodenal or intravenous AV or ND of either VEGF or PDGF for transducing the genes in experimental duodenal ulcers induced by cysteamine in rats. Animals with confirmed duodenal ulcers were randomly divided into control and treatment groups. The controls received either intraduodenal injection of buffer or the ß-galactosidase-transducing AV. Rats treated with a single or double dose of AV or ND of VEGF or PDGF had significantly smaller ulcers than the controls (Figure 4.6). Histologic analysis demonstrated that reepithelized granulation tissue with prominent angiogenesis replaced the ulcers (Figure 4.7). Western blotting, immunohistochemistry, and ELISA of duodenal mucosa confirmed that the expression of VEGF or PDGF proteins was enhanced by the transgenes, whereas ß-galactosidase staining in multiple organs identified that the transgenes, especially after local administration, were only localized in the duodenum, stomach, and jejunum.[90] These results suggest that gene therapy with either VEGF or PDGF may be a rapid approach to achieving duodenal ulcer healing.

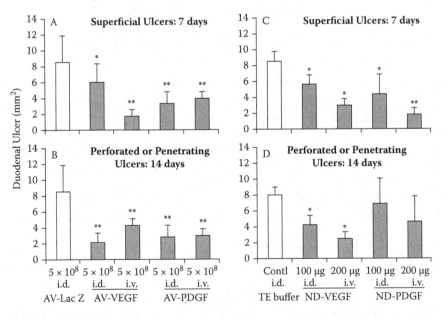

FIGURE 4.6 The size of duodenal ulcers 7 and 14 d after AV or ND of VEGF or PDGF compared with controls. *, $p < 0.05$, **, $p < 0.01$. $n = 6–12$. (From Deng, X., Szabo, S., Khomenko, T., Jadus, M.R., and Yoshida, M., *J. Pharmacol. Exp. Ther.*, 311, 982, 2004. With permission.)

MECHANISTIC IMPLICATIONS FOR THE FUTURE

Elucidation of the molecular mechanisms of ulcer healing bears a major impact on enhancing our understanding of accelerated ulcer healing; for example, unique gene therapy might be with genes encoding certain transcription factors. Our gene expression studies, initially by the Clontech DNA microarray (for about 3000 genes), and subsequently with Affymetrix gene chip (8000 genes) revealed a consistent activation of early growth response factor-1 (egr-1) and Sp1 as well as bFGF, PDGF, and VEGF genes after cysteamine administration.[2] One of the very early events in experimental duodenal ulceration is the rapid, local release of ET-1 followed by increased expression of transcription factor Egr-1 in the nuclear fraction of the duodenal mucosa.[29] We showed an overexpression of Egr-1 minutes after administration of duodenal ulcerogen cysteamine, and its expression remained elevated during 48 h. However, Sp1 expression was decreased as early as 30 min after cysteamine administration,[91] and this may be explained by replacement of Sp1 by Egr-1 in promoter regions of targeted genes.[92]

The zinc finger transcription factor Egr-1 is an immediate-early gene product that interacts with consensus GC-rich promoter regions to regulate the transcription of a diverse set of genes in response to mitogenic and nonmitogenic stimuli such as growth factors, shear stress, mechanical injury, or hypoxia. These stimuli induce the activation of ERK1/2, members of the MAPK family, leading to enhanced Egr-1 transcriptional function and activation of target genes. In turn, Egr-1 induces the expression of several target genes, including bFGF, PDGF-A, PDGF-B, and VEGF receptor-1/Flt-1.[92–94]

We hypothesized that egr-1 is a key mediator gene in the multifactorial mechanisms of duodenal ulcer development and healing because its protein, transcription factor product Egr-1, regulates the expression of angiogenic growth factors. An antisense oligonucleotide to egr-1 was used to inhibit the synthesis of Egr-1 and to determine its effect on ulcer formation in the rat model of cysteamine-induced duodenal ulceration. The egr-1 antisense oligonucleotide significantly aggravated experimental duodenal ulcers (Figure 4.8) and inhibited the expression of Egr-1 mRNA and protein as well as the synthesis of bFGF, PDGF, and VEGF in the rat duodenal mucosa.[95] Thus, the egr-1 gene and Egr-1 transcription factor seem to be important molecules in the etiology of experimental duodenal ulceration and can be a potential target for development of new therapeutics.

INFLAMMATORY BOWEL DISEASE (IBD)

Ulcerative colitis and Crohn's disease are chronic diseases of unknown etiology that are characterized by recurrent inflammation and ulcerations of intestinal or colonic mucosa and inappropriate healing.[96] The present basic therapy for IBD relies on classic anti-inflammatory and immunosuppressive drugs, such as glucocorticoids, mesalamine derivatives, azathioprine, and derivatives of the latter that vary in their ability to induce and maintain control of symptoms as well as in their tolerability and toxicities. Great progress has been made over the last decade in the development of targeted specific therapies for IBD, e.g., antibodies against TNF-α (infliximab) and

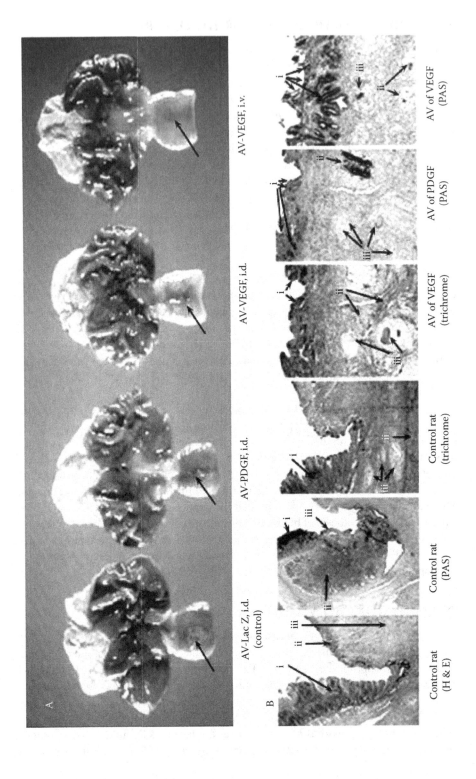

FIGURE 4.7 Panel A: The gross appearance of cysteamine-induced duodenal ulcers (arrows) after i.d. or i.v. administration of AV of LacZ (control), VEGF, or PDGF in 7 or 14 d the ulcers are reduced or not visible after gene therapy. Panel B: (a) duodenal ulcer in a control rat (hematoxylin and eosin)—the normal duodenal mucosa (i) is interrupted with a sharply demarcated ulcer crater that consists of necrotic (ii) and granulation (iii) tissue; (b) a control duodenal ulcer stained with the PAS method, which delineates mucus-secreting goblet cells (i) and Brunner glands (ii) in the submucosa. The new flat epithelium on the edge of the control ulcer does not contain mucus-secreting cells (iii); (c) another control ulcer stained with the trichrome technique—the epithelium (i) and liver (ii) (penetrating ulcer into the liver). The granulation tissue with dense collagen, (iii), and the loose granulation tissue, (iv); (d) trichrome staining of a healed duodenal ulcer after AV of VEGF—reepithelization (i) of granulation tissue that contains dense collagen (ii) and multinucleated giant cells (iii); (e) PAS staining of a completely healed duodenal ulcer 14 d after AV PDGF treatment—goblet cell-containing reepithelization (i) over the food particles (ii) and multinucleated giant cells (iii) trapped by dense granulation tissue; (f) similar, healed duodenal ulcer stained with PAS—new epithelium with mucus-secreting goblet cells (i), dense granulation tissue with food particles (ii), and multinucleated giant cells (iii). (From Deng, X., Szabo, S., Khomenko, T., Jadus, M.R., and Yoshida, M., *J. Pharmacol. Exp. Ther.*, 311, 982, 2004. With permission.)

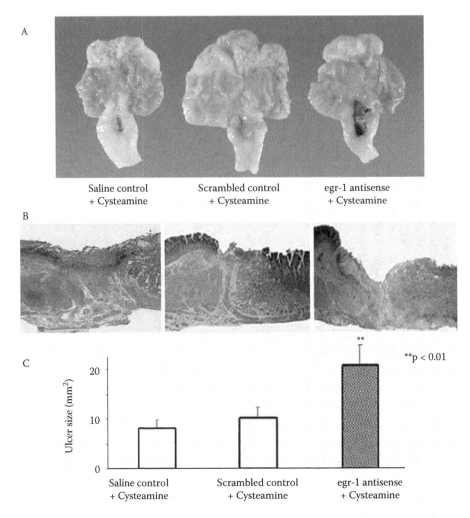

FIGURE 4.8 Effect of egr-1 antisense oligodeoxynucleotide on cysteamine-induced duodenal ulcers in rats. A: Gross appearance of duodenal ulcers 48 h after cysteamine administration. Animals were pretreated with either saline (S; $n = 10$), scrambled control oligodeoxynucleotide (SC; $n = 7$), or antisense egr-1 (phosphothioate-modified) oligodeoxynucleotide at 0.2 mg/rat (AS; $n = 10$). B: Light microscopy of duodenal ulcers in rats pretreated with either saline (showing superficial ulcer; *left*), scrambled control oligodeoxynucleotide (similar ulcer size; *middle*), or antisense egr-1 (phosphothioate-modified) oligodeoxynucleotide at 0.2 mg/rat before cysteamine (deep ulcer; *right*) with hematoxylin and eosin staining. C: Duodenal ulcer crater dimensions were measured in millimeters, and the ulcer areas were calculated using the ellipsoid formula. Sizes of duodenal ulcers are expressed as means ± SE. $**p < 0.01$. (From Khomenko, T., Szabo, S., Deng, X., Jadus, M.R., Ishikawa, H., Osapay, K., Sandor, Z., and Chen, L., *Am. J. Physiol. Gastrointest. Liver Physiol.*, 290, G1211, 2006. With permission.)

α-integrin (nataluzimab) molecules target the mechanisms of inflammation more narrowly by eliminating a specific major inflammatory cytokine or by disrupting accumulation of inflammatory cells at areas of mucosal lesions. Both strategies have added greatly to our ability to control IBD, but even this therapy is limited by lack or loss of efficacy and associated with significant complications, including fatal infections.[97,98] Thus, the main problem in IBD therapy is to avoid chronic steroid and immunosuppressant usage in an attempt to prevent disease progression and to eliminate the need for surgery, and, of course, to identify markers of disease susceptibility to prevent its expression.

Despite the historic research focus of our laboratory on ulceration and inflammation in the upper GI tract, during the last few years the molecular mechanisms of IBD have become the potential field of our research interest, especially in the context of comparative mechanisms in the pathogenesis of upper and lower GI tract lesions.

Based on our discovery of gastroprotective action of SH and other antioxidants, we developed a new animal model of IBD by intracolonic injection of SH blockers such as iodoacetamide or N-ethylmaleimide (e.g., 0.1 mL of 3–6% solutions by rubber tube, 7 cm from the anus).[99] Contrary to other chemically induced models of IBD, the SH alkylator alone is sufficient to induce these chronic inflammatory lesions, e.g., trinitrobenzene sulfonic acid (TNBS) needs to be administered in 30–50% ethanol, whereas ingestion of dextran sulfate results only in very mild and superficial colonic lesions. The spontaneously developing colitis in the IL-10 knockout mice is also a relatively mild mucosal inflammation, with very rare, if any, erosions or ulcers. The pathogenesis of SH alkylator-induced colonic lesions, in contrast to other models of colitis, is relatively well understood. It appears that iodoacetamide produces lesions by at least two pathways, involving both vascular and epithelial components: after depletion of endogenous protective glutathione in the mucosa, these chemicals alkylate cellular proteins, thus initiating irreversible cell damage.[99] Histologically, the initial colonic lesion (e.g., within 1–2 h after iodoacetamide) causes increased vascular permeability and massive mucosal edema, leading to erosions and ulcers, followed by extensive acute and chronic inflammation.[100]

PEPTIDE AND GENE THERAPY RELATED TO bFGF AND PDGF IN IBD

Recent clinical and animal studies suggest alteration and dysfunction of the intestinal microvasculature in IBD pathogenesis.[99,101–105] One of these studies showed increased microvascular density in IBD mucosa, upregulation of VEGF, bFGF, TGFβ, TNFα, and ET-1. We also detected differential changes in the serum and tissue levels of bFGF and PDGF in patients with Crohn's disease and ulcerative colitis.[106] More recently, we found a time-dependent increased expression of hypoxia-inducible transcription factor Egr-1 after iodoacetamide or TNBS administration into colonic mucosa in the rat models of IBD. Based on our previous strategy about successful treatment of gastroduodenal ulcers by increasing levels of angiogenic growth factors bFGF and PDGF, which stimulate wound healing without increasing vascular permeability, we tested this approach for IBD therapy.

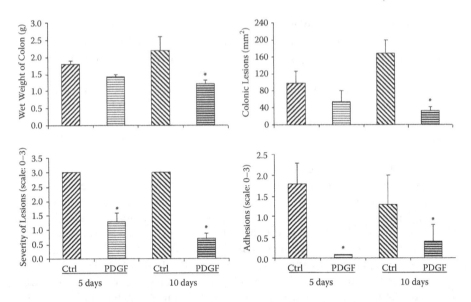

FIGURE 4.9   Effect of treatment with PDGF on the healing of ulcerative colitis induced by iodoacetamide in rats ($n = 10$–14). Results are expressed as means ± SEM. Statistical comparison is with appropriate vehicle.

The efficacy of therapy with **recombinant proteins bFGF and PDGF** was studied in the iodoacetamide-induced animal model of ulcerative colitis. Rats were given the SH alkylator iodoacetamide (6%) by rectum enema once to initiate IBD-like lesions. From the second day, bFGF or PDGF was given to the rats by rectal enema twice daily at doses of 100 ng/100 g or 500 ng/100 g. Both growth factors dose-dependently decreased the wet weight of the colon, the lesion area, and the severity of colonic ulcers as well as the number of adhesions after 10 d of treatment (Figure 4.9). Histologically, the ulcer size was smaller and the signs of inflammation were reduced, and extensive reepithelialization was seen in most cases. Our experiments demonstrated for the first time that intracolonic administration of bFGF or PDGF accelerated the healing of ulcerative colitis in rats.[45,107]

The development of colitis and rectosigmoiditis is a well-known side effect of abdominal irradiation for the treatment of certain malignant neoplasms. This so-called radiation colitis, enterocolitis, or rectosigmoiditis is a major clinical problem and relatively unresponsive to the usual therapy. We used a new animal model of radiation-induced enterocolitis for confirmation effectiveness growth factor therapy in lower GI lesions. One day after receiving abdominal 20 Gy gamma radiation, groups of rats were given sucralfate (10, 20, and 50 mg/100 g) or bFGF-CS23 (100 or 500 ng/100 g) by i.g. gavage twice daily. At autopsy on day 10, the area of the macroscopic and microscopic lesions in ileum and colon was estimated. All clinical signs and parameters were dose-dependently decreased by sucralfate. In the small doses given bFGF-CS23 exerted a beneficial effect on only a few parameters of enterocolitis. Light microscopic examination of inflamed bowel sections revealed that the lesions were reduced by both sucralfate and bFGF-CS23.[108]

Recently, our data were confirmed by two other animal models of experimental colitis: the dextran sulfate sodium model of colitis and the indomethacin–small bowel ulceration/inflammation model. Prophylactic and therapeutic administration of FGF significantly reduced the severity and extent of mucosal damage in both models. The mechanism of beneficial effect of FGF treatment was investigated *in vitro*. It demonstrated the elevated mRNA expression of COX-2, an intestinal trefoil factor; prostaglandin E_2 levels; restitution and stimulated growth in human intestinal epithelial cells; and enhanced growth of human intestinal fibroblasts after FGF administration.[109]

We also demonstrated that **gene therapy** with single or double doses of AV of PDGF accelerated the healing of ulcerative colitis induced by iodoacetamide. The area of colonic lesions, colon dilation, and thickness were statistically significantly smaller in the group with gene therapy versus control ($p < 0.05$). Histologic evaluation demonstrated marked reduction of colonic ulcers and extensive regeneration of mucosa, indicating increased healing in the AV-PDGF-treated group versus control (Figure 5.8). Gene therapy with AV-PDGF increased the levels of endogenous bFGF, PDGF, and VEGF. Thus, PDGF gene therapy seems to be a new option to achieve a rapid healing of experimental ulcerative colitis.[110]

NEW CLASS OF STEROIDS: ANGIOSTEROIDS FOR IBD

Another potential useful product of our focus on the vascular factors in ulcer pathogenesis was the discovery, in collaboration with Dr. Judah Folkman, of angiosteroids. When Folkman et. al. described in the early 1980s that cortisone plus heparin inhibited angiogenesis and experimental tumorgenesis in mice, and listed by surprise (but without explanation) that the much more potent synthetic glucocorticoid dexamethasone was less antiangiogenic than the natural cortisone,[111] we predicted that this would be yet another biphasic effect of glucocorticords and that angiogenisis modulation would be a novel, independent action of steroids. We suggested to him a brief structure–activity study with corticoids, estrogens, androgens, and other synthetic steroids (tetrahydro-S-21-Br, cortisol-21-I, cortisone-21-I, cortisone-21-Br), starting with 11-epi-cortisol (which is not a glucocorticoid, despite being almost identical to cortisol in its structure). After Folkman assigned these *in vitro* structure–activity tests to then-medical student Rosa Crum, we soon learned that 11-epi-cortisol, like cortisol or cortisone, *also* inhibited angiogenesis in the chicken egg assays. At the same time, cortisol-21-I exerted potent angiogenic activity, whereas cortisone-21-I was moderate, and cortisone-21-Br was mild. In the s.c.-implanted sponge (5 mg of steroids) assays for the quantitative assessment of granulation tissue in a week, tetrahydro-S-21-Br and cortisone-21-I increased the area of granulation tissue by 151.4 and 353.6%, respectively, and tetrahydro-S-21-Br doubled the density of blood vessels.

This led to a joint publication in *Science*,[112] reporting that steroids that lack glucocorticoid or mineralocorticoid activity inhibit angiogenesis in the presence of heparin or specific heparin fragments. This newly discovered steroid function appears to be governed by distinct structural configurations of the pregnane nucleus. Therefore, steroid compounds that influence angiogenesis independently of their endocrine effects could be classified as either angiostatic or angiogenic steroids. We subsequently demonstrated that the latter (angiogenic steroids) can accelerate the healing of cysteamine-induced chronic duodenal ulcer by stimulating angiogenesis and granulation tissue production.[113] This effect was not associated with glucocorticoid or mineralocorticoid potency or inhibition of gastric secretion. Therefore, angiogenic steroids represent a novel group of anti-ulcer agents. Because clinically there are many more anti-ulcer drugs for upper rather than lower GI tract ulcers, angiosteroids represent a novel type of antiulcer agent, especially for IBD. Furthermore, because these steroids, in addition to stimulating angiogenesis and healing, may also exert anti-inflammatory effect, they seem to be ideal candidates for the treatment of IBD.

VEGF VERSUS THE ANTI-ANGIOGENIC ANGIOSTATIN–ENDOSTATIN BALANCE: A NOVEL EXPLANATION WHY ULCERS DO NOT HEAL, AND PHARMACOLOGIC IMPLICATIONS

After our pharmacologic studies with angiogenic growth factor peptides (e.g., bFGF, PDGF, and VEGF), and with the related gene therapy, we postulated that the endogenous levels of these growth factors would be decreased in the early stages of ulcer development. To our surprise, just the opposite happened: numerous Western blot and gene expression studies revealed *elevated* levels of these growth factors in the rat duodenal mucosa after the administration of the duodenal ulcerogen cysteamine, and in the colonic mucosa during iodoacetamide-caused ulcerative colitis. This puzzle became more complicated after our most recent experiments demonstrating enhanced expression of angiogenesis inhibitors—angiostatin and endostatin—in the pathogenesis of both cysteamine-induced duodenal ulceration and iodoacetamide- or TNBS–induced ulcerative colitis.[101,114] In our pilot study we found that the colonic endostatin levels correlated with the severity of chemically induced ulcerative colitis (size of lesions) with Pearson correlation coefficient 0.85 ($p < 0.001$). Similar results were collected in the genetically determined animal model of IBD IL-10 knockout mice in whom rise of endostatin levels was also accompanied by enhanced disease progression. The biologic confirmation of our molecular biochemical results came from our pharmacologic experiments, when rats receiving daily injections of endostatin peptide (2 mg/100 g, s.c.) showed a markedly aggravated duodenal ulcer induced by cysteamine. Most rats with endostatin administration had a very extensive necrotic duodenal mucosa with large and deep ulcer craters that often perforated or penetrated into the adjacent organs such as liver or pancreas (unpublished data).

Angiostatin and endostatin are potent angiogenesis inhibitors. *In vitro*, both endostatin and angiostatin induced endothelial cell apoptosis and inhibited endothelial cell proliferation and migration.[115,116] It was shown that endostatin blocked VEGF-mediated signaling via direct interaction with VEGFR2 and thus interfered with VEGF actions.[117] Injection of angiostatin downregulated VEGF expression.[118] An *in vivo* study showed that the healing full-thickness skin wounds in mice were delayed by systemic administration of endostatin. Significant reduction in the number of functional vessels was observed in the wounds of endostatin-treated mice.[119]

Thus, we have described a new hypothesis of why ulcers develop and, especially, why they do not heal. It appears that the apparent "response to injury" resulting in the rapidly enhanced expression of angiogenic growth factors cannot achieve its goal of rapid healing because their effect is counteracted by the equally fast and robust expression of anti-angiogenic angiostatin and endostatin. Furthermore, the result is not only poor or inadequate healing, but the nonangiogenic effects of VEGF seem to become manifested, such as the increased vascular permeability (i.e., VEGF is also called "vascular permeability factor"—VPF) as well as the effect of VEGF on inflammatory cells that contain VEGF receptors, leading to enhanced and persistent inflammation, especially in ulcerative colitis. If these findings on the imbalance of angiogenic and anti-angiogenic peptides will be confirmed by other GI ulcerogens, these data will provide a truly novel explanation of the old clinical observation: the problem is not one of ulcer causation but why chemically induced ulcers do not heal (i.e., in contrast to mechanical lesions, such as after biopsy).

REFERENCES

1. Vogelstein, B. and Kinzler, K.W., Cancer genes and the pathways they control, *Nat. Med.*, 10, 789, 2004.
2. Szabo, S. et al., Gene expression and gene therapy in experimental duodenal ulceration, *J. Physiol.*, 95, 325, 2001.
3. Khabele, D. et al., Drug-induced inactivation or gene silencing of class I histone deacetylases suppresses ovarian cancer cell growth: Implications for therapy, *Cancer Biol. Ther.*, 6, Epub ahead of print, 2007.
4. Zheng, X.L., Yuan, S.G., and Peng, D.Q., Phenotype-specific inhibition of the vascular smooth muscle cell cycle by high glucose treatment, *Diabetologia*, 50, 881, 2007.
5. Houghton, J. et al., Gastric cancer originating from bone marrow-derived cells, *Science*, 306, 1568, 2004.
6. Sandler, R.S. et al., The burden of selected digestive diseases in the United States, *Gastroenterology*, 122, 1500, 2002.
7. Selye, H., A syndrome produced by diverse noxious agents, *Nature*, 138, 32, 1936.
8. Robert, A. et al., Cytoprotection by prostaglandins in rats. Prevention of gastric necrosis produced by alcohol, HCl, NaOH, hypertonic NaCl, and thermal injury, *Gastroenterology*, 77, 433, 1979.
9. Guth, P.H., Aures, D., and Paulsen, G., Topical aspirin plus HCl gastric lesions in the rat. Cytoprotective effect of prostaglandin, cimetidine, and probanthine, *Gastroenterology*, 76, 88, 1979.
10. Szabo, S., Trier, J.S., and Frankel, P.W., SH compounds may mediate gastric cytoprotection, *Science*, 214, 200, 1981.
11. Szabo, S. et al., Role of the adrenal cortex in gastric mucosal protection by prostaglandins, sulfhydryls and cimetidine in the rat, *Gastroenterology*, 85, 1384, 1983.
12. Meister, A., New aspects of glutathione biochemistry and transport: selective alteration of glutathione metabolism, *Fed. Proc.*, 43, 3031, 1984.
13. Szabo, S. et al., Protection against aspirin-induced hemorrhagic gastric erosions and mucosal vascular injury by co-administration of SH drugs, *Gastroenterology*, 88, 1604, 1985.
14. Pihan, G., Rogers, C., and Szabo, S., Vascular injury in acute gastric mucosal damage: Mediatory role of leukotrienes, *Dig. Dis. Sci.*, 33, 625, 1988.
15. Leung, F.W., Robert, A., and Guth, P.H., Gastric mucosal blood flow in rats after 16,16-dimethyl PGE_2 given at a cytoprotective dose, *Gastroenterology* 88, 1948, 1985.
16. Konturek, S.J., GI physiology and pharmacology of prostaglandins, in *New Pharmacology of Ulcer Disease*, Szabo, S. and Mozsik, G., Eds., Elsevier, New York, 1987, pp. 304–322.

17. Takeuchi, K. et al., Gastric motility changes in capsaicin-induced cytoprotection in the rat stomach, *Jpn. J. Pharmacol.*, 55, 147, 1991.
18. Szabo, S., Nagy, L., and Plebani, M., Glutathione, protein sulfydryls and cysteine proteases in gastric mucosal injury and protection, *Clin. Chim. Acta.*, 206, 95, 1992.
19. Hauser, J. and Szabo, S., Extremely long protection by pyrazole derivatives against chemically induced gastric mucosal injury, *J. Pharmacol. Exp. Ther.*, 256, 592, 1991.
20. Szabo, S., Pyrazole and thiazole derivatives for treatment of gastrointestinal ulcer disease, U.S. Patent 5,223,521, 1993.
21. Filaretova, L. et al., Various ulcerogenic stimuli are potentiated by glucocorticoid deficiency in rats, *J. Physiol. Paris*, 95, 59, 2001.
22. Filaretova, L. et al., Gastroprotective action of glucocorticoids during the formation and the healing of indomethacin-induced gastric erosions in rats, *J. Physiol. Paris*, 95, 201, 2001.
23. Im, E., Venkatakrishnan, A., and Kazlauskas, A., Cathepsin B regulates the intrinsic angiogenic threshold of endothelial cells, *Mol. Biol. Cell*, 16, 3488, 2005.
24. Kruszewski, W.J. et al., Overexpression of cathepsin B correlates with angiogenesis in colon adenocarcenoma, *Neoplasma*, 51, 38, 2004.
25. Nagy, L. et al., Role of cysteine proteases and protease inhibitors in gastric mucosal damage induced by ethanol or ammonia in the rat, *J. Clin. Inv.*, 98, 1047, 1996.
26. Nagy, L. et al., Characterization of proteases and protease inhibitors in the rat stomach, *Am. J. Physiol.*, 272, G1151, 1997.
27. Swarnakar, S. et al., Curcumin regulates expression and activity of matrix metalloproteinases 2 and 9 during prevention and healing of indomethacine-induced gastric ulcers, *J. Biol. Chem.*, 280, 9409, 2005.
28. Hellmig, S. et al., Genetic variants in matrix metalloproteinase genes are associated with development of gastric ulcer in *H. pylori* infection, *Am. J. Gastroenterol.*, 101, 29, 2006.
29. Szabo, S. et al., Vascular approach to gastroduodenal ulceration: new studies with endothelins and VEGF, *Dig. Dis. Sci.*, 43, 40S, 1998.
30. Hassan, M. et al., Gastric mucosal injury induced by local ischemia-reperfusion in rats. Role of endogenous endothelin-1 and free radical, *Dig. Dis. Sci.*, 42, 1375, 1997.
31. Michida, T. et al., Role of endothelin 1 in hemorrhagic shock-induced gastric mucosal injury in rats, *Gastroenterology*, 106, 988, 1994.
32. Matsumaru, K. et al., Bosentan, a novel synthetic mixed-type endothelin receptor antagonist, attenuates acute gastric mucosal lesions induced by indomethacin and HCI in rat: role of endothelin-1, *J. Gastoenterol.*, 32, 164, 1997.
33. Masuda, E. et al., Plasma and gastric mucosal endothelin-1 concentration in patients with peptic ulcer, *Dig. Dis. Sci.*, 42, 314, 1997.
34. Morales, R.E., Johnson, B.R., and Szabo, S., Endothelin induces vascular and mucosal lesions, enhances the injury by HCl/ethanol and the antibody exerts gastroprotection, *FASEB J.*, 6, 2354, 1992.
35. Pribnow, D. et al., Endothelin induces transcription of fos/jun family genes: a prominent role for calcium ion, *Mol. Endocrinol.*, 6, 1003, 1992.
36. Yada, Y., Higuchi, K., and Imokawa, G., Effects of endothelins on signal transduction and proliferation in human melanocytes, *J. Biol. Chem.*, 266, 18352, 1991.
37. Simonson, M.S. and Herman, W.H., Protein kinase C and protein tyrosine kinase activity contribute to mitogenic signaling by endothelin-1, *J. Biol. Chem.*, 268, 9347, 1993.
38. Hudson, N. et al., Enhanced gastric mucosal leukotriene B4 synthesis in patients taking non-steroidal anti-inflammatory drugs, *Gut*, 34, 742, 1993.
39. Kobayashi, K. and Arakawa, T., Arachidonic acid cascade and gastric mucosal injury, protection, and healing: topics of this decade, *J. Clin. Gastroenterol.*, 21, S12, 1995.
40. Folkman, J. et al., Duodenal ulcer: discovery of a new mechanism and development of angiogenic therapy that accelerates healing, *Ann. Surg.*, 214, 414, 1991.

41. Szabo, S., Pharmacological modulation of cellular, vascular and motility factors, in *Drug Therapy of GI Ulceration*, Garner, A. and Whittle, B.J.R., Eds., John Wiley, London, 1989, pp. 205–220.
42. Szabo, S. et al., Duodenal ulcerogens: Effect of FGF on cysteamine-induced duodenal ulcer, in *Mechanisms of Peptic Ulcer Healing*, Halter, F., Garner, A., and Tytgat, G.N.J., Eds., Kluwer Academic, London, 1991, pp. 139–150.
43. Guglietta, A. et al., Effect of platelet-derived growth factor–BB on indomethacin-induced gastric lesions in rats, *Scand. J. Gastroenterol.*, 27, 673, 1992.
44. Szabo, S. and Sandor, Z., Basic fibroblast growth factor and PDGF in GI diseases, in *Baillier's Clinical Gastroenterology*, Goodlad, R.A. and Wright, N.A., Eds., WB Saunders, London, 1996, pp. 97–112.
45. Szabo, S., Gombos, Z., and Sandor, Z., Growth factors in GI diseases, *BioDrugs*, 12, 27, 1999.
46. Satoh, H. et al., Effect of rhbFGF mutein CS23 (TGP-580) on the healing of gastric ulcers induced by acetic acid in rats, *Gastroenterology*, 100, A155, 1991.
47. Satoh, H. et al., Role of endogenous and exogenous bFGF in the healing of gastric ulcers in rats, *Gastroenterology*, 102, A159, 1992.
48. Satoh, H. et al., Role of endogenous and exogenous bFGF in the healing of gastric ulcers in rats, *Jpn. J. Pharmacol.*, 73, 59, 1997.
49. Pihan, G. and Szabo, S., Protection of gastric mucosa against hypertonic sodium chloride by 16,16-dimethyl prostaglandin E_2 or sodium thiosulfate in the rat: evidence for decreased mucosal penetration of damaging agent, *Dig. Dis. Sci.*, 34, 1865, 1989.
50. Lippe, I.T. and Szabo, S., New mechanism of mucosal protection: Gastroprotective prostaglandin and SHs delay the absorbtion of ethanol and 14C-aspirin from the rat stomach, *Gastroenterology*, 98, A79, 1990.
51. Gospodarowitz, D., Bialecki, H., and Greenburg, G.J., Purification of the fibroblast growth factor activity factor from bovine brain, *J. Biol. Chem.*, 253, 3736, 1978.
52. Shing, Y. et al., Heparin affinity: purification of a tumor-derived capillary endothelial call growth factor, *Science*, 223, 1296, 1984.
53. Szabo, S. et al., Accelerated healing of duodenal ulcers by oral administration of a mutein of basic fibroblast growth factor in rats, *Gastroenterology*, 106, 1106, 1994.
54. Folkman, J., Szabo, S., and Shing, Y., Sucralfate affinity for fibroblast growth factor, *J. Cell Biol.*, 111, 223a, 1990.
55. Szabo, S., The mode of action of sucralfate the 1×1×1 mechanism of action, *Scand. J. Gastroenterol.*, 26, 7, 1991.
56. Szabo, S. et al., Molecular and cellular basis of ulcer healing, *Scand. J. Gastroenterol.*, 208, 3, 1995.
57. Kusstatscher, S. et al., Additive effect of basic fibroblast growth factor (bFGF) and cimetidine on chronic duodenal ulcer healing in rats, *Gastroenterology*, 102, A104, 1993.
58. Hull, M.A. et al., Basic fibroblast growth factor treatment for non-steroidal anti-inflammatory drug associated gastric ulceration, *Gut*, 37, 610, 1995.
59. Kusstatscher, S. et al., Different molecular forms of basic fibroblast growth factor (bFGF) accelerates duodenal ulcer healing in rats, *J. Pharmacol. Exp. Ther.*, 275, 456, 1995.
60. Vincze, A. et al., ELISA and Western blot studies with basic fibroblast growth factor (bFGF) and platelet-derived growth factor (PDGF) in experimental duodenal ulceration and healing, *Inflammopharmacology*, 4, 261, 1996.
61. Nakamura, M. and Ishii, H., Increased effector site of platelet-derived growth factor-BB and basic fibroblast growth factor during acetic acid-induced gastric ulcer healing, *GI Forum—Molecular Biological Approach to Peptic Ulcer Disease*, 13, 1995.
62. Kusstatscher, S. et al., Inhibition of endogenous basic fibroblast growth factor (bFGF) delays duodenal ulcer healing in rats: implications for a physiologic role of bFGF, *Gastroenterology*, 106, A113, 1994.

63. Szabo, S. et al., Inactivation of basic fibroblast growth factor (bFGF) by gastric helico-bacters and not by *E. coli*, *Gastroenterology*, 106, A190, 1994.

64. Szabo, S. et al., *In vitro* effects of vac+ and vac- *H. pylori* on the cell proliferative response to bFGF and PDGF, in *Molecular Biological Approach to Peptic Ulcer Disease*, Terano, A., Kitajima, M., and Szabo, S., Eds., Axel Springer, Tokyo, 1996, pp. 57–60.

65. Vince, A. et al., Inhibition of cell proliferation by *Helicobacter pylori* supernatants interfere with growth and proliferative response of fibroblasts to bFGF and PDGF, *Gastroenterology*, 110, A286, 1996.

66. Vince, A. et al., Inhibition of cell proliferation by *Helicobacter pylori* supernatants and lysates might be a mechanism in delayed ulcer healing, *Gastroenterology*, 112, A323, 1997.

67. Piotrowski, J., Slomiany, A., and Slomiany, B.L., Supression of *Helicobacter pylori* protease activity towards growth factors by sulglycotide, *J. Physiol. Pharmacol.*, 48, 345, 1997.

68. Koutroubakis, I.E. et al., Role of angiogenesis in inflammatory bowel disease, *Inflamm. Bowel Dis.*, 12, 515, 2006.

69. Antoniades, H.N. and Hunkapiller, M.W., Human platelet-derived growth factor (PDGF): amino terminal amino acid sequence, *Science*, 220, 963, 1983.

70. Johnson, A. et al., The c-sis gene encodes a precursor of beta chain of platelet derived growth factor, *EMBO J.*, 3, 921, 1984.

71. Heldin, C.H. and Westermark, B., Platelet-derived growth factor: mechanism of action and possible *in vivo* function, *Cell Reg.*, 1, 555, 1990.

72. Dvorak, H.F., Angiogenesis: update, *J. Thromb. Haemost.*, 3, 1835, 2005.

73. Khachigian, L.M., Owemsby, D.A., and Chesterman, C.N., A tyrosinated peptide representing the alternatively spliced exon of platelet-derived growth factor A-chain binds specifically to cultured cells and interferes with binding to several growth factors, *J. Biol. Chem.*, 267, 1660, 1992.

74. Owen, A., Gayer, R., and Antoniades, H.N., Human platelet-derived growth factor stimulates amino acid transport and protein synthesis by human diploid fibroblast in plasma-free media, *Proc. Nat. Acad. Sci. U.S.A.*, 79, 3203, 1982.

75. Chait, A. et al., Platelet-derived growth factor stimulates activity of low density lipo-protein receptors, *Proc. Nat. Acad. Sci. U.S.A.*, 77, 4084, 1980.

76. Deuel, T.F. and Kawahara, R.S., Platelet-derived growth factor, in *Human Cytokines: Handbook for Basic and Clinical Research*, Aggarwal, B.B. and Gutterman, R.S., Eds., Blackwell, Boston, 1992, pp. 301–327.

77. Vattay, P. et al., Effect of orally administered platelet-derived growth factor (PDGF) on healing of chronic duodenal ulcers and gastric secretion in rats, *Gastroenterology*, 100, A180, 1991.

78. Vincze, A. et al., The role of endogenous basic fibroblast growth factor and platelet-derived growth factor in healing of experimental duodenal ulcers, *Z. Gastroenterol.*, 36, 452, 1998.

79. Neufeld, G. et al., Similarities and differences between the vascular endothelial growth factor (VEGF) splice variants, *Cancer Mets. Rev.*, 15, 153, 1996.

80. Dvorak, H.F. et al., Vascular permeability factor/vascular endothelial growth factor, microvascular hyperpermeability, and angiogenesis, *Am. J. Pathol.*, 146, 1029, 1995.

81. Dor, Y. and Keshet, E., Ischemia-driven angiogenesis, *Trends Cardiovasc. Med.*, 7, 289, 1997.

82. Ferrara, N. and Davis-Smyth, T., The biology of vascular endothelial growth factor, *Endocrinol. Rev.*, 18, 4, 1997.

83. Nash, A.D. et al., The biology of vascular endothelial growth factor-B (VEGF-B), *Pulm. Pharmacol. Ther.*, 19, 61, 2006.

84. Szabo, S., Folkman, J., and Vincze, A., Modulation of vascular factors by VEGF/ VPF (vascular endothelial growth factor/vascular permeability factor) is sufficient for chronic ulcer healing and acute gastroprotection, *Gastroenterology*, 112, A303, 1997.

85. Szabo, S. and Vincze, A., Growth factors in ulcer healing: lessons from recent studies, *J. Physiol.* (Paris), 94, 77, 2000.

86. Szabo, S. and Cho, C.H., From cysteamine to MPTP: structure-activity studies with duodenal ulcerogens, *Toxicol. Pathol.*, 16, 205, 1988.

87. Isner, J.M. et al., Clinical evidence of angiogenesis after arterial gene transfer of phVEGF$_{165}$ in patient with ischemic limb, *Lancet*, 348, 370, 1996.

88. Tsurumi, Y. et al., Direct intramuscular gene transfer of naked DNA encoding vascular endothelial growth factor augments collateral development and tissue perfusion, *Circulation*, 94, 3281, 1996.

89. Losordo, D.W. et al., Gene therapy for myocardial angiogenesis initial clinical results with direct myocardial injection of phVEGF$_{165}$ as sole therapy for myocardial ischemia, *Circulation*, 98, 2800, 1998.

90. Deng, X. et al., Gene therapy with adenoviral plasmids or naked DNA of vascular endothelial growth factor and platelet-derived growth factor accelerates healing of duodenal ulcer in rats, *J. Pharmacol. Exp. Ther.*, 311, 982, 2004.

91. Khomenko, T. et al., The distribution of the transcription factors Egr-1 and Sp1 in nuclear and cytoplasmic fractions of duodenal mucosa after cysteamine, *Proc. Soc. Exp. Biol. Med.*, 225, 166, 2000.

92. Silverman, E.S. and Collins, T., Pathways of Egr-1-mediated gene transcription in vascular biology, *Am. J. Pathol.*, 154, 665, 1999.

93. Khachigian, L.M. et al., Egr-1-induced endothelial gene expression: a common theme in vascular injury, *Science*, 271, 1427, 1996.

94. Vidal, F. et al., Up-regulation of vascular endothelial growth factor receptor Flt-1 after endothelial denudation: role of transcription factor Egr-1, *Blood*, 95, 3387, 2000.

95. Khomenko, T. et al., Suppression of early growth response factor-1 with egr-1 antisense ligodeoxynucleotide aggravates experimental duodenal ulcers, *Am. J. Physiol. Gastrointest. Liver Physiol.*, 290, G1211, 2006.

96. Strober, W., Fuss, I., and Mannon, P., The fundamental basis of inflammatory bowel disease, *J. Clin. Invest.*, 117, 514, 2007.

97. Chang, J.T. and Lichtenstein, G.R., Drug insight: antagonists of tumor-necrosis factor-alpha in the treatment of inflammatory bowel disease, *Nat. Clin. Pract. Gastroenterol. Hepatol.*, 3, 220, 2006.

98. Targan, S., Current limitations of IBD treatment: where do we go from here? *Ann. N. Y. Acad. Sci.*, 1072, 1, 2006.

99. Satoh, H. et al., New ulcerative colitis model induced by SH blockers in rats and the effects of antiinflammatory drugs on the colitis, *Jpn. J. Pharmacol.*, 73, 299, 1997.

100. Sandor, Z. et al., New animal model of ulcerative colitis associated with depletion of glutathione and protein SH alkylation, *Gastroenterology*, 106, A766, 1994.

101. Sandor, Z. et al., Altered angiogenic balance in ulcerative colitis: a key to impaired healing? *Biochem. Biophys. Res. Commun.*, 350, 147, 2006.

102. Danese, S. et al., Angiogenesis as a novel component of inflammatory bowel disease pathogenesis, *Gastroenetrology*, 130, 2060, 2006.

103. Chidlow, J.H. et al., Differential angiogenic regulation of experimental colitis, *Am. J. Pathol.*, 169, 2014, 2006.

104. Kanazawa, S. et al., VEGF, basic-FGF, and TGF-beta in Crohn's disease and ulcerative colitis: a novel mechanism of chronic intestinal inflammation, *Am. J. Gastroenterol.*, 96, 822, 2001.

105. Deniz, M., Cetinel, S., and Kurtel, H., Blood flow alterations in TNBS-induced colitis: role of endothelin receptors, *Inflamm. Res.*, 53, 329, 2004.

106. Sandor, Z. et al., Differential changes of basic fibroblast growth factor (bFGF) and platelet-derived growth factor (PDGF) in human serum and biopsy samples in ulcerative colitis and Crohn's disease, *Gastroenterology*, 114, G4402, 1998.

107. Sandor, Z. et al., The effect of platelet-derived growth factor on experimental inflammatory bowel disease, *Orvosi Hetilap.*, 136, 1059, 1995.

108. Szabo, S. et al., Radiation-induced enterocolitis: basic and applied science, *Eur. J. Surg.*, 582, 85, 1998.

109. Jeffers, M. et al., A novel human fibroblast growth factor treats experimental intestinal inflammation, *Gastroeterology*, 123, 1151, 2002.

110. Deng, X.M. et al., Gene therapy with adenoviral vector of PDGF increased levels of endogenous VEGF and PDGF, and accelerates healing of experimental UC in rats, *Gastroenterology*, 204, A243, 2002.

111. Folkman, J. et al., Angiogenesis inhibition and tumor regression caused by heparin or a heparin fragment in the presence of cortisone, *Science*, 221, 719, 1983.

112. Crum, R., Szabo, S., and Folkman, J., A new class of steroids inhibits angiogenesis in the presence of heparin or a heparin fragment, *Science*, 230, 1375, 1985.

113. Szabo, S. et al., Effect of a new class of steroids on healing of experimental chronic duodenal ulcers, *Gastroenterology*, 100, A171, 1991.

114. Deng, X.M. et al., The role of angiostatin and endostatin in cysteamine-induced duodenal ulceration in rats, *Gastroenterology*, 5, A976, 2005.

115. Claesson-Welsh, L. et al., Angiostatin induces endothelial cell apoptosis and activation of focal adhesion kinase independently of the integrin-binding motif RGD, *Proc. Natl. Acad. Sci. U.S.A.*, 95, 5579, 1998.

116. Taddei, L. et al., Inhibitory effect of full-length human endostatin on *in vitro* angiogenesis, *Biochem. Biophys. Res. Commun.*, 263, 340, 1999.

117. Kim, Y.-M. et al., Endostatin blocks vascular endothelial factor-mediated signaling via direct interaction with KDR/Flk-1, *J. Biol. Chem.*, 277, 27872, 2002.

118. Sima, J. et al., The effect of angiostatin on vascular leakage and VEGF expression in rat retina, *FEBS Lett.*, 564, 19, 2004.

119. Schmidt, A. et al., Endostatin influences endothelial morphology via the activated Erk ½-kinase endothelial morphology and signal transduction, *Microvasc. Res.*, 71, 152, 2006.

5 Stress Activates Corticotropin Releasing: Factor Signaling Pathways
Implication in Functional Bowel Disorders

Yvette Taché

CONTENTS

FROM HANS SELYE TO BIOCHEMICAL CODING OF STRESS RESPONSE

Hans Selye pioneered the concept of biological stress, borrowing the word *stress* from the terminology that defines stress as the interaction between a deforming force and the resistance to it. His initial publication in 1936 provided experimental evidence that the suprarenal glands, the immune system, and the gut are target organs of the stress response as shown by the hypertrophy of the adrenals, involution of the lymphatic organs, and the occurrence of gastric erosions in rats exposed to various noxious chemical or physical stimuli.[1,2] Subsequent seminal contributions by Geoffrey Harris in the 1950s established that stress-induced adrenocorticotropic hormone (ACTH) secretion involves "neural control via the hypothalamus and the hypophyseal portal vessels of the pituitary stalk."[3,4]

Biochemical support for this pathway came 15 years later when Guillemin[5] (a former Ph.D. student of Selye) and Schally's group[6] independently demonstrated the existence of hypothalamic factors that elicited ACTH release from the pituitary gland in intact rats. This substance was named corticotropin-releasing factor (CRF), in line with its stimulatory effect on ACTH release and the fact that its chemical structure was still to be identified. In 1981, Vale and his group contributed major milestones through the characterization of CRF as a 41-amino acid (aa) peptide isolated from ovine hypothalami, subsequently cloning CRF_1 and CRF_2 receptors, and developing specific peptide CRF receptor antagonists.[7] These seminal contributions opened a new era in the understanding of the biochemical coding of stress.[8] The role of CRF and CRF_1 activation in hypothalamic–pituitary–adrenal (HPA) stimulation induced by stress was further characterized using CRF antibody and CRF_1 receptor antagonists that inhibited the HPA responses to various psychological, immune, and physical stressors.[9]

In keeping with the insightful concepts Selye developed in the 1950s related to a "first mediator" that integrated the adaptive bodily response to stress,[10] further investigations revealed that CRF biological actions extended far beyond the HPA axis. CRF injected into the brain recapitulated the overall behavioral (anxiety and/or depression, feeding alterations), autonomic (sympathetic and sacral parasympathetic activation), immune, metabolic, and visceral adaptive changes induced by various systemic or cognitive stressors largely independently from the HPA stimulation.[11–16]

In particular, earlier reports in rats and dogs showed that exogenous administration of CRF into the brain or peripherally mimicked Walter Cannon's experimental observations that stress suppresses gastric acid secretion and transit.[15,17–20] In addition, we showed that blockade of CRF receptors by the central injection of a peptide CRF antagonist, α-helical CRF_{9-41}, prevented the inhibition of gastric acid and emptying in rats exposed to restraint or abdominal surgery supporting a role of CRF receptors in the alteration of gastric functions induced by these stressors.[20,21]

The central actions of CRF to suppress gastric function were mediated by the autonomic nervous system independently from the activation of HPA axis; they were prevented by autonomic blockade while hypophysectomy had no effect.[19,22] These initial findings paved the way to a number of investigations to delineate the central and peripheral sites of CRF actions, autonomic effectors, and physiological relevance of CRF receptors in the alterations of gut function by stress. They also provided a potential new venue for pharmacological interventions of stress-related development of functional bowel disorders.[23,24]

Irritable bowel syndrome (IBS) is a highly prevalent functional disorder that can affect up to 20% of the population in North America. It is characterized by chronic abdominal pain or discomfort associated with changes in bowel habits in the absence of detectable organic diseases.[25–27] The etiology of IBS reflects a multifactorial disorder with physiological and psychological components contributing to the expression of symptoms. For instance, highly stressful life events and psychosocial trauma, alone or combined with previous episodes of bowel infection/inflammation, have been identified as important risk factors in the development, severity, and/or maintenance of IBS symptoms.[28–30] Psychiatric disorders related to anxiety and depression are also common comorbidities among patients with IBS, reaching 40 to 90% in tertiary care centers.[31] Because various forms of emotional distress are important cofactors that lead to IBS symptoms,[26,32] the understanding of stress signaling pathways may have relevance for unraveling some underlying mechanisms of IBS and new modalities of interventions.

This chapter first summarizes the state of the art of the CRF signaling system including the expanding numbers of mammalian CRF-related peptides, their pharmacological characteristics on cloned CRF_1 and CRF_2 receptors and their isoforms, and the growing interests of pharmaceutical companies to develop orally active, brain penetrant, selective CRF_1 receptor antagonists. The role of CRF_1 receptors in stress-related alterations of colonic motor function and the development of visceral hypersensitivity in experimental animals and phase I clinical studies will be also presented.

CRF SIGNALING PATHWAYS: LIGANDS AND RECEPTORS

CRF LIGANDS

Within the past decade, seminal contributions to the identification of the CRF ligands and receptors greatly expanded our understanding of the biochemical coding of stress-related processes. They relate first to the discovery of CRF and related peptide members called urocortin 1 (Ucn 1), urocortin 2 (Ucn 2), and urocortin 3 (Ucn 3). In mammals, CRF is a well-conserved 41-aa peptide with identical primary structures in humans, primates, dogs, horses, and rodents.[33] In non-mammalian vertebrates, CRF-like peptides include sauvagine, a 40-aa peptide isolated from amphibian *Phyllomedua sauvagei* serous glands,[34] and urotensin-I, a 41-aa residue peptide from teleost neurosecretory systems (urophysis).[35,36]

Originally, these peptides were considered amphibian and teleost CRF orthologs in line with their 50% homologies with CRF. However, the cloning of closer homologues of CRF peptides in these species[37,38] raised the quest for possible mammalian homologs of these peptides. In 1995, Ucn 1[39] was isolated from rat midbrain as a 40-aa peptide displaying a 45% sequence identity with human and rat CRF, 44 to 63% with urotensin-I sequenced from various fish species,[33,39,40] and 35% with sauvagine.[33] The structure of Ucn 1 is also highly conserved across mammalian species: 100% identity with rat, mouse, and sheep primary sequences and 95% with rodents.[40–42]

The mismatches between the brain distribution of CRF and Ucn 1 and distribution of CRF receptors[43] led to additional searches for endogenous CRF-related agonists. The genetic approach involving screening of human and mouse genome databases for sequence homology with mammalian CRF and Ucn 1 and non-mammalian urotensin-I and sauvagine culminated in the simultaneous cloning of two novel putative CRF-related peptides by two groups.[44,45] They developed divergent nomenclatures that were recently addressed in UPHARM guidelines.[39] Mouse mUcn 2 is a 38-aa peptide that shares 34% homology with r/hCRF and 42% with r/mUcn

1,[46] while mUcn 3 and hUcn 3[45] (also named human stresscopin[44]) are more distantly related to r/hCRF, r/hUcn 1, and mUcn 1 (18 and 21% homologies, respectively).[45]

CRF RECEPTORS: CLONING AND PHARMACOLOGICAL CHARACTERISTICS

CRF ligands bind to CRF_1 and/or CRF_2 receptors cloned from two distinct genes that share 70% identity at the aa level within species homologs.[11,34] Both CRF receptor subtypes belong to the B1 subfamily of seven transmembrane domain receptors that signal largely, but not exclusively, by coupling to Gs, resulting in the stimulation of adenylate cyclase.[47] CRF_1 and CRF_2 receptors exist as multiple isoforms produced by alternative splicing of mRNA.[48,49]

Functional characterization of these isoforms shows that only $CRF_{1\alpha}$ is coupled directly to adenylate cyclase, while the majority of other variants lack ligand binding and/or signaling domains.[48,50] Soluble isoforms may play a modulatory role. CRF_{1e} decreased and CRF_{1h} amplified $CRF_{1\alpha}$-coupled cAMP production induced by Ucn 1 in transfected COS cells.[50] CRF_2 receptors exist in three functional splice variants, 2α, 2β, and 2γ, in humans and two, 2α and 2β, in rodents. They are structurally distinct in their N-terminal extracellular domains.[39] In the final translated proteins, the 34-aa N-terminal extracellular region of the $CRF_{2\alpha}$ (involved in ligand receptor interaction) is replaced by a 61-aa region in $CRF_{2\beta}$ and a 20-aa region in $CRF_{2\gamma}$. The C-terminus is common to all CRF_2 receptor splice variants.[49,51,52]

Sequence comparison revealed high level homologies of rat and mouse $CRF_{2\alpha}$ (94%) and mouse and human (92%).[51] The presence of $CRF_{2\alpha}$ in amphibian species suggests an extremely early occurrence of this splice variant during vertebrate evolution. $CRF_{2\beta}$ is less conserved, appears to be evolutionarily younger, and is found only in mammals.[34] Chen et al.[53] reported a novel $CRF_{2\alpha}$ splice variant in the mouse brain that includes the first extracellular domain of the $CRF_{2\alpha}$ receptor and acts as a soluble binding protein ($sCRF_{2\alpha}$) capable of modulating CRF family ligand activity. In rat esophagus, $CRF_{2\beta}$ wild-type transcript is predominantly expressed and in addition, several new CRF_2 splice variants including six $CRF_{2\alpha}$ isoforms were identified.[49]

CRF_1 and CRF_2 receptors differ considerably in their binding characteristics, as established by radioreceptor and functional assays. CRF_1 receptor displays high affinity to CRF and Ucn 1, but shows no appreciable binding affinity to Ucn 2 and Ucn 3.[39,54] In contrast, Ucn 1, Ucn 2, and Ucn 3 bind CRF_2 receptor with greater affinity than CRF, making this receptor subtype highly selective for Ucn signaling.[40,46,51,54,55] Although CRF is 100-fold more potent for binding CRF_1 compared to CRF_2 receptor,[56] no natural ligands with selectivity for CRF_1 receptors have yet been identified. Cortagine and stressin_1-A peptides have been developed as selective CRF_1 agonists.[57,58]

The binding characteristics of CRF receptor variants, $CRF_{2\beta}$, $CRF_{2\alpha}$, and $CRF_{2\gamma}$ isoforms, are almost identical; they show high affinity for Ucn 1, Ucn 2, and Ucn 3, and lower affinity for r/hCRF.[45,59-61] Interestingly, the mouse $sCRF_{2\alpha}$ receptor displays very low affinity for Ucn 2 and Ucn 3, while binding Ucn 1 (Ki 6.6 nM), and to lesser extent CRF (23 nM), and inhibiting the cAMP and ERK1/2-p42,44 responses to Ucn 1 and CRF.[53] In contrast, rat $CRF_{2\alpha-tr}$ binds with low affinity to CRF (Kd 12.7 nM) and does not bind to Ucn 1.[62]

CRF RECEPTOR ANTAGONISTS

Key to the assessment of the role of endogenous CRF ligands and CRF receptors in the stress response was the development of specific CRF antagonists. Earlier studies relied on the use of specific peptide CRF antagonists subtypes such as α-helical CRF_{9-41}, D-Phe$^{12}CRF_{12-41}$, and the longer acting more potent, astressin and astressin-B.[63–65] These peptides are competitive antagonists that bind to both CRF_1 and CRF_2 receptors and equally to the α, β, and γ variants of CRF_2 receptors and $sCRF_{2\alpha}$.[51,53,59,60] Recently, the antisauvagine-30 and K41498 peptides and the more potent, longer acting astressin$_2$-B analog were developed as selective CRF_2 receptor competitive antagonists.[66,67]

As a whole, these peptide antagonists generally displayed poor penetrance to the brain. We found that an intravenous (iv) injection of astressin at a dose that blocked iv CRF-induced delayed gastric emptying did not influence intracisternal (ic) CRF-induced inhibition of gastric emptying.[68] Likewise, α-helical CRF_{9-41} infused intravenously did not influence intracerebroventricular (icv) CRF-induced suppression of pentagastrin-stimulated acid secretion. When injected icv, the antagonist blocked the icv CRF action.[69]

In the context of targeting the CRF_1 system to alleviate various human pathologies including anxiety disorders, pharmaceutical companies developed a number of small hydrophobic, orally active CRF_1 antagonists that cross the blood–brain barrier (Table 5.1). They include the first developed CP-154,526[70,71] and several others.[72] The newly developed CRF_1 antagonists known as NBI 30775 (or R 121919), NBI 30545, and NBI 35965 have the distinct advantage of being water-soluble,[73–75] compared to previously developed antagonists that were not water-soluble.[72] These compounds have been largely used to establish the role of brain CRF and CRF_1 signaling path-

TABLE 5.1
Non-Peptide CRF$_1$ Antagonists: Brain Penetration upon Peripheral Administration

Name	References
CP-154,526	70, 124, 125
NBI 30775/R121919	74, 126
DMP695, DMP696, DMP904	127, 128, 129
Antalarmin	130
SSR125543A	131
NBI 27914, NBI 35965, NBI 30545	71, 73, 132
SN003	133
R278995/CRA0450	134
CRA1000, CRA1001	135

ways in stress-related activation of the HPA axis, anxiogenic behavior,[11] and visceral responses.[76]

BRAIN CRF$_1$ RECEPTORS AND STRESS-INDUCED STIMULATION OF COLONIC MOTOR FUNCTION

CRF peptides and receptors are located in the brain, the gut, and other organs. Their central nervous system distribution involves specific regions linked with anxiogenic and digestive behaviors and regulation of autonomic nervous system activity innervating the gastrointestinal tract.[72] Consistent with this distribution, convergent preclinical data point to the role of central CRF$_1$ signaling pathways in stress-related processes including those related to stimulation of colonic function.[76,77] This was established by the mimicry between the colonic propulsive response to centrally administered CRF/Ucn 1 and exposure to various stressors and the blockade of the colonic responses to stress by CRF receptor antagonists including selective CRF$_1$ antagonists.[78]

Numerous experimental studies established that CRF and Ucn 1 injected icv leads to colonic spike burst activity, acceleration of colonic transit, and defecation in freely moving rats, mice, and gerbils.[20,22,79–87] Brain nuclei responsive to CRF to stimulate colonic motor function (increased tonic and phasic colonic motility, decreased colonic transit time, and induction of watery fecal output) have been localized in the paraventricular nucleus of the hypothalamus (PVN) and pontine areas (locus coeruleus or LC and Barrington's nucleus complex),[86,88–90] that are also brain nuclei involved in central CRF-induced anxiety and depression.[91,92]

Colonic motor stimulation in response to central CRF is independent of activation of the HPA axis and brought about through the activation of sacral parasympathetic outflow and related enteric nerve activity.[22,80] This was shown by pharmacological and surgical approaches whereby chlorisondamine (a ganglionic blocker) prevented and vagotomy attenuated the stimulation of colonic transit, phasic and tonic contractions, and defecation induced by CRF injected icv or into the PVN.[22,88,89,93] Bretylium (a noradrenergic blocker) had no effect.[22] Effector mechanisms within the colon that activate propulsive colonic motor function involve parasympathetic cholinergic-mediated activation of colonic serotonin (5-HT) interacting with 5-HT$_3$ and 5-HT$_4$ receptors. This is supported by the demonstration that icv injection of CRF increased the 5-HT content in the feces of the rat proximal colon.[94] 5-HT$_3$ antagonists, granisetron, ramosteron, ondansetron, and azasetron; 5-HT$_4$ antagonist, SB-204070; and muscarinic antagonist, atropine prevented icv CRF-induced colonic motor stimulation.[85,94,95]

CRF receptor antagonists are able to block the central CRF and stress-related stimulation of colonic motor function.[76] α-Helical CRF$_{9-41}$ injected into brain cerebrospinal fluid abolished icv CRF and restrained stress-induced stimulation of colonic transit in rats.[20,85,96,97] α-Helical CRF$_{9-41}$, D-Phe^{12}CRF$_{12-41}$ and astressin injected icv inhibited defecatory response to icv CRF and wrap restraint and the increased frequency of colonic spike-bursts induced by conditioned fear stress.[20,79,80,83,98] Furthermore, a number of selective CRF$_1$ antagonists that cross the blood–brain barrier, namely CP-154,526, CRA 1000, NBI 27914, NBI 35965, antalarmin, and JTC-017 injected centrally or peripherally dampened icv CRF and various stressor-induced stimulation (via restraint, water avoidance, elevation plus maze, social intruder) of colonic motor function.[76,84,97] By contrast, the selective CRF_2 antagonist, astressin$_2$-B injected centrally at doses that blocked CRF_2 receptor-mediated inhibition of gastric emptying did not alter stress-related defecation in mice.[84,99]

Further anatomical support for the role of pontine and PVN CRF signaling pathways in the regulation of pelvic organ functions came from tracing studies showing that a proportion of CRF immunoreactive neurons in the Barrington's nucleus and PVN are linked transynaptically to the colon.[100,101] The centrally administered CRF receptor antagonists alone do not affect basal or postprandial colonic transit and motility or defecation events, indicating that central CRF receptors do not play a role in the regulation of colonic motor function under non-stress conditions.[78]

BRAIN CRF$_1$ RECEPTORS AND STRESS-RELATED HYPERSENSITIVITY

In addition to the stress-related stimulation of colonic motor function, stress-induced visceral hyperalgesia also involves the activation of central CRF_1 receptors. An initial report by Gué et al.[102] established that icv injection of CRF mimicked the stress-induced colonic hyperalgesia in male rats submitted to colorectal distention (CRD). Similarly, selective CRF_1 antagonists,CP-154,526, antalarmin, NBI 35965, and JTC-017 were effective in various models of acute[78] or delayed colonic hypersensitivity to CRD in naïve or chronically stressed rats.[75,103] Of interest is the observation that microinjection of α-helical CRF_{9-41} into the hippocampus or peripheral injection of the CRF_1 antagonist, JTC-017 reduced visceral pain induced by noxious tonic CRD along with the anxiety response to CRD in rats.[103]

The central mechanisms through which blockade of CRF_1 receptor activation influences the development of stress-related visceral hyperalgesia may involve the dampening of brain noradrenergic pathways. Recent electrophysiologic reports of anesthetized rats showed that injection of a DPhe^{12}CRF$_{12-41}$ or astressin peptide antagonist and the selective CRF_1 antagonist, NBI 35965, into the cerebrospinal fluid or directly into the LC prevented the activation of LC noradrenergic neurons induced by central injection of CRF and CRD.[104]

Consistent with electrophysiological demonstrations of LC silencing by CRF receptor antagonists, the JTC-017 CRF_1 antagonist reduced noradrenaline release in the hippocampus induced by CRD.[103] The release of noradrenaline in the cortical and limbic rostral efferent projections from the LC led to arousal and anxiogenic responses along with hypervigilance to visceral input—a common characteristic of IBS patients.[105] In the periphery, central CRF and stress may contribute to colonic hypersensitivity by activating colonic mast cells that release neurotransmitters sensitizing visceral afferents. It has been shown that central injection of CRF induced a rapid release of rat mast cell protease II, prostaglandin E_2, and histamine in the colon. In addition doxantrazole, a colonic mast cell stabilizer, prevented icv CRF and stress-induced colonic hypersensitivity to a second set of CRD, and α-helical CRF_{9-41} injected icv prevented both icv CRF and stress-induced enhancement of colonic mast cell content of histamine.[102,106,107]

CRF₁ RECEPTORS AND PATHOPHYSIOLOGY OF IBS SYMPTOMS

Experimental studies showed that exogenous CRF and Ucn 1 act in the brain to reproduce features of IBS symptoms including anxiogenic behavior, enhanced visceral pain to CRD, stimulation of colonic secretion and propulsive motility, development of watery stool/diarrhea, and increased permeability of colonic mucosa with translocation of luminal bacteria into the colonic tissue.[76,108] In addition, CRF and Ucn 1 can act directly within the colon to stimulate endocrine (serotonin containing enterochromaffin cells), immune (mucosal mast cells), and neuronal (enteric nervous system) cell types regulating colonic functions.[77] Peripheral administration of CRF or a CRF₁ selective agonist increases mucus and watery secretion from colonic epithelial cells and activates the enteric nervous system, promoting propulsive colonic motility, defecation, and diarrhea.[20,106,109–112]

Peripheral administration of CRF stimulates mast cells to release enzymes, histamine, prostaglandins, and cytokines that play a role in enhancing permeability, bacterial uptake from the lumen to the colonic tissue, and pain response.[106] Our recent studies established that colonic myenteric neurons have a distinct responsiveness to peripheral injection of CRF from those in other segments of rat gastrointestinal tract[113] based on the selective activation of cholinergic neurons in both the proximal and distal regions and nitrergic neurons more prominently in the distal colon, while enteric neurons located in the stomach and small intestine were not activated. Laser capture combined with reverse transcriptase polymerase chain reaction or immunohistochemical detection revealed that CRF₁ receptors are primarily expressed at the gene and protein levels on colonic myenteric neurons compared with other layers of the proximal colon under basal conditions in rats. The presence of functional CRF₁ receptor within the colonic myenteric neurons along with CRF₁ receptor peptide agonists, Ucn 1 and CRF, in the large intestines of several species including rats, guinea pigs, and humans supports the presence of a Ucn 1 or CRF/CRF₁ receptor signalling system in the colon that may have physiological relevance. Indeed, peripherally restricted peptide CRF antagonists blocked the stress-related alterations of colonic motility and permeability.[76,77] Preliminary evidence indicates that selective activation of CRF₁ receptors enhanced visceral pain to CRD. The issue of how alterations of the autonomic nervous system induced by stress recruit CRF ligands and CRF₁ receptor signaling in the colon remains to be resolved.

Clinical studies indicate that systemic administration of CRF induces colonic motility responses more prominently in IBS compared with controls that include clustered contractions in the descending and sigmoid colons.[114] The induction of abdominal pain and discomfort in IBS patients were not observed in healthy controls.[114] Other groups showed that the iv administration of CRF to healthy human subjects decreased visceral pain thresholds to repetitive rectal distensions and enhanced the intensity of discomfort sensation to CRD.[115,116] Recent reports indicate that CRF administered iv activates subepithelial mast cells and stimulates transcellular uptake of protein antigens in the mucosa in colonic biopsies of healthy subjects[117] as previously reported in experimental animals.[77,118] These changes may be consistent with enhanced colonic CRF signaling because increased uptake of antigen-sized macromolecules is associated with inflammation. Increasing evidence indicates that IBS patients display low grade colonic inflammation, including plasmatic cytokines (interleukin-6), intraepithelial lymphocytes, mast cell degranulation, and increased permeability. These changes may be consistent with enhanced colonic CRF signaling.

Other support for CRF signaling pathways in the pathophysiology of IBS came from a report that the peripheral injection of α-helical CRF_{9-41} prevented electrical stimulation-induced enhanced sigmoid colonic motility, visceral perception, and anxiety in IBS patients compared to healthy controls, without altering the HPA axis.[121] CRF antagonists almost normalized the altered EEG activities in IBS patients under basal and in response to CRD.[122] Consistent with preclinical studies, anatomical and biochemical studies of colonic biopsies from IBS patients support a possible role of mast cells as effectors in visceral nociception.[123]

CONCLUSIONS

Experimental investigations support the concept that activation of CRF_1 pathways recaptures cardinal features of IBS symptoms (stimulation of colonic motility, defecation and/or watery diarrhea, visceral hypersensitivity, anxiogenic or hypervigilance behavior, and mast cell activation). Further key preclinical evidence arises from pharmacological interventions whereby blockade of CRF_1 receptors prevents or attenuates the development of the above CRF- or stress-related functional or cellular alterations.[1] Selective CRF_1 antagonists abolished or reduced exogenous CRF and stress-induced anxiogenic and/or depressive behavior, stimulation of colonic motility, mucus secretion, mast cell activation, defecation, diarrhea, and pain related to colonic hypersensitivity (Table 5.2). Therefore, the conceptual framework that sustained activation of the CRF_1 system at central and/or peripheral sites may be one component underlying IBS symptoms. Targeting these mechanisms by CRF_1 receptor antagonists may provide a novel therapeutic approach for IBS patients.

TABLE 5.2
CRF_1 Receptor Agonists as Potential Approaches to Curtailing IBS Manifestations

Characteristics in IBS Patients	Preclinical Study Results
Comorbidity with anxiety and depression	CRF_1 antagonists blocked stress-related anxiety and depression
Changes in autonomic functions	Autonomic and endocrine responses
Increased bowel movements	Stimulation of colonic motility/defecation
Ion transport dysfunction	Diarrhea
Change in mast cells	Activation of mast cells
Increase barrier permeability	Increase barrier permeability
Lower pain threshold to colorectal distention	Hypersensitivity to colorectal distention

ACKNOWLEDGMENTS

The author's work was supported by the National Institute of Arthritis, Metabolism and Digestive Diseases, Grants R01 DK-33061, R01 DK-57236, DK-41301, P50 AR-049550 and Veterans' Administration Merit and Senior Scientist Awards.

REFERENCES

1. Selye, H., Syndrome produced by diverse nocuous agents, *Nature*, 138, 32, 1936.
2. Selye, H. and Collip, J.B., Fundamental factors in the interpretation of stimuli infuencing endocrine glands, *Endocrinology*, 20, 667, 1936.
3. De Groot, J. and Harris, G.W., Hypothalmic control of the anterior pituitary gland and blood lymphocytes, *J. Physiol.*, 3, 335, 1950.
4. Harris, G.W., The hypothalamus and endocrine glands, *Br. Med. Bull.*, 6, 345, 1950.
5. Guillemin, R. and Rosenberg, B., Humoral hypothalamic control of anterior pituitary: a study with combined tissue cultures, *Endocrinology*, 57, 599, 1955.
6. Saffran, M., Schally, A.V., and Benfey, B.G., Stimulation of the release of corticotropin from the adenohypophysis by a neurohypophysial factor, *Endocrinology*, 57, 439, 1955.
7. Perrin, M.H. and Vale, W.W., Corticotropin releasing factor receptors and their ligand family, *Ann. NY Acad. Sci.*, 885, 312, 1999.
8. Hillhouse, E.W. and Grammatopoulos. D.K., The molecular mechanisms underlying the regulation of the biological activity of corticotropin-releasing hormone receptors: implications for physiology and pathophysiology, *Endocr. Rev.*, 27, 260, 2006.
9. Herman, J.P. et al., Central mechanisms of stress integration: hierarchical circuitry controlling hypothalamo–pituitary–adrenocortical responsiveness, *Front. Neuroendocrinol.*, 24, 151, 2003.
10. Selye, H., Theories, in *Stress in Health and Disease*, Selye, H., Ed., Butterworths, Boston, 1976, p. 928.
11. Bale, T.L. and Vale, W.W., CRF and CRF receptor: role in stress responsivity and other behaviors. *Annu. Rev. Pharmacol. Toxicol.*, 44, 525, 2004.
12. De Souza, E.B., Corticotropin-releasing factor receptors: physiology, pharmacology, biochemistry and role in central nervous system and immune disorders, *Psychoneuroendocrinology*, 20, 789, 1995.
13. Dunn, A.J. and Berridge, C.W., Physiological and behavioral response to corticotropin-releasing factor administration: is CRF a mediator of anxiety or stress responses? *Brain Res. Rev.*, 15, 71, 1990.
14. Habib, K.E. et al., Oral administration of a corticotropin-releasing hormone receptor antagonist significantly attenuates behavioral, neuroendocrine, and autonomic responses to stress in primates, *Proc. Natl. Acad. Sci. USA*, 97, 6079, 2000.
15. Taché, Y. et al., Inhibition of gastric acid secretion in rats and in dogs by corticotropin-releasing factor, *Gastroenterology*, 86, 281, 1984.
16. Taché, Y. et al., Stress and the gastrointestinal tract III. Stress-related alterations of gut motor function: role of brain corticotropin-releasing factor receptors, *Am. J. Physiol. Gastrointest. Liver Physiol.*, 280, G173, 2001.
17. Cannon, W.B., *Bodily Changes in Pain, Hunger, Fear And Rage*, Boston, Branford, CT, 1953.
18. Taché, Y. et al., Central nervous system action of corticotropin-releasing factor to inhibit gastric emptying in rats, *Am. J. Physiol.*, 253, G241, 1987.
19. Taché, Y. et al., Inhibition of gastric acid secretion in rats by intracerebral injection of corticotropin-releasing factor, *Science*, 222, 935, 1983.

20. Williams, C.L. et al., Corticotropin-releasing factor directly mediates colonic responses to stress, *Am. J. Physiol.*, 253, G582, 1987.
21. Stephens, R.L. el al., Intracisternal injection of CRF antagonist blocks surgical stress-induced inhibition of gastric secretion in the rat, *Peptides*, 9, 1067, 1988.
22. Lenz, H.J. et al., Central nervous system effects of corticotropin-releasing factor on gastrointestinal transit in the rat, *Gastroenterology*, 94, 598, 1988.
23. Gué, M. and Buéno, L., Involvement of CNS corticotropin-releasing factor in the genesis of stress-induced gastric motor alterations, in *Nerves and the Gastrointestinal Tract*, Singer, M.V. and Goebell, H., Eds., Kluwer, Dordrecht, 1988, p. 417.
24. Taché, Y., Gunion, M.M. and Stephens, R., CRF: central nervous system action to influence gastrointestinal function and role in the gastrointestinal response to stress, in *Corticotropin-Releasing Factor: Basic and Clinical Studies of a Neuropeptide*, De Souza, E.B. and Nemeroff, C.B., Eds., CRC Press, Boca Raton, 1989, p. 300.
25. Camilleri, M., Management of the irritable bowel syndrome, *Gastroenterology*, 120, 652, 2001.
26. Halpert, A. and Drossman, D., Biopsychosocial issues in irritable bowel syndrome, *J. Clin. Gastroenterol.*, 39, 665, 2005.
27. Talley, N.J., Irritable bowel syndrome: disease definition and symptom description, *Eur. J. Surg. Suppl.*, 24, 1998.
28. Chang, L., The association of functional gastrointestinal disorders and fibromyalgia, *Eur. J. Surg. Suppl.*, 32, 1998.
29. Dunlop, S.P. et al., Relative importance of enterochromaffin cell hyperplasia, anxiety, and depression in postinfectious IBS, *Gastroenterology*, 125, 1651, 2003.
30. Kurland, J.E. et al., Prevalence of irritable bowel syndrome and depression in fibromyalgia, *Dig. Dis. Sci.*, 51, 454, 2006.
31. North, C.S., Hong, B.A. and Alpers, D.H., Relationship of functional gastrointestinal disorders and psychiatric disorders: implications for treatment, *World J. Gastroenterol.*, 13, 2020, 2007.
32. Mayer, E.A. et al., Stress and the gastrointestinal tract. V. Stress and irritable bowel syndrome, *Am. J. Physiol. Gastrointest. Liver Physiol.*, 280, G519, 2001.
33. Lovejoy, D.A. and Balment, R.J., Evolution and physiology of the corticotropin-releasing factor (CRF) family of neuropeptides in vertebrates, *Gen. Comp. Endocrinol.*, 115, 1, 1999.
34. Dautzenberg, F.M. et al., Molecular biology of the CRH receptors in the mood, *Peptides*, 22, 753, 2001.
35. Lederis, K. et al., Complete amino acid sequence of urotensin I, a hypotensive and corticotropin-releasing neuropeptide from Catostomus, *Science*, 218, 162, 1982.
36. Pallai, P.V. et al., Structural homology of corticotropin-releasing factor, sauvagine, and urotensin I: circular dichroism and prediction studies, *Proc. Natl. Acad. Sci. USA*, 80, 6770, 1983.
37. Okawara, Y. et al., Cloning and sequence analysis of cDNA for corticotropin-releasing factor precursor from the teleost fish *Catostomus commersoni*, *Proc. Natl. Acad. Sci. USA*, 85, 8439, 1988.
38. Stenzel-Poore, M.P. et al., Characterization of the genomic corticotropin-releasing releasing factor (CRF) gene from *Xenopus laevis*: two members of the CRF family exist in amphibians, *Mol. Endocrinol.*, 6, 1716, 1992.
39. Hauger, R.L. et al., International Union of Pharmacology. XXXVI. Current status of the nomenclature for receptors for corticotropin-releasing factor and their ligands, *Pharmacol. Rev.*, 55, 21, 2003.
40. Vaughan, J. et al., Urocortin, a mammalian neuropeptide related to fish urotensin I and to corticotropin-releasing factor, *Nature*, 378, 287, 1995.

41. Cepoi, D. et al., Ovine genomic urocortin: cloning, pharmacologic characterization, and distribution of central mRNA, *Brain Res. Mol. Brain Res.*, 68, 109, 1999.

42. Zhao, L. et al., The structures of the mouse and human urocortin genes (Ucn and UCN), *Genomics*, 50, 23, 1998.

43. Bittencourt, J.C. et al., Urocortin expression in rat brain: evidence against a pervasive relationship of urocortin-containing projections with targets bearing type 2 CRF receptors, J. *Comp. Neurol.*, 415, 285, 1999.

44. Hsu, S.Y. and Hsueh, A.J., Human stresscopin and stresscopin-related peptide are selective ligands for the type 2 corticotropin-releasing hormone receptor, *Nat. Med.*, 7, 605, 2001.

45. Lewis, K. et al., Identification of urocortin III, an additional member of the corticotropin-releasing factor (CRF) family with high affinity for the CRF2 receptor, *Proc. Natl. Acad. Sci. USA*, 98, 7570, 2001.

46. Reyes, T.M. et al., Urocortin II: A member of the corticotropin-releasing factor (CRF) neuropeptide family that is selectively bound by type 2 CRF receptors, *Proc. Natl. Acad. Sci. USA*, 98, 2843, 2001.

47. Grammatopoulos, D.K. and Chrousos, G.P., Functional characteristics of CRH receptors and potential clinical applications of CRH receptor antagonists, *Trends Endocrinol. Metab.* 13, 436, 2002.

48. Pisarchik, A. and Slominski, A.T., Alternative splicing of CRH-R1 receptors in human and mouse skin: identification of new variants and their differential expression, *FASEB J.*, 15, 2754, 2001.

49. Wu, S.V. et al., Identification and characterization of multiple corticotropin-releasing factor type 2 receptor isoforms in the rat esophagus, *Endocrinology*, 148, 1675, 2007.

50. Pisarchik, A. and Slominski, A., Molecular and functional characterization of novel CRFR1 isoforms from the skin, *Eur. J. Biochem.*, 271, 2821, 2004.

51. Dautzenberg, F.M. et al., Molecular cloning and functional expression of the mouse CRF2(a) receptor splice variant, *Regul. Pept.*, 121, 89, 2004.

52. Miyata, I. et al., Localization and characterization of a short isoform of the corticotropin-releasing factor receptor type 2 alpha (CRF(2) alpha-tr) in the rat brain, *Biochem. Biophys. Res. Commun.*, 280, 553, 2001.

53. Chen, A.M. et al., A soluble mouse brain splice variant of type 2 alpha corticotropin-releasing factor (CRF) receptor binds ligands and modulates their activity, *Proc. Natl. Acad. Sci. USA*, 102, 2620, 2005.

54. Dautzenberg, F.M. and Hauger, R.L., The CRF peptide family and their receptors: yet more partners discovered, *Trends Pharmacol. Sci.*, 23, 71, 2002.

55. Chalmers, D.T. et al., Corticotropin-releasing factor receptors: from molecular biology to drug design, *Trends Pharmacol. Sci.*, 17, 166, 1996.

56. Smart, D. et al., Characterisation using microphysiometry of CRF receptor pharmacology, *Eur. J. Pharmacol.*, 379, 229, 1999.

57. Rivier, J. et al., Stressin1-A, a potent corticotropin releasing factor receptor 1 (CRF1)-selective peptide agonist, *J. Med. Chem.*, 50, 1668, 2007.

58. Tezval, H. et al., Cortagine, a specific agonist of corticotropin-releasing factor receptor subtype 1, is anxiogenic and antidepressive in the mouse model, *Proc. Natl. Acad. Sci. USA*, 101, 9468, 2004.

59. Ardati. A. et al., Human CRF_2 α and β splice variants: pharmacological characterization using radioligand binding and a luciferase gene expression assay, *Neuropharmacology*, 38, 441, 1999.

60. Kostich, W.A. et al., Molecular identification and analysis of a novel human corticotropin-releasing factor (CRF) receptor: the CRF_2y receptor, *Mol. Endocrinol.*, 12, 1077, 1998.

61. Suman-Chauhan, N. et al., Expression and characterisation of human and rat CRF$_2$ alpha receptors, *Eur. J. Pharmacol.*, 379, 219, 1999.
62. Miyata, I. et al., Cloning and characterization of a short variant of the corticotropin-releasing factor receptor subtype from rat amygdala, *Biochem. Biophys. Res. Commun.*, 256, 692, 1999.
63. Hernandez, J.F. et al., Synthesis and relative potency of new constrained CRF antagonists, *J. Med. Chem.*, 36, 2860, 1993.
64. Rivier, J., Rivier, C. and Vale, W., Synthetic competitive antagonists of corticotropin-releasing factor: effect on ACTH secretion in the rat, *Science*, 224, 889, 1984.
65. Rivier, J.E. et al., Constrained corticotropin releasing factor antagonists (astressin analogues) with long duration of action in the rat, *J. Med. Chem.*, 42, 3175, 1999.
66. Rivier, J. et al., Potent and long-acting corticotropin releasing factor (CRF) receptor 2 selective peptide competitive antagonists, *J. Med. Chem.*, 45, 4737, 2002.
67. Ruhmann, A. et al., Structural requirements for peptidic antagonists of the corticotropin- releasing factor receptor (CRFR): development of CRFR2β selective antisauvagine-30, *Proc. Natl. Acad. Sci. USA*, 95, 15264, 1998.
68. Martinez, V., Rivier, J. and Taché, Y., Peripheral injection of a new corticotropin-releasing factor (CRF) antagonist, astressin, blocks peripheral CRF-and abdominal surgery-induced delayed gastric emptying in rats, *J. Pharmacol. Exp. Ther.*, 290, 629, 1999.
69. Druge, G. et al., Pathways mediating CRF-induced inhibition of gastric acid secretion in rats, *Am. J. Physiol.*, 256, G214, 1989.
70. Keller, C. et al., Brain pharmacokinetics of a nonpeptidic corticotropin-releasing factor receptor antagonist, *Drug Metab. Dispos.*, 30, 173, 2002.
71. McCarthy, J.R., Heinrichs, S.C. and Grigoriadis, D.E. Recent advances with the CRF1 receptor: design of small molecule inhibitors, receptor subtypes and clinical indications, *Curr. Pharm. Des.*, 5, 289, 1999.
72. Zorrilla, E.P. and Koob, G.F., The therapeutic potential of CRF1 antagonists for anxiety, *Expert Opin. Investig. Drugs*, 13, 799, 2004.
73. Chen, C. et al., Optimization of 3-phenylpyrazolo[1,5-a]pyrimidines as potent corticotropin-releasing factor-1 antagonists with adequate lipophilicity and water solubility, *Bioorg. Med. Chem. Lett.*, 14, 3669, 2004.
74. Heinrichs, S.C. et al., Brain penetrance, receptor occupancy and antistress *in vivo* efficacy of a small molecule corticotropin releasing factor type I receptor selective antagonist, *Neuropsychopharmacology*, 27, 194, 2002.
75. Million, M. et al., A novel water-soluble selective CRF$_1$ receptor antagonist, NBI 35965, blunts stress-induced visceral hyperalgesia and colonic motor function in rats, *Brain Res.*, 985, 32, 2003.
76. Taché, Y. and Bonaz, B., Corticotropin-releasing factor receptors and stress-related alterations of gut motor function, *J. Clin. Invest.*, 117, 33, 2007.
77. Taché, Y. and Perdue, M.H., Role of peripheral CRF signaling pathways in stress-related alterations of gut motility and mucosal function, *Neurogastroenterol. Motil.*, 16 (Suppl. 1), 1, 2004.
78. Taché, Y. et al., CRF$_1$ receptor signaling pathways are involved in stress-related alterations of colonic function and viscerosensitivity: implications for irritable bowel syndrome, *Br. J. Pharmacol.*, 141, 1321, 2004.
79. Bonaz, B. and Taché, Y., Water-avoidance stress-induced c-fos expression in the rat brain and stimulation of fecal output: role of corticotropin-releasing factor, *Brain Res.*, 641, 21, 1994.
80. Gué, M. et al., Conditioned emotional response in rats enhances colonic motility through the central release of corticotropin-releasing factor, *Gastroenterology*, 100, 964, 1991.

81. Gué, M. et al., Cholecystokinin blockade of emotional stress- and CRF-induced colonic motor alterations in rats: role of the amygdale, *Brain Res.*, 658, 232, 1994.

82. Jiménez, M. and Buéno, L., Inhibitory effects of neuropeptide Y (NPY) on CRF and stress-induced cecal motor response in rats, *Life Sci.*, 47, 205, 1990.

83. Martinez, V. et al., Central injection of a new corticotropin-releasing factor (CRF) antagonist, astressin, blocks CRF- and stress-related alterations of gastric and colonic motor function, *J. Pharmacol. Exp. Ther.*, 280, 754, 1997.

84. Martinez, V. et al., Central CRF, urocortins and stress increase colonic transit via CRF1 receptors while activation of CRF2 receptors delays gastric transit in mice, *J. Physiol.*, 5561, 221, 2004.

85. Miyata, K., Ito, H. and Fukudo, S., Involvement of the 5-HT3 receptor in CRH-induced defecation in rats, *Am. J. Physiol.*, 274, G827, 1998.

86. Mönnikes, H. et al., Microinfusion of corticotropin releasing factor into the locus coeruleus/subcoeruleus stimulates colonic motor function in rats, *Brain Res.*, 644, 101, 1994.

87. Okano, S. et al., Effects of TAK-637, a novel neurokinin-1 receptor antagonist, on colonic function in vivo, J. *Pharmacol. Exp. Ther.*, 298, 559, 2001.

88. Mönnikes, H., Schmidt, B.G. and Taché, Y., Psychological stress-induced accelerated colonic transit in rats involves hypothalamic corticotropin-releasing factor, *Gastroenterology*, 104, 716, 1993.

89. Mönnikes, H. et al., CRF in the paraventricular nucleus mediates gastric and colonic motor response to restraint stress, *Am. J. Physiol.*, 262, G137, 1992.

90. Mönnikes, H. et al., CRF in the paraventricular nucleus of the hypothalamus stimulates colonic motor activity in fasted rats, *Peptides*, 14, 743, 1993.

91. Mönnikes, H., Heymann-Monnikes, I. and Taché, Y., CRF in the paraventricular nucleus of the hypothalamus induces dose-related behavioral profile in rats, *Brain Res.*, 574, 70, 1992.

92. Weiss, J.M. et al., Depression and anxiety: role of the locus coeruleus and corticotropin-releasing factor, *Brain Res. Bull.*, 35, 561, 1994.

93. Tebbe, J.J. et al., Excitatory stimulation of neurons in the arcuate nucleus intitiates central CRF-dependent stimulation of colonic propulsion in rats, *Brain Res.*, 1036, 130, 2005.

94. Nakade, Y. et al., Restraint stress stimulates colonic motility via central corticotropin-releasing factor and peripheral 5-HT3 receptors in conscious rats, *Am. J. Physiol. Gastrointest. Liver Physiol.*, 292, G1037, 2007.

95. Ataka, K. et al., Wood creosote prevents CRF-induced motility via 5-HT3 receptors in proximal and 5-HT4 receptors in distal colon in rats, *Auton Neurosci.*, 133, 136, 2007.

96. Lenz, H.J. et al., Stress-induced gastrointestinal secretory and motor responses in rats are mediated by endogenous corticotropin-releasing factor, *Gastroenterology*, 95, 1510, 1988.

97. Martinez, V. and Taché, Y., Role of CRF receptor 1 in central CRF-induced stimulation of colonic propulsion in rats, *Brain Res.*, 893, 29, 2001.

98. Million, M. et al., Differential Fos expression in the paraventricular nucleus of the hypothalamus, sacral parasympathetic nucleus and colonic motor response to water avoidance stress in Fischer and Lewis rats, *Brain Res.*, 877, 345, 2000.

99. Taché, Y. et al., Role of corticotropin releasing factor receptor subtype 1 in stress-related functional colonic alterations: implications in irritable bowel syndrome, *Eur. J. Surg.*, 168 (Suppl. 587), 16, 2002.

100. Valentino, R.J. et al., Transneuronal labeling from the rat distal colon: Anatomic evidence for regulation of distal colon function by a pontine corticotropin-releasing factor system, *J. Comp. Neurol.* 417, 399, 2000.

101. Rouzade-Dominguez, M.L. et al., Convergent responses of Barrington's nucleus neurons to pelvic visceral stimuli in the rat: A juxtacellular labelling study, *Eur. J. Neurosci.*, 18, 3325, 2003.
102. Gué, M. et al., Stress-induced visceral hypersensitivity to rectal distension in rats: role of CRF and mast cells, *Neurogastroenterol. Motil.*, 9, 271, 1997.
103. Saito, K. et al., Corticotropin-releasing hormone receptor 1 antagonist blocks brain–gut activation induced by colonic distention in rats, *Gastroenterology*, 129, 1533, 2005.
104. Kosoyan, H., Grigoriadis, D. and Taché, Y., The CRF1 antagonist, NBI-35965 abolished the activation of locus coeruleus by colorectal distention and intracisternal CRF in rats, *Brain Res.* 1056, 85, 2004.
105. Naliboff, B.D. et al., Evidence for two distinct perceptual alterations in irritable bowel syndrome, *Gut*, 41, 505, 1997.
106. Castagliuolo, I. et al., Acute stress causes mucin release from rat colon: role of corticotropin releasing factor and mast cells, *Am. J. Physiol.*, 271, G884, 1996.
107. Eutamene, H. et al., Acute stress modulates the histamine content of mast cells in the gastrointestinal tract through interleukin-1 and corticotropin-releasing factor release in rats, *J. Physiol.*, 553, 959, 2003.
108. Martinez, V. and Taché, Y., CRF_1 receptors as a therapeutic target for irritable bowel syndrome, *Curr. Pharm. Des.*, 12, 1, 2006.
109. Maillot, C. et al., Peripheral corticotropin-releasing factor and stress-stimulated colonic motor activity involve type 1 receptor in rats, *Gastroenterology*, 119, 1569, 2000.
110. Martinez, V. et al., Differential actions of peripheral corticotropin-releasing factor (CRF), urocortin II, and urocortin III on gastric emptying and colonic transit in mice: role of CRF receptor subtypes 1 and 2, *J. Pharmacol. Exp. Ther.*, 301, 611, 2002.
111. Miampamba, M. et al., Peripheral CRF activates myenteric neurons in the proximal colon through CRF_1 receptor in conscious rats, *Am. J. Physiol Gastrointest. Liver Physiol.*, 282, G857, 2002.
112. Saunders, P.R. et al., Peripheral corticotropin-releasing factor induces diarrhea in rats: role of CRF_1 receptor in fecal watery excretion, *Eur. J. Pharmacol.*, 435, 231, 2002.
113. Yuan, P.-Q. et al., Peripheral corticotropin releasing factor (CRF) and a novel CRF1 receptor agonist, stressin1-A activate CRF1 receptor expressing cholinergic and nitrergic myenteric neurons selectively in the colon of conscious rats, *Neurogastroenterol. Motil.*, 19, 923, 2007.
114. Fukudo, S. et al., Impact of corticotropin-releasing hormone on gastrointestinal motility and adrenocorticotropic hormone in normal controls and patients with irritable bowel syndrome, *Gut*, 42, 845, 1998.
115. Lembo, T. et al., Effects of the corticotropin-releasing factor (CRF) on rectal afferent nerves in humans, *Neurogastroenterol. Motil.*, 8, 9, 1996.
116. Nozu, T. and Kudaira, M., Corticotropin-releasing factor induces rectal hypersensitivity after repetitive painful rectal distention in healthy humans, J. Gastroenterol., 41, 740, 2006.
117. Wallon, C. et al., Corticotropin releasing hormone (CRH) regulates macromolecular permeability via mast cells in normal human colonic biopsies *in vitro*, *Gut*, 2007 (Epub ahead of print).
118. Barreau, F. et al., Pathways involved in gut mucosal barrier dysfunction induced in adult rats by maternal deprivation: corticotrophin-releasing factor and nerve growth factor interplay, *J. Physiol.*, 580, 347, 2007.
119. Chadwick, V.S. et al., Activation of the mucosal immune system in irritable bowel syndrome, *Gastroenterology*, 122, 1778, 2002.

120. Dinan, T.G. et al., Hypothalamic-pituitary-gut axis dysregulation in irritable bowel syndrome: plasma cytokines as a potential biomarker? *Gastroenterology*, 130, 304, 2006.

121. Sagami, Y. et al., Effect of a corticotropin releasing hormone receptor antagonist on colonic sensory and motor function in patients with irritable bowel syndrome, Gut, 53, 958, 2004.

122. Tayama, J. et al., Effect of alpha-helical CRH on quantitative electroencephalogram in patients with irritable bowel syndrome, *Neurogastroenterol. Motil.*, 19, 471, 2007.

123. Barbara, G. et al., Functional gastrointestinal disorders and mast cells: implications for therapy, *Neurogastroenterol. Motil.*, 18, 6, 2006.

124. Chen, Y.L. et al., Synthesis and oral efficacy of a 4-(butylethylamino)pyrrolo[2,3-d]pyrimidine: a centrally active corticotropin-releasing factor1 receptor antagonist, *J. Med. Chem.*, 40, 1749, 1997.

125. Schulz, D.W. et al., CP-154,526: a potent and selective nonpeptide antagonist of corticotropin releasing factor receptors, *Proc. Natl. Acad. Sci. USA*, 93, 10477, 1996.

126. Chen, C. et al., Design of 2,5-dimethyl-3-(6-dimethyl-4-methylpyridin-3-yl)-7-dipropylaminopyrazolo[1,5-a]pyrimidine (NBI 30775/R121919) and structure--activity relationships of a series of potent and orally active corticotropin-releasing factor receptor antagonists, *J. Med. Chem.*, 47, 4787, 2004.

127. Gilligan, P.J. et al., Corticotropin releasing factor (CRF) receptor modulators: progress and opportunities for new therapeutic agents, *J. Med. Chem.*, 43, 1641, 2000.

128. Lelas, S. et al., Anxiolytic-like effects of the corticotropin-releasing factor1 (CRF1) antagonist DMP904 [4-(3-pentylamino)-2,7-dimethyl-8-(2-methyl-4-methoxyphenyl)-pyrazolo-[1,5-a]-pyrimidine] administered acutely or chronically at doses occupying central CRF1 receptors in rats, *J. Pharmacol. Exp. Ther.*, 309, 293, 2004.

129. Millan, M.J. et al., Anxiolytic properties of the selective, non-peptidergic CRF(1) antagonists, CP154,526 and DMP695: a comparison to other classes of anxiolytic agent, *Neuropsychopharmacology*, 25, 585, 2001.

130. Webster, E.L. et al., *In vivo* and in vitro characterization of antalarmin, a nonpeptide corticotropin-releasing hormone (CRH) receptor antagonist: suppression of pituitary ACTH release and peripheral inflammation, *Endocrinology*, 137, 5747, 1996.

131. Griebel, G. et al., 4-(2-Chloro-4-methoxy-5-methylphenyl)-N-[(1S)-2-cyclopropyl-1-(3-fluoro-4-   methylphenyl)ethyl]5-methyl-N-(2-propynyl)-1,   3-thiazol-2-amine hydrochloride (SSR125543A), a potent and selective corticotrophin-releasing factor(1) receptor antagonist. II. Characterization in rodent models of stress-related disorders, *J. Pharmacol. Exp. Ther.*, 301, 333, 2002.

132. Maciejewski-Lenoir, D. et al., Selective impairment of corticotropin-releasing factor1 (CRF1) receptor-mediated function using CRF coupled to saporin, *Endocrinology*, 141, 498, 2000.

133. Zhang, G. et al., Pharmacological characterization of a novel nonpeptide antagonist radioligand, (+/-)-N-[2-methyl-4-methoxyphenyl]-1-(1-(methoxymethyl) propyl)-6-methyl-1H-1,2,3-triazolo[4,5-c]pyridin-4-amine ([3H]SN003) for corticotropin-releasing factor1 receptors, *J. Pharmacol. Exp. Ther.*, 305, 57, 2003.

134. Chaki, S. et al., Anxiolytic- and antidepressant-like profile of a new CRF1 receptor antagonist, R278995/CRA0450, *Eur. J. Pharmacol.*, 485, 145, 2004.

135. Okuyama, S. et al., Receptor binding, behavioral, and electrophysiological profiles of nonpeptide corticotropin-releasing factor subtype 1 receptor antagonists CRA1000 and CRA1001, *J. Pharmacol. Exp. Ther.*, 289, 926, 1999.

Part II

Classical Medicinal Chemistry

6 Design, Synthesis, and Pharmacological Evaluation of High-Affinity and Selectivity Sigma-1 and Sigma-2 Receptor Ligands

Jacques H. Poupaert

CONTENTS

SUMMARY

Sigma-1 ligands have demonstrated potentials as neuroprotective, antiepileptic, and antiamnesic agents in the modulation of opioid analgesia, and in the modification of cocaine-induced locomotor activity. On the other hand, sigma-2 agonists induce apoptosis in drug-resistant cancer cells and may be useful in the treatment of drug-resistant cancers. Sigma-2 antagonists might prevent the irreversible motor side effects of typical neuroleptics.

In an effort to produce new pharmacological probes with high and selective affinity either for the sigma-1 or sigma-2 receptors and that could be tested *in vivo*, we synthesized original 2(3*H*)-benzothiazolones, taking compound **1** [3-(1-piperidinoethyl)-6-propylbenzothiazolin-2-one] as lead compound, and their receptor binding affinity at sigma-1 and sigma-2 receptor was determined. The best sigma-1 ligand (**4**, K_i 0.56 nM, $K_i(\sigma_2)/K_i(\sigma_1) > 1000$) was obtained with an azepine side-chain. When tested on a wide battery of receptors, including $5HT_{2A}(h)$, $5HT_3(h)$, α_1, α_2, β_1, β_2, H_1, H_2, opioids μ, δ, and κ, D_1 (h), D_2 (h), 5-HT uptake, DOPA uptake, etc., compound **4** showed submicromolar affinity only for α_2 ($K_i = 205$ nM) and H_1 ($K_i = 311$ nM).

Based on a similar strategy using the 2(3H)-benzothiazolone template as pivotal structure, we obtained compound **7** (K_i 2.0 nM, $K_i(\sigma_1)/K_i(\sigma_2)$ 70), a compound devoid of affinity for the receptors mentioned in the above battery, except for H_1 ($K_i = 9$ nM).

Benzothiazolinone-based low molecular weight and achiral ligands selective for the sigma-1 (compound **4**) and sigma-2 (compound **7**) receptors were obtained that can be used as pharmacological probes to investigate in particular the problem of the multiplicity of the sigma-1 receptor. Their physicochemical characteristics make them also compatible with *in vivo* evaluations.

INTRODUCTION

Sigma-binding sites[1] recognize a wide array of structurally unrelated compounds exhibiting potent antipsychotic, neuroprotective, and immunoregulatory effects. These ligands include, e.g., haloperidol, 1,3-di-o-tolylguanidine (DTG), (+)-3-(3-hydroxyphenyl)-N-(1-propyl) piperidine [(+)-3PPP], and (+)-benzomorphans such as (+)-pentazocine and (+)-N-allylnormetazocine (Figure 6.1 and Figure 6.2). According to biochemical and radioligand binding data criteria, recognition sites have been classified thus far into two types, termed sigma-1 and sigma-2. Whereas the sigma-1 receptor has been characterized in molecular terms, the sigma-2 receptor has not yet been cloned. Sigma-1 is a 26 kDa protein related to the yeast *Saccharomyces cerevisiae* Δ8–Δ7-sterol isomerase encoded by the erg2 gene, and both proteins share 30% sequence homology. Despite this substantial homology, sigma-1 does not exhibit any sterol isomerase activity.

Although it is now fairly well established that sigma receptors represent unique binding sites in the brain and peripheral organs, it has not always been so. The existence of a sigma receptor was initially proposed by Martin et al. to account for the psychotomimetic effects of N-allylnormetazocine ((±)-SKF-10,047) in the morphine-dependent chronic spinal dog model and, as a consequence, sigma receptors were associated with opioid receptors. It became, however, rapidly evident that most of the behaviors elicited by the drug were resistant to blockade by classical opioid antagonists such as naloxone or naltrexone. The sigma receptors were also confused with the high-affinity phencyclidine (PCP) binding sites, located within the ion

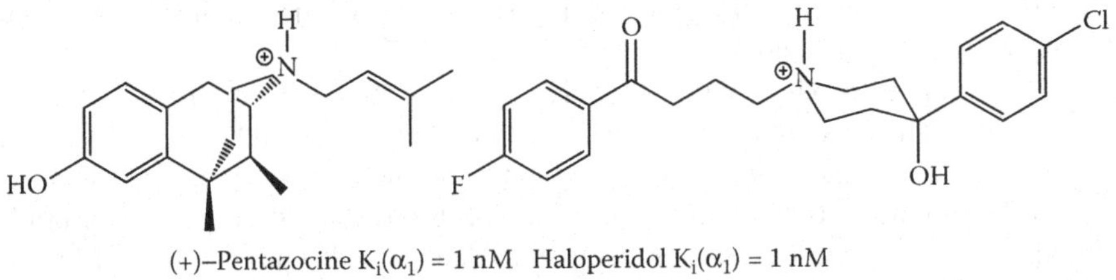

(+)–Pentazocine $K_i(\alpha_1) = 1$ nM   Haloperidol $K_i(\alpha_1) = 1$ nM

FIGURE 6.1   Structurally dissimilar compounds bind with equal affinity at sigma-1 receptors.

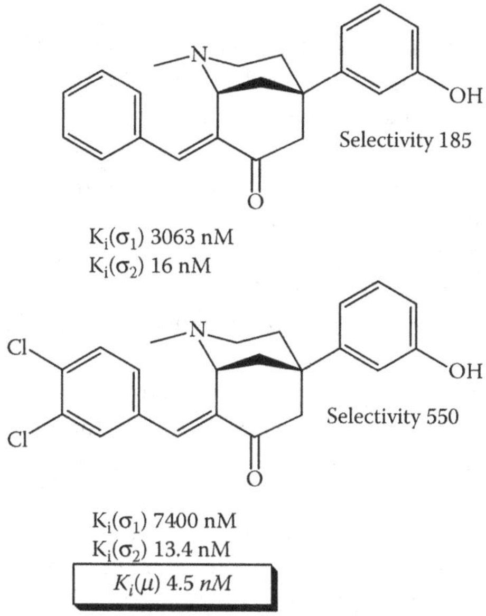

$K_i(\sigma_1)$ 3063 nM
$K_i(\sigma_2)$ 16 nM

Selectivity 185

Selectivity 550

$K_i(\sigma_1)$ 7400 nM
$K_i(\sigma_2)$ 13.4 nM
$K_i(\mu)$ 4.5 nM

FIGURE 6.2 (+)-1R,5R-(E)-8-benzylidene-5-(3-hydroxyphenyl)-2-methylmorphan-7-one and its 3,4-dichloro derivative (CB64D and CB184) are selective for sigma-2. Unfortunately, CB184 binds with high affinity at the μ opioid receptor.

channel associated with the NMDA glutamate receptor, because of similar affinities of these sites for several compounds, e.g., PCP and (+)-SKF-10,047. This confusion was cleared up by the availability of more selective drugs, including dizocilpine or thienylcyclidine for the PCP site and DTG, (+)-pentazocine, ((+)-3-PPP), igmesine (JO-1784), and (+)-cis-N-methyl-N-[2-(3,4-dichlorophenyl)ethyl]-2-(1-pyrrolidinyl) cyclohexylamine (BD737), among others, for the sigma sites.

The sigma-1 and sigma-2 sites can be distinguished based on their different drug selectivity patterns. Some examples of potent and highly selective sigma-1 ligands are the cis-(+)-N-substituted N-normetazocine derivatives (+)-pentazocine, (+)-N-allyl-normetazocine (SKF 10,047) and (+)-N-benzyl-N-normetazocine. Examples of sigma-2 selective ligands, relative to 1, are (+)-1R,5R-(E)-8-benzylidene-5-(3-hydroxyphenyl)-2-methylmorphan-7-one and its 3,4-dichloro derivative (CB64D and CB184, respectively; see Figure 6.2). However, these last two ligands exhibit high opioid receptor affinity. DTG, (+)-3-PPP, and haloperidol are nondiscriminating ligands with high affinity on both subtypes. Moreover, sigma-1 sites are allosterically modulated by phenytoin and sensitive to pertussis toxin and to the modulatory effects of guanosine triphosphate. It also has been shown that several drugs, such as haloperidol, α-(4-fluorophenyl)-4-(5-fluoro-2-pyrimidinyl)-1-piperazine butanol (BMY-14,802), rimcazole, and N,N-dipropyl-2-(4-methoxy-3-(2-phenylethoxy)phenyl)ethylamine (NE-100) act as antagonists in several physiological and behavioral tests relevant to the sigma-1 pharmacology. Unfortunately, most of these ligands are nonselective and bind to other pharmacological targets.

The sigma receptors are widely expressed throughout the body and have been reported in a variety of tissues, including regions of the central nervous system (CNS) and in the gastrointestinal (GI) tract, the liver, and the kidney. The sigma-2 receptors are overexpressed in tumor cells as compared to normal cells, including breast, neural, lung, prostate, and melanoma tumors. The sigma-1 receptor is involved in a direct modulation of intracellular calcium ($[Ca^{2+}]i$) mobilizations through a complex mechanism. The sigma-1 receptor plays a key role of sensor-modulator for the neuronal intracellular Ca^{2+} traffic and consequently for the extracellular Ca^{2+} influx. This cellular role may explain the effective but nonselective neuromodulation mediated by the sigma-1 receptor on several neurotransmitter systems, including dopaminergic pathways.

The sustained interest for sigma ligands stems from the possibility of developing clinical agents for the treatment of psychiatric and motor disorders, tumor diagnosis, cocaine abuse, neuroprotection, and several neurological diseases. More specifically, sigma-1 ligands have demonstrated significant potentials as neuroprotective, anti-epileptic, and anti-amnesic agents in the modulation of opioid analgesia and in the modification of cocaine-induced locomotor activity. Many classes of drugs such as antihistaminics (H_1) and antipsychotic drugs produce important side effects due to their interaction with sigma receptors.

DESIGN OF SIGMA-1 LIGANDS

We had previously reported the discovery of a series of 2(3H)-benzoxazolone and 2(3H)-benzothiazolone derivatives endowed with good affinity both for sigma-1 and sigma-2 receptor subtypes. Several of these derivatives showed preferential selectivity for sigma-1 binding sites, and compound 1 [3-(1-piperidinoethyl)-6-propylbenzothiazolin-2-one] (Figure 6.3) came out as a potent sigma-1 receptor ligand (Ki = 0.6 nM) but displayed rather poor selectivity over the sigma-2 receptor subtype (Ki for sigma-2/Ki for sigma-1 = 30).[2]

Compound 1 demonstrated remarkable anticonvulsant activity, being as potent as phenytoin in the maximal electroshock seizure test (subcutaneous administration, mice). Compounds 2 [3-(1-piperidinopropyl)-6-propanoylbenzothiazolin-2-one] and

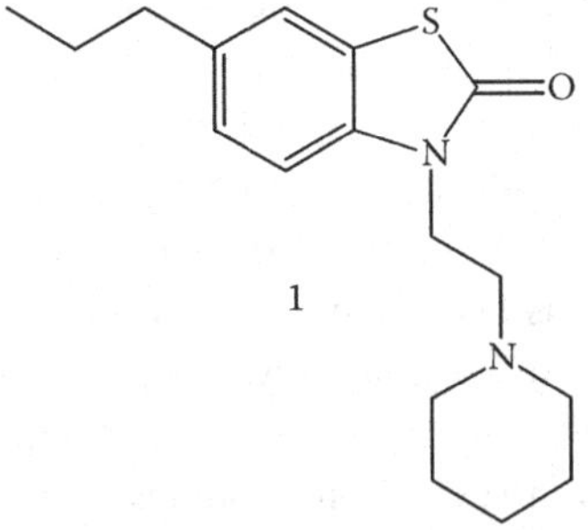

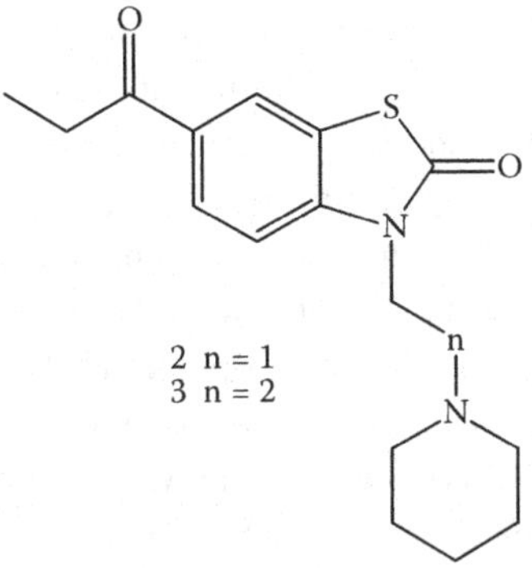

FIGURE 6.3   Structure of benzothiazolinone-based sigma-1 lead structures 1–3.

3 [3-(1-piperidinopropyl)-6-propanoyl benzoxazolin-2-one] (Figure 6.3), although maintaining rather high affinities at sigma-1 binding sites (Ki values of 2.3 and 8.5 nM, respectively) were endowed with relatively higher selectivity (K_i for sigma-2/Ki for sigma-1 = 87 and 58, respectively). The study concluded that the benzothiazolinones were more potent than the benzoxazolinones, that a 6-alkyl substitution was more efficacious than a 6-acyl, and that a piperidino side-chain along with a linker with two or three methylene represented an optimum for affinity. In an effort to further study the structure–activity relationship in this class of benzothiazolinones and improve their selectivity for sigma-1 receptors, we now report on analogues of our lead compound **1** with the main structure variations affecting either the nature of the tertiary amine residue or the link between the tertiary amine and heterocyclic nitrogen. The affinity of benzothiazolinone derivatives at sigma-1 and sigma-2 receptor subtypes was measured classically by competition-binding experiments. $[^3H](+)$-pentazocine was used determining affinity at sigma-1 receptor. DTG was used in the presence of 100 nM (+)-N-allylnormetazocine (NANM) (to saturate sigma-1 binding sites) in guinea-pig brain membranes.

As already observed, the 6-alkyl substitution was found better than the 3-acyl one. The 6-n-propyl substituent was found superior to 6-n-pentyl. The optimal number of methylenes for the link between the amine and heterocycle nitrogen was found equal to two. Interestingly enough, lengthening of this linker with $n = 4$ produced a ligand with appreciable affinity for sigma-2 receptor (Ki = 13 nM).

We previously observed that the nature of the surrounding of the tertiary amine nitrogen had a profound impact on affinity. For example, when associated to a 6-propanoyl substitution, the morpholino derivative was found inactive, whereas the pyrrolidino and piperidino terms gave ligands with increasing affinity and selectivity. Consequently, the synthesis of the azepino derivative (**4**) was undertaken (Figure 6.4). This compound not only had very high affinity (Ki 0.56 nM) but was also endowed with high selectivity for the sigma-1 receptor (>1000).[3] Compound **4** compares well with recently published monoamine molecules in terms of affinity and selectivity, many sigma-1/2 ligands being indeed generally diamine systems. Marazzo et al.[4]

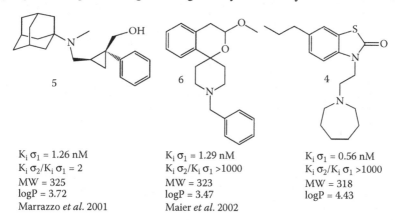

5

$K_i\ \sigma_1 = 1.26$ nM
$K_i\ \sigma_2/K_i\ \sigma_1 = 2$
MW = 325
logP = 3.72
Marrazzo *et al.* 2001

6

$K_i\ \sigma_1 = 1.29$ nM
$K_i\ \sigma_2/K_i\ \sigma_1 >1000$
MW = 323
logP = 3.47
Maier *et al.* 2002

4

$K_i\ \sigma_1 = 0.56$ nM
$K_i\ \sigma_2/K_i\ \sigma_1 >1000$
MW = 318
logP = 4.43

FIGURE 6.4   Structures of sigma-1 selective ligands.

have reported high affinity ligands for sigma-1, e.g., (+)-*cis*-(2-{[1-adamantyl(meth-yl)amino]methyl}-1-phenylcyclopropyl) methanol (5) had Ki = 1.26, unfortunately associated to a low selectivity ratio (=2) for sigma-1. Maier[5] has also published spi-ropiperidine structures with high affinity for the sigma-1 receptor. Among the best compounds obtained, 6 had a Ki = 1.29 with a selectivity ratio >1000 for sigma-1. Although devoid of affinity for various receptors or transport systems, compound 6 revealed submicromolar affinity for kappa opioid receptors (Ki = 418 nM). Our compound 4 has been tested on a similar battery of affinity tests including $5HT_{2A}$(h), $5HT_3$(h), α_1, α_2, β_1, β_2, H_1, H_2, opioids, D_1 (h), D_2 (h), 5HT uptake, DOPA uptake, GABA-A, etc. Compound 4 showed submicromolar affinity only for α_2 (Ki = 205 nM) and H_1 (Ki = 311 nM) receptors.

DESIGN OF SIGMA-2 LIGANDS

The design of these ligands was based on the following arguments: (1) CoMFA investigations showed that most sigma ligands are characterized by a central basic nitrogen flanked by two aromatic hydrophobic groups; (2) it is well established that the *N*-phenylpiperazine moiety is a versatile platform in drug design; (3) 2(3*H*)-ben-zothiazolones have been successfully used as a template in the sigma axis.[1,2] There were, however, indications in the literature that, to obtain sigma-2 ligands, it was necessary to differentiate in an important manner the electron density at both hydro-phobic aromatic ends. Consequently, in an effort to produce new pharmacological probes with preferential affinity for the sigma-2 receptor, we have synthesized new 6-piperidino- or piperazino-alkyl-2(3*H*)-benzothiazolones, and their receptor-bind-ing profile (sigma-1, sigma-2, $5-HT_{1A}$, $5-HT_{2A}$, $5-HT_3$, D_2, H_1, and M_1) was deter-mined (Figure 6.5).

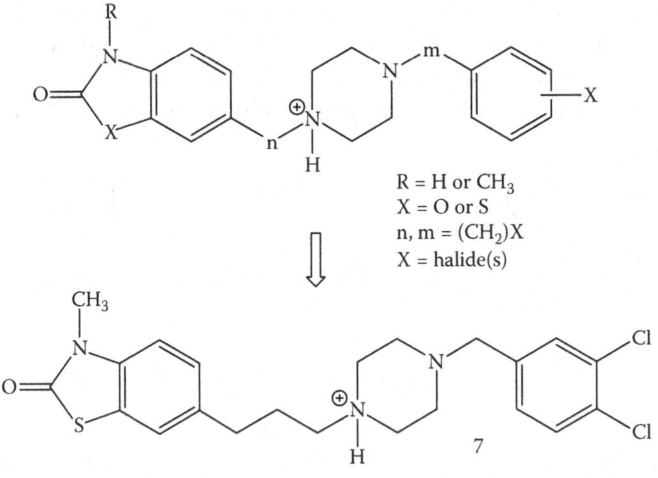

R = H or CH_3
X = O or S
n, m = $(CH_2)X$
X = halide(s)

$K_i(\sigma_1)$ = 140 nM; $K_i(\sigma_2)$ = 2 nM, Selectivity ratio = 70

FIGURE 6.5 General structure of the σ-2 ligands (n = number of methylenes, R = H or CH_3, X = N or CH, Y = phenyl, benzyl, 2,4- or 3,4-dichlororobenzyl) and design of compound 7.

Most of these compounds exhibited very high affinity (in the nanomolar range) at sigma and/or 5-HT$_{1A}$ receptors. Within the piperidine series, binding studies showed that, for sigma affinity, 4-phenyl substitution was superior to 4-benzyl irrespective of the length (n) of the methylene linker between the benzothiazolinone and piperidine moiety, whereas, for 5-HT$_{1A}$ affinity, 4-phenyl and 4-benzyl substitution gave nearly equal scores, however with a superiority for $n = 2$. Within the piperazine series, binding studies indicated that, for sigma affinity, 4-benzyl and 3,4-dichlorobenzyl substitutions were superior to 2,4-dichlorobenzyl with low incidence of the methylene number n, and a selective sigma-2 ligand (**7**)[6] was obtained with 3,4-dichlorobenzyl associated to $n = 4$ (Ki 2 nM) (Figure 6.5). For this compound, the selectivity ratio Ki σ1/Ki σ2 was 70. This situation is reminiscent of CB-184 (1R,5R-(E)-8-(3,4-dichlorobenzylidene)-(3-hydroxyphenyl)-2-methylmorphan-7-one and ibogaine, reported by Bowen et al.[7] as σ2-selective ligands (see Figure 6.2).

CONCLUSION

Conclusively, in the course of this research, we have synthesized low-molecular-weight achiral ligands selective for the sigma-1 (compound **4**) and sigma-2 (compound **7**) receptors that can be used as pharmacological probes to investigate in particular the problem of the multiplicity of the sigma-1 receptor. Their physicochemical characteristics are also compatible with *in vivo* evaluations.

ACKNOWLEDGMENTS

The author thanks Santi Spampinato (University of Bologna, Italy), Saïd Yous, Daniel Lesieur (University of Lille, France), Giuseppe Ronsisvalle (University of Catania, Italy), and Christopher McCurdy for their collaboration or useful suggestions during the realization of this work.

REFERENCES

1. For a comprehensive review on sigma receptors, see Maurice, T. et al., Sigma(1) (sigma(1)) receptor antagonists represent a new strategy against cocaine addiction and toxicity, *Neurosci. Biobehav. Rev.*, 26, 499, 2002; see also Bowen, W.D., Sigma receptors: recent advances and new clinical potentials, *Pharm. Acta. Helv.*, 74, 211, 2000.
2. Ucar, H. et al., 2(3*H*)-benzoxazolone and 2(3*H*)-benzothiazolone derivatives: Novel, potent and selective sigma1 receptor ligands, *Eur. J. Pharmacol.*, 335, 267, 1997; see also Ucar, H. et al., Synthesis and anticonvulsant activity of 2(3*H*)-benzoxazolone and 2(3*H*)-benzothiazolone derivatives, *J. Med. Chem.*, 41, 1138, 1998; Ucar, H. et al., Synthesis and anticonvulsant activity of 2(3*H*)-benzoxazolone and 2(3*H*)-benzothiazolone derivatives, *J. Med. Chem.*, 41, 3102, 1998; for an account of the discovery of the initial lead, see Yous, S. et al., Synthesis and pharmacological evaluation of analgesic 6-substituted 2(3*H*)-benzothiazolones, *Drug Des. Discov.* 17, 331, 2001.
3. Mouithys-Mickalad, A. et al., Synthesis and pharmacological evaluation of 6-piperidino- and 6-piperazinoalkyl-2(3*H*)-benzothiazolones as mixed sigma/5-HT(1A) ligands, *Bioorg. Med. Chem. Lett.*, 12, 1149, 2002.

4. Marrazzo, A. et al., Synthesis of (+)- and (−)-*cis*-2-[(1-adamantylamino)-methyl]-1-phenylcyclopropane derivatives as high affinity probes for sigma1 and sigma2 binding sites, *Farmaco*, 57, 45, 2002; see also Marrazzo, A. et al., Synthesis and pharmacological evaluation of potent and enantioselective sigma 1, and sigma 2 ligands, *Farmaco*, 56, 181, 2001.
5. Maier, C.A. and Wunsch, B., Novel sigma receptor ligands. Part 2. SAR of spiro[[2]benzopyran-1,4′-piperidines] and spiro[[2]benzofuran-1,4′-piperidines] with carbon substituents in position 3, *J. Med. Chem.*, 45, 4923, 2002; Maier, C.A. and Wunsch, B., Novel spiropiperidines as highly potent and subtype selective sigma-receptor ligands. Part 1, *J. Med. Chem.*, 45, 438, 2002.
6. Poupaert, J. et al., 2(3*H*)-benzoxazolone and bioisosters as "privileged scaffold" in the design of pharmacological probes, *Curr. Med. Chem.*, 12, 877, 2005.
7. Bertha, C.M. et al., (*E*)-8-benzylidene derivatives of 2-methyl-5-(3-hydroxyphenyl) morphans: highly selective ligands for the sigma 2 receptor subtype, *J. Med. Chem.*, 38, 4776, 1995.

7 Synthesis of Biologically Active Taxoids

K. C. Nicolaou and R. K. Guy

CONTENTS

INTRODUCTION

Taxol® (Figure 7.1; **1**), a diterpene originally isolated from the American Pacific yew *Taxus brevifolia*,[1] has emerged as one of the most effective chemotherapeutic agents for the treatment of breast and ovarian cancers.[2] The compound is also in development for non-small-cell lung cancers,[3] and head and neck cancers.[4] Taxol's development has included periods of both unusual languor and speed.[5] It acts by blocking the normal depolymerization of microtubules, a process that is crucial for a number of cellular functions including mitosis.[6] The elucidation of this unique mechanism by Horwitz's group in the early 1980s spurred the drug's development.[7]

Since the publication of Taxol's structure in 1971, the compound has interested synthetic chemists who have followed both semisynthetic and totally synthetic approaches to the assembly of Taxol and simpler mimetics.[8,9] This ongoing synthetic effort produced, early in 1994, two separate total syntheses of Taxol.[10,11] Our own route[11–15] (Figure 7.2) involved the convergent joining of A- and C-ring synthons **2** and **3**, formed themselves by Diels–Alder reactions, to give intermediate **4**. Closure of the B-ring by McMurry ketyl coupling then elaborated this intermediate to baccatin III derivative **5**. Subsequent manipulations then converted this compound to Taxol. Although this route was unsuitable for the commercial production of Taxol, knowledge gained about Taxol's chemistry during its execution enabled the design and production of novel taxoids.[16,17]

As with all chemotherapeutics, the activity and selectivity of Taxol could be improved by applying a thorough understanding of Taxol's structure–activity relationship (SAR) to the design of next generation analogues. When we began our synthetic endeavors, most areas of Taxol's SAR had not been carefully probed. Due to the ready availability of side-chain analogues, the contributions of the side-chain

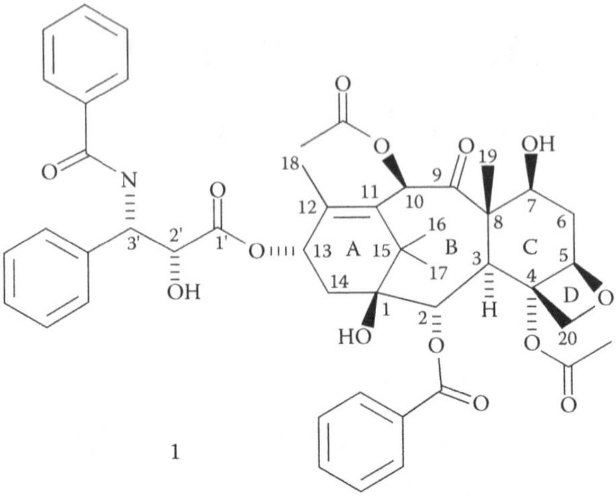

FIGURE 7.1   The structure and numbering scheme of Taxol.

moieties were fairly well assessed.[18–22] It was generally believed that these domains were crucial to Taxol's activity. The oxetane was likewise believed crucial,[23] whereas the C-7 hydroxyl[24,25] and C-10 acyl[25–27] were thought unimportant. Additionally, although Taxol has established itself as an important pharmaceutical, it does suffer from formulation problems.[28] These problems are rooted in its low aqueous solubility, a property that could be ameliorated by chemical means. For these reasons, when planning our synthetic endeavors, we included deliberate probing of certain areas of Taxol's pharmacology and SAR.

C2-ESTER DERIVATIVES OF TAXOL

During our total synthesis of Taxol, one of the crucial steps involved the treatment of a C1-C2 cyclic carbonate derivative of baccatin III (Figure 7.3; **6**) with phenyllithium to provide, regioselectively, hydroxy-benzoate **8**.[12,29] At the time this transformation was discovered, almost no concrete information existed concerning the importance of the C2-ester to Taxol's SAR.[8,30,31] This void prompted us to examine the suitability of this process for producing derivatives of Taxol termed *taxoids*.

Gratifyingly, the addition of organolithium reagents to baccatin carbonates proved to be quite general and allowed the production of a number of C2-taxoids (Table 7.1).[16,17,29,32] After the initial addition of the organolithium reagent, the same methodology used for the total synthesis of Taxol transformed the resulting baccatin derivative to the corresponding taxoid. Although almost all organolithium reagents would add to the carbonate in the desired manner, some of the derivatives, such as the 2-pyridyl ester, proved quite difficult to carry through to the taxoid.

The method produced three classes of taxoids: 5-membered heteroaromatic esters, 6-membered aromatic esters, and linear aliphatic esters. Careful biological analysis by the standard methods of tubulin depolymerization inhibition and cytotoxicity assays allowed the ordering of the relative activities of these derivatives. These tests revealed that all of the monocyclic unsubstituted esters had an activity

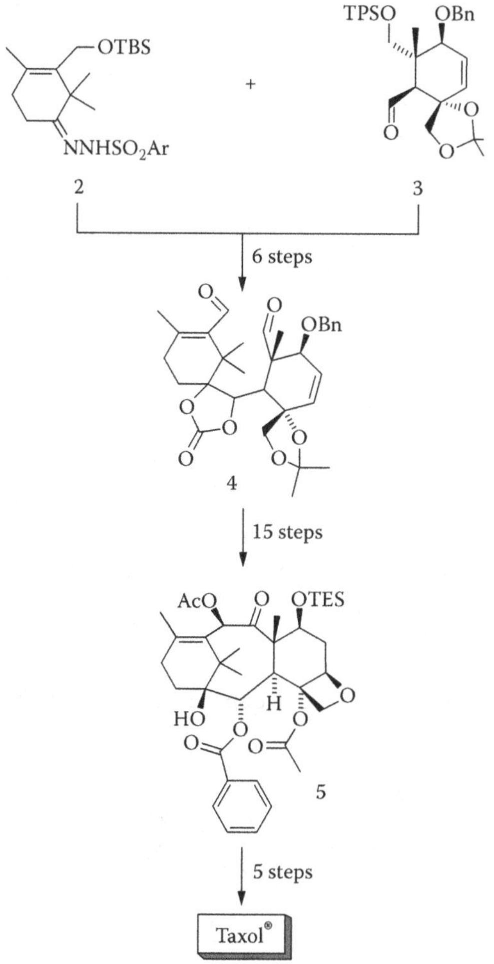

FIGURE 7.2 The total synthesis of Taxol.

within two orders of the magnitude of Taxol. The other derivatives (**8f–h**) had very low activity. Further evaluation of the two most promising taxoids, thiophenes **8b** and **8c**, using distal tumor models in nude mice revealed that both compounds possessed pharmaceutical and therapeutic profiles very similar to those of Taxol.

These derivatives revealed some general rules about the contribution of the C2-ester to Taxol's SAR. First, the position is tolerant of any simple aromatic ester. This implies that receptor binding is not completely specific because the "void" caused by a smaller 5-membered aromatic ester does not significantly perturb binding. On the other hand, the *p*-position of the ester is quite intolerant of substitution, perhaps indicating a negative steric interaction. Finally, the receptor does require a cyclic group immediate to the carbonyl of the C2-ester. These observations represented, at the time, the first concrete information about this region of Taxol's receptor. Since then other groups have confirmed these results.[10,30,31,33,34]

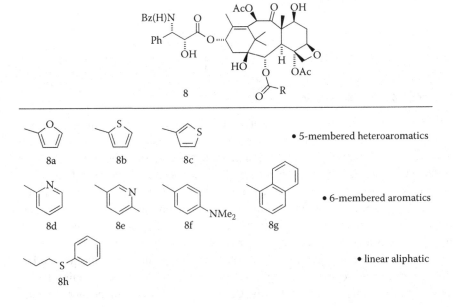

FIGURE 7.3   The opening of baccatin carbonates to provide C2-substituted taxoids.

TABLE 7.1

Synthetic Taxoids Produced By Nucleophilic Carbonate Opening

8a	• 5-membered heteroaromatics
8b	
8c	
8d	• 6-membered aromatics
8e	
8f	
8g	
8h	• linear aliphatic

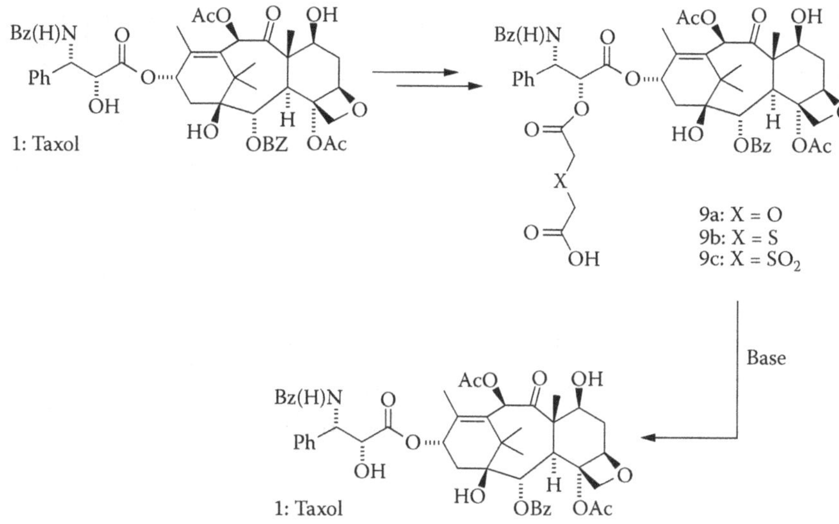

FIGURE 7.4 First-generation proTaxols.

PRODRUGS FOR TAXOL

One of the major problems associated with Taxol is the difficulty of administering this highly insoluble compound. Although the literature presents a range of values for the solubility of Taxol, clearly it is insoluble to the degree of <0.01 mg/mL in water. This property forces Taxol's formulation in a mixture of Cremaphor® and ethanol,[35] a vehicle that itself may cause hypersensitivity reactions in patients.[36] During clinical trials, workers developed a complex regimen involving the pretreatment of patients with anti-inflammatories and the slow infusion of a suspension of formulated Taxol in saline.[37] The nursing literature has noted this regimen as particularly difficult to administer.[38] Clearly, an active water-soluble form of Taxol would prove advantageous. However, drugs effective in the treatment of solid tumors usually exhibit high lipophilicity. Thus, an ideal derivative would possess two seemingly opposite solubility characteristics. The use of a prodrug, a compound that is water-soluble for administration but would discharge Taxol *in vivo*, might solve this dilemma.[39–42]

Our first efforts to produce such a prodrug for Taxol,[42] or proTaxol, hinged upon the observation that cancer cells exhibit a higher pH in their cytosol than normal cells.[43] Derivatives such as **9a–c** that contained an acidic proton whose abstraction would provoke the self-immolative discharge of Taxol in a basic environment were targeted for synthesis. These compounds did indeed posses a higher solubility than Taxol and could release Taxol at a controlled rate in basic solution. However, these first-generation proTaxols exhibited several undesired characteristics (see Figure 7.4). First, although reasonably stable, they would degrade in solution over the period of hours to days, making their formulation more problematic. Second, and more seriously, although quite active in cytotoxicity assays, these compounds exhibited no activity in distal tumor models *in vivo*. Radiolabeling experiments allowed tracing this lack of activity to rapid excretion of the water-soluble compound via the kidney.

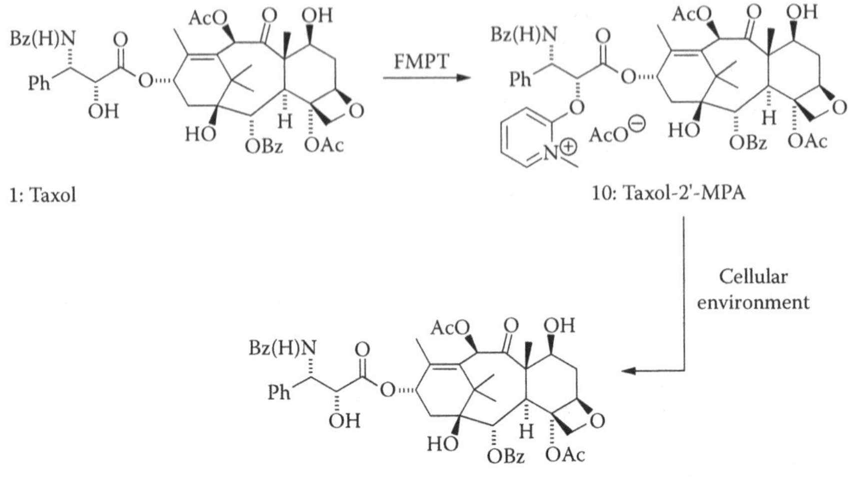

1: Taxol   10: Taxol-2'-MPA

1: Taxol

FIGURE 7.5   Second-generation proTaxols.

Our second generation of proTaxols fell, literally, out of our habit of testing synthetic intermediates for biological activity (see Figure 7.5).[44,45] The methylpyridinium salt of Taxol **10** was prepared as an intermediate expected to allow nucleophilic substitution at the 2′ position. Upon routine testing, it revealed cytotoxicity equitable with Taxol. Further investigation revealed that the material was functioning as a highly stable prodrug of Taxol. Although extremely (>1 month) stable in solution, the material almost immediately discharges Taxol when exposed to human plasma. *In vivo* tumor models showed that the compound was not only as effective as Taxol but also less toxic.

CONCLUSION

The studies outlined in this article show how a guided synthetic program, ultimately aimed at the total synthesis of a given drug, can simultaneously provide important information about the pharmacology of that substance. In the case of Taxol, a method designed to allow the regioselective substitution of a diol provided the ability to synthesize a novel class of taxoids that eventually revealed crucial information about Taxol's SAR. In a similar manner, explorations of the degradative chemistry of Taxol gave, unexpectedly, a novel class of water-soluble prodrugs that have proved more useful than any of those designed specifically for that purpose. Both of these studies prove the crucial and wide-reaching role of total synthesis in the development of pharmaceuticals.

REFERENCES

1. Wani, M.C. et al., Plant antitumor agents. VI. The isolation and structure of Taxol, a novel antileukemic and antitumor agent from Taxus brevifolia, *J. Am. Chem. Soc.*, 93, 2325, 1971.

2. Valero, V. et al., Phase II trial of docetaxel: a new, highly effective antineoplastic agent in the management of patients with anthracycline-resistant metastatic breast cancer, *J. Clin. Oncol.*, 13, 2886, 1995.

3. Ettinger, D.S., Overview of paclitaxel (Taxol) in advanced lung cancer, *Semin. Oncol.*, 20, 46, 1993.

4. Forastiere, A.A., Paclitaxel (Taxol) for the treatment of head and neck cancer, *Semin. Oncol.*, 21, 49, 1994.

5. Suffness, M., Taxol: From discovery to therapeutic use, *Annu. Rep. Med. Chem.*, 28, 305, 1993.

6. Horwitz, S.B., Mechanism of action of Taxol, *Trends Pharmacol. Sci.*, 13, 134, 1992.

7. Wall, M.E. and Wani, M.C., Camptothecin and Taxol: discovery to clinic—Thirteenth Bruce F. Cain Memorial Award Lecture, *Cancer Res.*, 55, 753, 1995.

8. Nicolaou, K.C., Dai, W.M., and Guy, R.K., Chemistry and biology of Taxol, *Angew. Chem., Int. Ed. Engl.*, 33, 15, 1994.

9. Kingston, D.G.I., Molinero, A.A., and Rimoldi, J.M., The taxane diterpenoids, *Prog. Chem. Org. Nat. Prod.* 61, 1, 1993.

10. Holton, R.A. et al. The total synthesis of paclitaxel starting with camphor, *ACS Symp. Ser.*, 583, 288, 1995.

11. Nicolaou, K.C. et al., Total synthesis of Taxol, *Nature*, 367, 630, 1994.

12. Nicolaou, K.C. et al., Total synthesis of Taxol. 1. Retrosynthesis, degradation, and reconstitution, *J. Am. Chem. Soc.*, 117, 624, 1995.

13. Nicolaou, K.C. et al., Total synthesis of Taxol. 2. Construction of A- and C-ring intermediates and initial attempts to construct the ABC ring system, *J. Am. Chem. Soc.*, 117, 634, 1995.

14. Nicolaou, K.C. et al., Total synthesis of Taxol. 3. Formation of Taxol's ABC ring skeleton, *J. Am. Chem. Soc.*, 117, 645, 1995.

15. Nicolaou, K.C. et al., Total synthesis of Taxol. 4. The final stages and completion of the synthesis, *J. Am. Chem. Soc.*, 117, 653, 1995.

16. Nicolaou, K.C. et al., Chemical synthesis and biological evaluation of C-2 taxoids, *J. Am. Chem. Soc.,* 117, 2409, 1995.

17. Nicolaou, K.C. et al., Synthesis of C-2 Taxol analogues, *Angew. Chem. Int. Ed. Engl.* 33, 1581, 1994.

18. Swindell, C.S. et al., Characterization of two Taxol photoaffinity analogues bearing azide and benzophenone-related photoreactive substituents in the A-ring side chain, *Med. Chem.*, 37, 1446, 1994.

19. Burkhart, C.A. et al., Relationship between the structure of Taxol and other taxanes on induction of tumor necrosis factor-alpha gene expression and cytotoxicity, *Cancer Res.*, 54, 5779, 1994.

20. Li, L. et al., Synthesis and biological evaluation of C-3'-modified analogs of 9(R)-dihydrotaxol, *J. Med. Chem.*, 37, 2655, 1994.

21. Kant, J. et al. Studies toward structure–activity relationships of Taxol®: synthesis and cytotoxicity of Taxol® analogues with C-2' modified phenylisoserine side chains, *Bioorg. Med. Chem. Lett.*, 3, 2471, 1993.

22. Jayasinghe, L.R. et al., Structure–activity studies of antitumor taxanes: synthesis of novel C-13 side chain homologated Taxol and taxotere analogs, *J. Med. Chem.*, 37, 2981, 1994.

23. Samaranayake, G. et al., Modified Taxols. 5. Reaction of Taxol with electrophilic reagents and preparation of a rearranged Taxol derivative with tubulin assembly activity, *J. Org. Chem.*, 56, 5114, 1991.

24. Chaudhary, A.G., Rimoldi, J.M., and Kingston, D.G.I., Modified Taxols. 10. Preparation of 7-deoxyTaxol, a highly bioactive Taxol derivative, and interconversion of Taxol and 7-epi-Taxol, *J. Org. Chem.*, 58, 3798, 1993.

25. Chen, S.H. et al., Synthesis of 7-deoxy- and 7,10-dideoxytaxol via radical intermediates, *J. Org. Chem.*, 58, 5028, 1993.
26. Chaudhary, A.G and Kingston, D.G.I., Synthesis of 10-deacetoxytaxol and 10-deoxytaxotere, *Tetrahedron Lett.*, 34, 4921, 1993.
27. Chen, S.H. et al., Taxol structure–activity relationships: synthesis and biological evaluation of 10-deoxytaxol, *J. Org. Chem.*, 58, 2927, 1993.
28. Goldspiel, B.R., Pharmaceutical issues: preparation, administration, stability, and compatibility with other medications, *Ann. Pharmacother.*, 28, S23, 1994.
29. Nicolaou, K.C. et al., Novel chemistry of Taxol. Retrosynthetic and synthetic studies, *J. Chem. Soc., Chem. Commun.*, 295, 1994.
30. Holton, R.A. and Tao, C., Florida State University, U.S. PCT Int. Appl. WO 9417052 A1 940804WO 94-US1223 940128 US 93-10798 930129 US 93-34852 930322 US 93-94715 930720.
31. Chen, S.H. et al., Structure–activity relationships of Taxol®: synthesis and biological evaluation of C2 Taxol analogs, *Bioorg. Med. Chem. Lett.*, 4, 479, 1994.
32. Nicolaou, K.C. et al., US 5274137A 931228 US 92-902390 920623.
33. Sengupta, S. et al., Probing the environment of tubulin-bound paclitaxel using fluorescent paclitaxel analogues, *Biochemistry*, 36, 5179, 1997.
34. Boge, T.C. et al., The effect of the aromatic rings of Taxol on biological activity and solution conformation: synthesis and evaluation of saturated Taxol and taxotere analogues, *J. Med. Chem.*, 37, 3337, 1994.
35. Adams, J.D. et al., Taxol: a history of pharmaceutical development and current pharmaceutical concerns, *J. Natl. Cancer Inst. Monogr.*, 141, 1993.
36. Dorr, R.T., Pharmacology and toxicology of Cremophor EL diluent, *Ann. Pharmacother.*, 28, S11, 1994.
37. Finley, R.S. and Rowinsky, E.K., Patient-care issues: the management of paclitaxel-related toxicities, *Ann. Pharmacother.*, 28, S27.
38. Lubejko, B.G. and Sartorius, S.E., Nursing considerations in paclitaxel (Taxol) administration, *Semin. Oncol.*, 20, 26, 1993.
39. Skwarczynski, M., Hayashi, Y., and Kiso, Y., Paclitaxel prodrugs: toward smarter delivery of anticancer agents, *J. Med. Chem.*, 49, 7253, 2006.
40. Greenwald, R.B. et al., Highly water-soluble Taxol derivatives: 2′-polyethyleneglycol esters as potential prodrugs, *Bioorg. Med. Chem. Lett.*, 4, 2465, 1994.
41. Hostetler, K.Y., Kumar, R., and Sridhar, N.C., Vestar, Inc., U.S. PCT Int. Appl. WO 9413324 A1 940623 WO 93-US12241 931215 US 92-991166 921216.
42. Nicolaou, K.C. et al., Design, synthesis and biological activity of protaxols, *Nature*, 364, 464, 1993.
43. Keizer, H.G. and Joenje, H., Increased cytosolic pH in multidrug-resistant human lung tumor cells: Effect of verapamil, *J. Natl. Cancer Inst.*, 82, 706, 1989.
44. Nicolaou, K.C. et al., A water-soluble prodrug of Taxol with self-assembling properties, *Angew. Chem. Int. Ed. Engl.*, 33, 1583, 1994.
45. Gomez-Paloma, L.G. et al., Conformation of a water-soluble derivative of Taxol in water by 2D-NMR spectroscopy, *Chem. Biol.*, 1, 107, 1994.

8 From the Molecular Pharmacology to the Medicinal Chemistry of Cannabinoids

Didier M. Lambert

CONTENTS

INTRODUCTION

The aim of this chapter is to briefly review the molecular pharmacology of cannabinoids and to present thereafter our main research interests: (1) the synthesis and the pharmacological characterization of new cannabinoid-receptor antagonists and (2) the synthesis and biological evaluation of *N*-palmiloylethanolamine derivatives as new endocannabinoids interfering with the metabolism of anandamide.

CANNABIS, CANNABINOIDS, AND ENDOCANNABINOID SIGNALING

Among the plants known by humans since ancient times, the hemp *Cannabis sativa* has been widely used both for therapeutic, ritual, and recreational purposes. According to Chinese legend, the emperor Sheng Nung discovered the medicinal properties of three plants: ephedra, ginseng, and cannabis. The Chinese pharmacopoeia already described, in 200 B.C., the beneficial effects of cannabis for human health as well as the psychotropic properties of abuse. However, the isolation of the psychoactive ingredient of the plant, Δ^9-tetrahydrocannabinol abbreviated here as THC, and its correct structure assignation were successfully achieved only in the mid-1960s by Professor Mechoulam at the Hebrew University.[1] At that time, typical C21 terpenes isolated from the plant were named cannabinoids.

Nowadays, the term "cannabinoid" has been extended, at least by pharmacologists, to include all the compounds that bind to the cannabinoid receptors or other cannabinoid proteins. However, the isolation of THC did not solve the question of how THC and cannabis preparations act. Indeed, the sites of action, i.e., cannabinoid G-protein-coupled receptors (GPCR), were identified only in the late-1980s,[2] first by biochemical evidence and finally by the cloning and sequencing of their respective cDNAs.[3–5] One of the reasons for such a delay was the hydrophobic nature of THC, which does not facilitate its use as a radioligand.

Today, two different receptors are known, the cannabinoid CB_1 receptor found in the brain, the uterus, the lungs, and the testis, and the cannabinoid CB_2 receptor, which is restricted to the membranes of cells from the immune system. The CB_1 cannabinoid receptor was first evidenced by autoradiography and radioligand binding studies using [^{3}H]-CP55940, a THC analogue less hydrophobic than its parent compound, and was cloned from rat, humans, and mouse. It is expressed in the brain and some peripheral tissues including testis, small intestine, urinary bladder, and *vas deferens*. An alternative spliced form of CB_1, christened CB_{1A}, has also been described, but thus far no peculiar property in terms of ligand recognition and receptor activation has been shown for this variant.[6] The CB_2 was discovered by sequence homology and is predominantly found in the immune system[5] (spleen, tonsils, and

	CB_1 Cannabinoid receptor	CB_2 Cannabinoid receptor
Amino acid sequence	472 (human) 473 (mouse) 473 (rat)	360 (human) 347 (mouse) 360 (rat)
Localization	Brain, testis, uterus, lungs....	Immune system
Receptor type	G protein coupled receptor	G protein coupled receptor
Main transduction mechanism	Adenylate cyclase, AMPc	Adenylate cyclase, AMPc
Homologies	Human-rat: 98% total, 100% TM Human mouse: 97% total	Human-rat: 81% total Human-mouse: 83%
Therapeutical interests	Pain Appetite Asthma Emesis	Pain Inflammation Immunity

FIGURE 8.1 Comparison of the properties of the two subtypes of cannabinoid receptors (abbreviations: aa = amino acids; GPCR = G-protein-coupled receptor; TM = transmembrane

immune cells). A comparative table illustrates the main characteristics of both receptors (Figure 8.1).

The discovery of cannabinoid receptors led to the identification of endogenous lipid compounds such as anandamide and 2-arachidonylglycerol, which bind to cannabinoid receptors. In 1993, Raphael Mechoulam, almost 30 years after his discovery of THC, proposed, together with William Devane, that an endogenous lipid, arachidonoylethanolamide, is the endogenous ligand of cannabinoid receptors.[7] They christened the compound anandamide, as *ananda* in Sanskrit means internal bliss. Congeners have been identified, differing by the degree of unsaturation or the length of the arachidonoyl moiety. Some years later, almost at the same time, Mechoulam's team again, and the group of Sugiura in Japan, identified another endogenous lipid— not an amide, but an ester: the 2-arachidonoylglycerol. This compound has been proposed as the preferred endogenous CB_2 agonist.

From a structural point of view, until now, agonists for these receptors belong to several distinct chemical classes:[8] molecules derived from THC (usually named classical cannabinoid, and the close derivatives have been named nonclassical cannabinoids), the aminoalkylindoles derived from pravadoline, and the fatty acid amides and esters derived from anandamide, the first described endogenous ligand (Figure 8.2). The diversity of structures is paralleled with the variety of origins: THC was isolated from a plant, pravadoline is a synthetic molecule, and anandamide was isolated from mammalian brain. Three classes of antagonists have been described thus far: SR 141716A[9] (Figure 8.2) and SR 144528[10] (Figure 8.2) are diarylpyrazoles, LY-320135 is an arylbenzofuran,[11] and AM630 is an aminoalkylindole.[12] Point mutation studies as well as chimera CB_1 and CB_2 receptors provide new information on how these different ligands bind to the receptors. If agonists and antagonists have distinct sites inside the receptor (which is the case with most GPCR receptors), the point mutation studies indicated that some residues are crucial for some families of agonists but not for others (see the review by Reggio.[13]).

With the discovery of anandamide (arachidonoylelhanolamide) and, later, 2-arachidonylglycerol, a new signaling system has been identified.[14,15] Proteins have now been described either to trigger an intracellular signaling or to degrade the endocannabinoids. The first group of proteins involve cannabinoid receptors (CB_1 and CB_2), which are metabotropic receptors, and probably the vanilloid VR_1 receptor, a calcium ligand-gated channel. The proteins involved in the degradation of anandamide neurotransmission include an active uptake protein that removes the endocannabinoids, and a degrading enzyme—the fatty acid amide hydrolase (FAAH) (Figure 8.3).

3-ALKYL-(5,5'-DIPHENYL)IMIDAZOLIDINEDIONES AS NEW CANNABINOID RECEPTOR ANTAGONISTS

Huffman et al.[16,17] at Clemson University found that simplified derivatives of aminoalkylindoles such as 1-alkyl-3-(1-naphthoyl)pyrroles also exhibit a significant affinity for cannabinoid receptors.[17,18] The more potent ligand was the *n*-pentyl derivative (JWH-030, Scheme 8.1) with a Ki of 87 nM on brain cannabinoid receptors. On analyzing the structures of 1-alkyl-3-(1-naphthoyl)pyrroles and of diarylpyrazoles antagonists, and considering that in the past our laboratory was involved in hydantoin

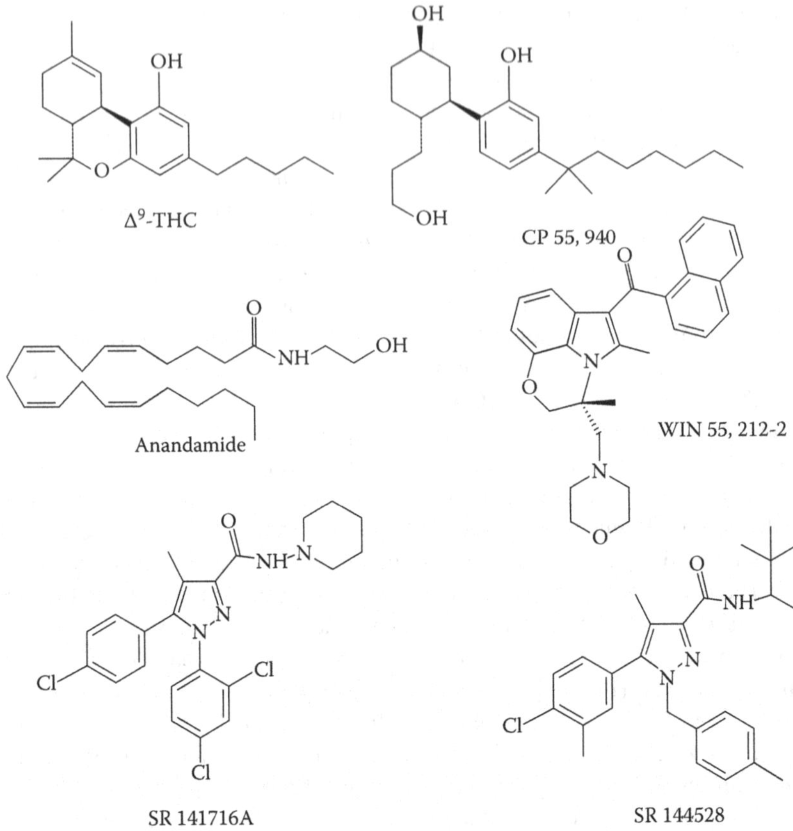

FIGURE 8.2 Different classes of ligands of cannabinoid receptors. Anandamide is the endogenous compound found originally in mammalian brain, Δ^9-THC is the major psychoactive compound isolated from the hemp *Cannabis sativa* L.; CP-55,940 was designed by Pfizer by ring opening and development of THC analogues; Win 55,212-2 is derived from pravadoline, an analgesic and anti-inflammatory compound; and SR141716A and SR144528 have been synthesized by Sanofi (nowadays Sanofi-Synthélabo) and are cannabinoid antagonists, selective for CB_1 and CB_2 receptors, respectively.

chemistry, we decided to investigate whether a 5,5′-diphenylhydantoin nucleus may constitute a new template for CB recognition by screening a very small library of hydantoins we had left. Comparing the structures, the diphenyl rings of the diphenylhydantoin may mimic the phenyl rings in the reference molecules, and the N_3 nitrogen of hydantoin would be useful for further N-alkylations. After a first hit with our small series, a set of 3-alkylated 5,5′-diphenylimidazolidinediones (**1–24**) was prepared[14] and tested by radioligand binding assay on transfected cells stably expressing the human CB_1.

At first glance, 3-alkyl 5.5′-diphenylimidazolinediones were readily obtained in two steps, summarized in Scheme 8.1. The first step, known as the Biltz synthesis, needs benzil or the corresponding substituted benzils, urea and KOH. The reagents were stirred in refluxing ethanol for 2 h. After cooling and washing with

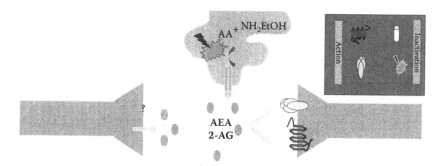

FIGURE 8.3 Putative and schematic representation of a cannabinoid synapse. The secretion step is still controversial; some authors proposed that anandamide and 2-arachidoylglycerol are cleaved upon demand from phospholipid precursors in the membrane. Once the endocannabinoids are in the synapse, they may activate the targets (GPCR: CB_1 and CB_2, ligand-gated channel receptor: the vanilloid VR_1) or be degraded by the inactivation system consisting of an active uptake process and a fatty acid amide hydrolase (FAAH).

0.5 N NaOH, the glycolureide product was removed by filtration. After the addition of acetic acid to the filtrate, a precipitate was obtained providing **25–29** in 65–88% yields. In a second step, the alkylation of **25–29** (1 mmol) by an excess (1.2 mmol) of bromo- or chloro-alkyl chains was carried out in anhydrous dimethylformamide in the presence of K_2CO_3 at room temperature overnight. However, looking carefully at the side-products of the reactions, we noticed that some amounts of bi-alkylated (N1–N3) or, surprisingly, even (N-O) bi-alkylated compounds were also obtained, giving both purification problems and low yields (Scheme 8.2). The nature and the amount of the side-products depend on the alkylating agent, the temperature, and the stoichiometry. An alternative synthesis is under investigation.

Apart from the diphenylhydantoin ring, we were also interested in having bulkier substituents in position 5, and we prepared the tryptophan hydantoin from the amino acid and potassium isocyanate. As D- and L-tryptophan are commercially available, both enantiomers have been prepared.

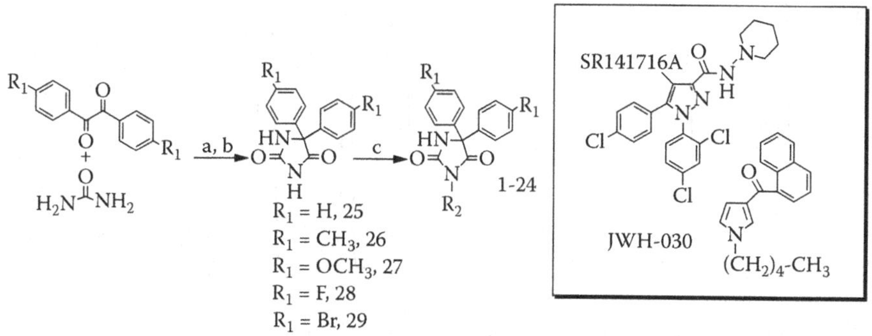

SCHEME 8.1 Synthesis of structures of target compounds (**1–24**) compared to SR141716A and JWH030. The conditions and reagents were: (a) KOH, C_2H5OH, reflux (b) CH_3COOH, and (c) R_2-Cl or R_2-Br, K_2CO_3, DMF.

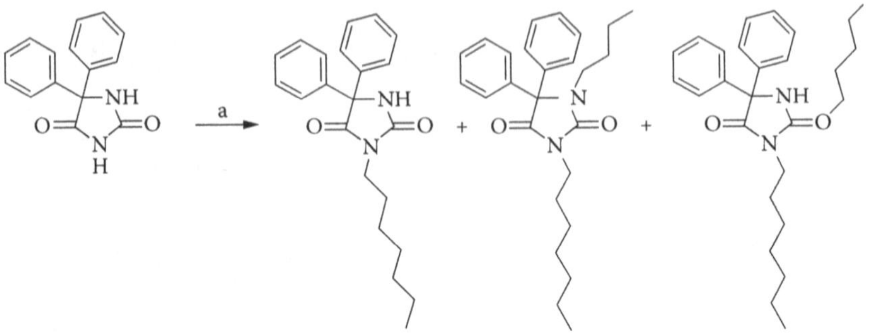

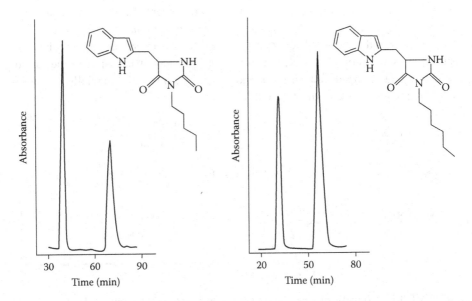

SCHEME 8.2 Alkylation of phenytoin by chloroheptane: desired compounds and side products. Reagents (a) chloroheptane, K_2CO_3, DMF.

Alkylation using bromo- or chloroalkanes has been achieved at room temperature in dimethylformamide in quite low yields. This reaction has some drawbacks. If there is a chiral center on the starting hydantoin (such as tryptophan hydantoins), racemization occurs during this step. A liquid chromatography system using chiral columns has been developed to resolve the enantiomers. Three most widely used polysaccharide-type column stationary phases, such as cellulose tris(4-methylbenzoate) (Chiralcel OJ), cellulose tris(3.5-dimelhylphenylcarbamate) (Chiralcel OD), and amylose tris(3,5-dimethylphenalcarbamate) (Chiralpak AD) have been tested for the analytical scale enantioseparations of the newly synthesized chiral hydantoin derivatives. An example of the resolution is given in Figure 8.4. Details of the analytical procedures have been published elsewhere. [20]

FIGURE 8.4 Separation of the enantiomers of the hydantoins on Chiralcel OD column. Mobile phase *n*-hexane/2-propanol 90/10 (v/v). Flow rate is 1 ml/min.

TABLE 8.1

Displacement of [³H]-SR 141716A Binding to CB_1 CHO Cell Membranes By Compounds 1–24

Compounds	R_1	n	R_2	Displacement at 10 (µM (%)	Displacement at 1 (µM (%)
1	H	2	-N(CH₂CH₂)₂O	<5	—
2	H	2	-N(CH₂)₅	<15	—
3	H	2	-N(CH₃)₂	<5	—
4	H	2	-CH₃	<20	—
5	H	3	-CH₃	25.1 ± 2.2	—
6	H	4	-CH₃	35.4 ± 2.9	—
7	H	5	-CH₃	35.6 ± 1.5	—
8	H	7	-CH₃	61.2 ± 4.7	—
9	H	1	-C₆H₅	40.6 ± 3.9	—
10	H	0	-CH(CH₃)₂	<5	—
11	CH₃	2	-N(CH₂CH₂)₂O	23.9 ± 1.9	—
12	CH₃	5	-CH₃	46.8 ± 3.9	—
13	CH₃	6	-CH₃	51.3 ± 3.8	—
14	OCH₃	2	-N(CH₂CH₂)₂O	21.7 ± 1.7	—
15	OCH₃	5	-CH₃	66.6 ± 5.3	23.0 ± 1.9
16	F	2	-N(CH₂CH₂)₂O	30.3 ± 2.1	—
17	F	5	-CH₃	40.6 ± 3.1	—
18	F	6	-CH₃	51.4 ± 2.9	—
19	F	7	-CH₃	62.5 ± 5.7	15.3 ± 1.1
20	Br	2	-N(CH₂CH₂)₂O	91.2 ± 7.3	54.1 ± 4.5
21	Br	3	-OH	88.4 ± 6.7	50.2 ± 5.3
22	Br	5	-CH₃	72.1 ± 5.3	30.2 ± 1.9
23	Br	6	-CH₃	89.2 ± 7.6	51.5 ± 3.4
24	Br	7	-CH₃	80.0 ± 6.0	48.2 ± 3.8

Note: Results are expressed as the percentages of the displaced specific radioactivity (mean ± sem, $n = 3$–5).

The affinity of this family of molecules has been determined on human cannabinoid receptors expressed after stable transfection in CHO cells. Euroscreen S.A. (Brussels, Belgium) has provided the cells. CHO cells in our hands selectively express either the CB_1 cannabinoid receptors or the CB_2 cannabinoid receptors. Several radioligands are commercially available; for the study we decided to use the CB_1 cannabinoid receptors [³H]-SR 1417I6A, a high-affinity radioligand selective for the CB_1 cannabinoid receptor, and [³H]-CP55,940, also a high-affinity radioligand but devoid of selectivity. Our synthesized molecules including 1–24 and the hydantoin intermediates 25–29 were screened at a first dose of 10 µM, and when >60% displacement of the specific radioactivity bound was obtained, they were further tested at 1 µM (Table 8.1). None of the intermediates 25–29 showed a significant displacement of the radioligand at 10 µM (data not shown). This point is important as the imidazolidinedione ligands are derived from phenytoin, a major anti-epileptic drug

TABLE 8.2

Ki determination of 20, 21, 23 and Reference Compounds

Compounds	Ki (nM) against [^{3}H]SR 141716A
20	70.3 ± 4.3
21	103.2 ± 6.8
23	97.9 ± 5.5
HU210	0.82 ± 0.004
CP 55940	5.2 ± 0.3
SR 141716A	8.9 ± 0.4
Win55212-2	152.2 ± 9.3

Note: Ki values are expressed as means ± sem ($n = 3–5$).

acting through sodium and calcium channels. It is known that some hindrance at the N_3 abolishes the anticonvulsant effect of phenytoin.

Whatever the nature of the R_1 substituent, the affinity for CB_1 increased with an increase in the length of alkyl chains as for the case of l-alkyl-3-(l-naphthoyl)pyrroles. The dibromo derivatives showed the highest affinity.

Three compounds, i.e., the 3-ethylmorpholino-5,5′-di-*p*-bromophenylimidazoli-dinedion3, **20**; 3-(l-hydroxypropyl)-5,5′-di-*p*-bromophenyl-imidazolidinedione, **21**; and 3-heptyl-5.5′-di-*p*-bromophenyl imidazolidinedione, **23**, were selected for further pharmacological evaluations, and their Ki are shown in Table 8.2. Compared to reference cannabinoids, the 3-alkyl-5,5-di-*p*-bromophenylimidazolidinediones exhibited an affinity inferior to those of classical cannabinoids (HU210, CP55940) but superior to that of the reference aminoalkylindole tested (WIN55212-2). The Ki obtained for **20, 21**, and **23** were close to the value described in the literature for l-pentyl-3-naphthoylpyrrole.[18]

These three new cannabinoid ligands have been further characterized with regard to their interactions with CB_1 cannabinoid receptors to determine if they act as agonists, partial agonists, antagonists, or inverse agonists. The advantage to working on G-protein-coupled receptors is that this characterization might be achieved based on the recognition between the G-protein and the receptor. Indeed, guanylyl nucleotides are known to disrupt the functional coupling of G-protein-coupled receptors, resulting in decreased affinity of agonists. Nonhydrolyzable GTP analogues such as 5′-guanylylimidodiphosphate (Gpp(NH)p) and guanosine-5′-*O*-(3-thiotriphosphate) (GTP-γ-S) constitute useful tools to distinguish between agonists and antagonists, the latter being unaffected by such uncoupling. Using either CHO cells expressing the CB_1 receptor or a preparation of rat cerebellar membranes, a tissue known to be rich in cannabinoid receptors, we measure the dose-dependent displacement curve of [^{3}H]-SR141716A by reference agonists (HU210. CP55,940) in the presence and in the absence of 5′-guanylylimidodiphosphate (Gpp(NH)p) (50 μM), a nonhydrolyzable GTP analogue. The addition of 5′-guanylylimidodiphosphate (Gpp(NH)p) (50 μM) to the binding assay significantly reduced the affinity of CP 55940 (IC$_{50}$ values: agonist-Gpp(NH)p, 14.4 nM; +Gpp(NH)p, 27.5 nM) but was without effect on the binding of **20, 21**, and **23**, supporting the evidence for their antagonist properties at the CB_1 receptor.

A further step was to determine with a similar technique the activity of DML20, DML21, and DML23 at CB_1 cannabinoid receptors. As the radiolabeled $[^{35}S]$-GTPγS is commercially available, a derived binding technique is used to characterize the activity of ligands at G-protein-coupled receptors, including cannabinoid receptors. This assay constitutes a functional measure of the interaction of the receptor and the G-protein, the first step in activation of the G-protein-coupled receptors. In addition, whatever the transduction mechanism, it is possible to define the functional activity of ligands as agonist (positive intrinsic activity), partial agonist (partial positive intrinsic activity), antagonist (no intrinsic activity), and inverse agonist (negative intrinsic activity). We measured the modulation of $[^{35}S]$-GTPγS binding on the rat cerebellar membranes by several reference cannabinoids, such as classical and nonclassical terpene ($Δ^8$-THC, $Δ^9$-THC, HU-210, CP-55,940, CP-55244, CP-55243), aminoalkylindole (WIN 55,212-2), and diarylpyrazole cannabinoids (SR141716), and by DML20–23. In our hands, HU-210, CP-55,940, CP-55244, CT-55243, and WIN 55,212-2 are full agonists in this assay, the natural plant derivatives $Δ^8$-THC, $Δ^9$-THC were partial agonists, and SR141716 decreases the stimulation, indicating that it is an inverse agonist. DML**20**, **21**, and **23** do not modify the basal level of $[^{35}S]$-GTPγS binding and, moreover, they antagonize the stimulation of the agonist HU210 with pK_B values of $6.11 ± 0.14$, $6.28 ± 0.06$, and $5.74 ± 0.09$, respectively, suggesting that these molecules are true antagonists without inverse agonist activity.

These data indicate that the 3-alkyl-5,5-(di(bromophenyl))imidazolidinediones represent a new chemical template with true antagonist properties and could either constitute new tools for cannabinoid pharmacology or new lead compounds for medicinal chemistry development.

FATTY ACID AMIDES DERIVED FROM
N-PALMITOYLETHANOLAMINE INTERACTING
WITH ANANDAMIDE METABOLISM

As mentioned previously, anandamide, the amide of arachidonic acid with ethanolamine, has been first proposed to act as an endogenous ligand for cannabinoid receptors. It has been isolated from pig brain.[7] Later, two congeners of anandamide, i.e., dihomo-γ-linolenoylethanolamide (C20:3) and docosatetraenoylethanolamide (C22:4) have been discovered also in the pig brain.[21] Endogenous esters have been identified as well. Indeed, almost at the same time, Mechoulam[22] and Sugiura[23] identified the presence of a 2-monoglyceride—the 2-arachidonoylglycerol—in different tissues, i.e., the intestine and the brain, respectively; 2-arachidonoylglycerol binds to cannabinoids receptors, albeit with a weaker affinity than that of anandamide for the CB_1 receptor, but was found in higher amounts in the body. This compound has been proposed as the preferential CB_2 endocannabinoid ligand.[24] The fascinating discovery of anandamide and other polyunsaturated fatty acid ethanolamides as endogenous ligands for the cannabinoid receptors renewed the interest of the scientific community in the fatty acid amides, the N-acylethanolamines. In fact, the N-acylethanolamines, including N-palmitoylethanolamine (PEA), a shorter and fully saturated analogue of anandamide, have been known for almost 30 years.

PEA is an intriguing molecule; it is a C16:0 fatty acid derivative where the carboxylate function is amidated by the primary amine of ethanolamine. Its chemical name is N-(2-hydroxyethyl)hexadecanamide, and is termed by some authors as *palmidrol*. This endogenous compound has been shown to accumulate during ischemia and inflammation and, in rodent animal models, it exhibits analgesic and anti-inflammatory properties.[25] The exact mechanism of its action is still under debate.[26] It has been proposed that PEA was a potent, nanomolar agonist of the cannabinoid CB_2 receptors, but, after all, it is widely accepted that it is inactive to cannabinoid receptors, albeit some of its pharmacological actions being reversed in the presence of a CB_2 antagonist.[27]

Anandamide has, as a result of its actions upon cannabinoid and vanilloid receptors, a number of interesting pharmacological properties, including effects on nociception, memory processes, lung function, spasticity, and cell proliferation. It constitutes, therefore, a lead compound for medicinal chemists, and since its discovery, numerous studies about the influence of the chemical modifications in the anandamide structure on receptor recognition or enzyme degradation have been performed. However, anandamide is metabolized by an FAAH, and inhibition of this enzyme by phenylmethylsulfonyl fluoride and palmitylsulfonyl fluoride has been shown to potentiate the pharmacological actions of anandamide. Our drug design is based on the possibility that compounds interfering with the metabolism of AEA may be therapeutically useful. As lead compounds, we decided to investigate fatty acid amides differing from the anandamide structure. Several FAAH inhibitors have been described, often based on the structure of anandamide itself; but in such cases, the possibility to interact with cannabinoid receptors directly persists and may hamper the approach with the central psychotropic side effects of CB_1 agonists. An alternative approach is to prevent the metabolism of AEA with compounds related to N-palmitoylethanolamine because this fatty acid amide, although a substrate for FAAH,[28] is inactive at cannabinoid receptors.[29]

The fatty acid amides interacting with anandamide metabolism are amide derivatives of palmitic and oleic acids. Varying the chain length of the different ethanolamides or the amide head of palmitic acid, a series of compounds have been characterized in terms of cannabinoid receptor recognition, FAAH activity, and influence of the anandamide uptake (Table 8.3). Propanoylethanolamine (C3:0), butanoylethanolamine (C4:0), hexanoylethanolamine (C6:0), octanoylethanolamine (C8:0), decanoylethanolamine (C10:0), lauroylethanolamine (C12:0), myristoylethanolamine (C14:0), palmiloylethanolamine (C16:0), palmitoylbutylamide (C16:0), palmitoylisopropylamide (C16:0), palmitoylcyclohexamide (CI6:0), stearoylethanolamide (Cl 8:0), and oleoylethanolamide (C18:l) have been prepared either from the respective acyl chlorides and excess of ethanolamine, or from the corresponding acids, ethanolamine and 1,1′-carbonyldiimidazole. With the exception of oleoylethanolamide, the compounds, at concentrations of 100 μM, did not produce dramatic inhibition of the binding of [³H]WIN 55,212-2 to human CB_2 receptors expressed on CHO cells. Similarly, PEA and palmitoylisopropylamide had modest effects upon the binding of [³H]CP 55,940 to human CB_1 receptors expressed on CHO cells at 100 μM, whereas oleoylethanolamide produced 65% inhibition of binding.

TABLE 8.3
Affinity of N-Palmitoylethanolamide Derivatives on Cannabinoid Receptors

	Inhibition of Specific Binding (%)			
Cell line:	CHO-CB$_1$		CHO-CB$_2$	
Ligand:	[^{3}H]CP 55,940		[^{3}H]WIN 55,212-2	
Compound	10 μM	100 μM	10 μM	100 μM
Propanoylethanolamide (C3:0)	19.5 ± 8.8	7.8 ± 4.3	8.9 ± 0.2	2.9 ± 1.0
Butanoylethanolamide (C4:0)	19.7 ± 5.5	8.0 ± 2.3	4.4 ± 0.5	3.4 ± 2.4
Hexanoylethanolamide (C6:0)	31.6 ± 5.15	8.0 ± 2.3	1.6 ± 0.3	6.2 ± 1.3
Octanoylethanolamide (C8:0)	26.7 ± 2.29	17.6 ± 5.3	5.7 ± 0.8	12.7 ± 0.8
Decanoylethanolamide (C10:0)	11.0 ± 0.98	12.3 ± 2.3	10.0 ± 0.8	17.3 ± 2.6
Lauroylethanolamide (C12:0)	21.9 ± 1.84	37.7 ± 8.3	5.1 ± 0.6	16.0 ± 2.3
Myristoylethanolamide (C14:0)	23.6 ± 6.16	41.1 ± 7.4	5.3 ± 0.6	21.9 ± 3.0
Palmitoylethanolamide (C16:0)	23.8 ± 0.07	25.4 ± 5.3	13.9 ± 1.7	24.5 ± 0.5
Stearoylethanolamide (C18:0)	30.3 ± 9.8	21.9 ± 6.4	3.4 ± 0.6	11.6 ± 1.8
Palmitoylbutylamide (C16:0)	19.6 ± 3.4	35.4 ± 5.3	12.9 ± 0.4	21.8 ± 0.7
Palmitoylisopropylamide (C16:0)	25.4 ± 4.5	14.5 ± 6.0	12.3 ± 0.8	21.1 ± 0.5
Palmitoylcyclohexamide (C16:0)	34.2 ± 1.5	37.5 ± 2.1	8.1 ± 0.2	13.8 ± 0.8
Oleoylethanolamide (C18:1)	13.3 ± 2.7	63.4 ± 3.9	14.5 ± 0.5	64.6 ± 1.5

Note: Shown are means ± s.e. mean of quadruplicate determinations.

All compounds have been tested in Chris Fowler Laboratory at the Umea University in Sweden for their ability to inhibit the uptake and FAAH-catalyzed hydrolysis of [^{3}H]anandamide ([^{3}H]AEA). Regarding the FAAH activity,[30] PEA inhibited the metabolism of [^{3}H]AEA by rat brain homogenates with a pI$_{50}$ value of 5.30. The potencies of the homologues were roughly similar to PEA as the chain length was reduced to 14 and 12 carbon atoms, but was lower for the C10:0 compound. Homologues with chain lengths <8 carbon atoms gave <20% inhibition at 100 μM. Palmitoylisopropylamide inhibited the metabolism of [^{3}H]AEA with pI$_{50}$ values of 4.89 (mixed-type inhibition). Palmitoylbutylamide and palmitoylcyclohexamide produced <25% inhibition at 100 μM. Oleoylethanolamide was a mixed-type inhibitor with a pI$_{50}$ value of 5.33.[31]

Assuming that PEA may use its own transporter,[32] the analogues and homologues have been tested in an assay regarding the uptake of [^{3}H]AEA using two sources of cells, C6 glioma and RBL-2H3 cells. Modest effects of most of the compounds upon the uptake of [^{3}H]AEA were observed. However, palmitoylcyclohexamide (100 μM) and palmitoylisopropylamide (30 and 100 μM) produced more inhibition of [^{3}H]AEA uptake than expected to result from inhibition of [^{3}H]AEA metabolism alone.[31]

Finally, in a combined model using intact C6 cells, palmitoylisopropylamide and oleoylethanolamide inhibited formation of [^{3}H]ethanolamine from [^{3}H]AEA with pI$_{50}$ values of 5.08 and 5.14.[31]

Some years ago, Sugiura et al.[23] interestingly found the presence of a monoacylglycerol analogue of PEA, and they described 2-arachidonylglycerol as a possible endogenous cannabinoid in the brain. Although the relative abundance of the monopalmitoylglycerols was high in the rat brain—11.2% of the monoacylglycerols detected—no information was provided as to the cannabinoid-receptor-binding profile of this molecule in that study. The two positional isomers 1(3)-palmitoylglycerol and 2-palmitoylglycerol are present in rat brain, the 2 isomer being more abundant. Later, Ben-Shabat et al.[33] reported that 2-palmiloylglycerol does not bind to cannabinoid CB_1 and CB_2 receptors. However, despite the fact that this endogenous glycerol ester and 2-linoleoylglycerol are intrinsically inactive, they enhance both the binding and the activity of the other endocannabinoid 2-arachidonylglycerol at cannabinoid receptors as well as its pharmacological effects in the mouse "tetrad" of behavioral tests of cannabinomimetic activity. This novel route of molecular regulation of the activity of an endogenous ligand received the name of "entourage effect."

Our compounds, without interacting directly with the cannabinoid receptors but by inhibiting the metabolism or the uptake of anandamide, act as "entourage" compounds. New pharmacomodulations of this series are under progress in two directions: compounds with increasing solubility and compounds that may selectively recognize either the FAAH or the uptake protein.

ACKNOWLEDGMENTS

This work was supported by grants from the Université Catholique de Louvain (UCL) and from the Belgian National Fund for Scientific Research (Crédits aux Chercheurs, 1.5.133.97 and 1.5.207.99. Convention FRSM 3.4601.01). The author wishes to thank Sanofi-Research and Pfizer Inc. for their donations of SR 141716A and CP 55940, respectively. He is also indebted to Euroscreen for providing CHO cells expressing human CB_1 and CB_2 receptors. And last but not least, the author would like to specially thank Chris J. Fowler and his team (Umea University, Sweden) for the fruitful collaboration on the endocannabinoid research project.

REFERENCES

1. Gaoni, Y. and Mechoulam. R., Isolation, structure and partial synthesis of an active constituent of hashish, *J. Am. Chem. Soc.*, 86, 1646, 1964.
2. Devane, W.A. et al., Determination and characterization of a cannabinoid receptor in rat brain, *Mol. Pharmacol.*, 34, 605, 1998.
3. Matsuda, L.A. et al., Structure of a cannabinoid receptor and functional expression of the cloned cDNA, *Nature*, 346, 561, 1990.
4. Gerard, C.M. et al., Molecular cloning of a human cannabinoid receptor which is also expressed in testis, *Biochem. J.*, 279, 129, 1991.
5. Munro, S., Thomas, K.L., and Abu-Shaar, M., Molecular characterization of a peripheral receptor for cannabinoids, *Nature*, 365, 61, 1993.
6. Shire, D. et al., An amino-terminal variant of the central cannabinoid receptor resulting from alternative splicing, *J. Biol. Chem.*, 270, 3726, 1995.
7. Devane, W.A. et al., Isolation and structure of brain constituent that binds to the cannabinoid receptor, *Science*, 258, 1946, 1992.
8. Barth, F. and Rinaldi-Carmona, M., The development of cannabinoid antagonists, *Curr. Med. Chem.*, 6,745, 1999.
9. Rinaldi-Carmona, M. et al., SR141716A, a potent and selective antagonist of the brain cannabinoid receptor, *EEBS Lett.*, 350, 240, 1994.

10. Rinaldi-Carmona, M. et al., SR144528, the first potent and selective antagonist of the CB$_2$ cannabinoid receptor, *J. Pharmacol. Exp. Ther.*, 284, 644, 1998.

11. Felder, C.C. et al., LY320135, a novel cannabinoid CB1 receptor antagonist, unmasks coupling of the CB1 receptor to stimulation of cAMP accumulation, *J. Pharmacol. Exp. Ther.*, 284, 291, 1998.

12. Landsman, R.S. et al., AM630 is an inverse agonist at the human cannabinoid CB1 receptor, *Life Sci.*, 62, PL109, 1998.

13. Reggio, P.H., Ligand-ligand and ligand-receptor approaches to modeling the cannabinoid CB1 and CB2 receptors: achievements and challenges, *Curr. Med. Chem.*, 6, 665, 1999.

14. Di Marzo, V. et al., Endocannabinoids: endogenous cannabinoid receptor ligands with neuromodulatory action, *Trends Neurosci.*, 21, 521, 1998.

15. Piomelli, D. et al., The endocannabinoid system as a target for therapeutic drugs, *Trends Pharm. Sci.*, 27, 218, 2000.

16. Lainton, J.A.H. et al., *l*-Alkyl-3-(*l*-naphthoyl)pyrroles—a new class of cannabinoid, *Tetrahedron Lett.*, 36, 140, 1995.

17. Huffman, J.W. and Lainton, J.A.H., Recent developments in the medicinal chemistry of cannabinoids, *Curr. Med. Chem.*, 3, 101, 1996.

18. Willey, J.L. et al., Structure–activity relationships of indole- and pyrrole-derived cannabinoids, *J. Pharm. Exp. Ther.*, 285, 995, 1998.

19. Kanyonyo, M.R. et al., 3-Alkyl-(5,5′-diphenyl)imidazolidinediones as new cannabinoid receptor ligands, *Bioorg. Med. Chem. Lett.*, 9, 2233, 1999.

20. Kartozia, I. et al., Comparative HPLC enantioseparation of new chiral hydantoin derivatives on three different polysaccharide type chiral stationary phases, *J. Pharm. Biomed. Anal.*, 27, 457, 2002.

21. Hanus, L. et al., Two new unsaturated fatty acid ethanolamides in brain that bind to the cannabinoid receptor, *J. Med. Chem.*, 36, 3032, 1993.

22. Mechoulam, R. et al., Identification of an endogenous 2-monoglyceride present in canine gut, that binds to cannabinoid receptors, *Biochem. Pharmacol.*, 50, 83, 1995.

23. Sugiura. T. et al., 2-Arachidonoylglycerol: a possible endogenous cannabinoid receptor ligand in brain, *Biochem. Biophys. Res. Commun.*, 275, 89, 1995.

24. Sugiura, T. et al., Evidence that 2-arachidonoylglycerol but not *N*-palmitoylethanol-amine or anandamide is the physiological ligand for the cannabinoid CB$_2$ receptor. Comparison of the agonistic activities of various cannabinoid receptor ligands in HL-60 cells, *J. Biol. Chem.*, 275, 605, 2000.

25. Mazzari, S. et al., *N*-(2-hydroxyethyl)hexadecamide is orally active in reducing edema formation and inflammatory hyperalgesia by down-modulating mast cell activation, *Eur. J. Pharmacol.*, 200, 227, 1996.

26. Lambert, D.M. and Di Marzo, V., The palmitoylethanolamide and oleamide enigmas: Are these two fatty acid amides cannabimimetic? *Curr. Med. Chem.*, 6, 757, 1999.

27. Calignano, A. et al., Control of pain initiation by endogenous cannabinoids, *Nature*, 394, 277, 1998.

28. Cravatt, B.F. et al., Molecular characterization of an enzyme that degrades neuromodu-latory fatty-acid amides, *Nature*, 384, 83, 1996.

29. Lambert, D.M. et al., Analogues and homologues of N-palmitoylethanolamide, a puta-tive endogenous CB$_2$ cannabinoid, as potential ligands for the cannabinoid receptors, *Biochim. Biophys. Acta*, 1440, 266, 1999.

30. Tiger, G., Strenström, A., and Fowler, C.J., Pharmacological properties of rat brain fatty acid amidohydrolase in different subcellular fractions using palmitoylethanol-amide as substrate, *Biochem. Pharmacol.*, 59, 647, 2000.

31. Jonsson, K.O. et al., Effects of homologues and analogues of palmitoylethanolamide upon the inactivation of the endocannabinoid anandamide, *Br. J. Pharmacol.*, 133, 1263, 2001.

32. Jacobsson, S.O.P. and Fowler, C.J., Characterization of palmitoylethanolamide transport in mouse Neuro-2a neuroblastoma and rat RBL-2H3 basophilic leukaemia cells: comparison with anandamide, *Br. J. Pharmacol.*, 132, 1743, 2001.
33. Ben-Shabat, S. et al., An entourage effect: inactive endogenous fatty acid glycerol esters enhance 2-arachidonoyl-glycerol cannabinoid activity, *Eur. J. Pharmacol.*, 353, 23, 1998.

9 An Appraisal of Fomocaines: *Current Situation and Outlook*

Herbert Oelschläger and Andreas Seeling

CONTENTS

INTRODUCTION

Some groups of drugs, e.g., anti-infectives, are undergoing a process of continual development. In contrast, the local anesthetics that are indispensable to attain freedom from pain in diagnostic and therapeutic interventions tend to be a rather quiescent group. Compared to general anesthetics, there are major regional differences in their proportional share in anesthetic techniques that vary between 5% and 70%. Despite the introduction of medical hypnosis, dental medicine is absolutely dependent on local anesthetics. Temporary abolition of pain sensation by chemical substances was achieved thanks to the Vienna ophthalmologist Karl Koller, who experimented with cocaine at the suggestion of Sigmund Freud.

Today, about 40 local anesthetics are available to the physician. These comprise three groups of compounds. From the group of basic esters, the first useful molecule was procaine, the conduction and infiltration anesthetic synthesized by Alfred Einhorn in 1904. In 1931, it was followed by tetracaine, which is mainly suitable as a topical anesthetic but is also very toxic (Figure 9.1).

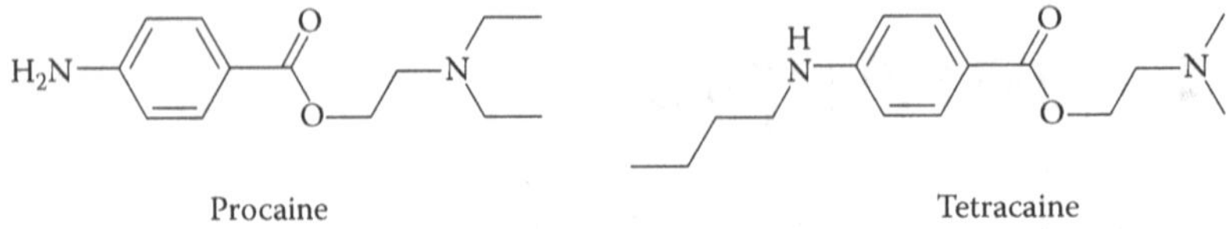

Procaine   Tetracaine

FIGURE 9.1   Important local anesthetics (type basic esters).

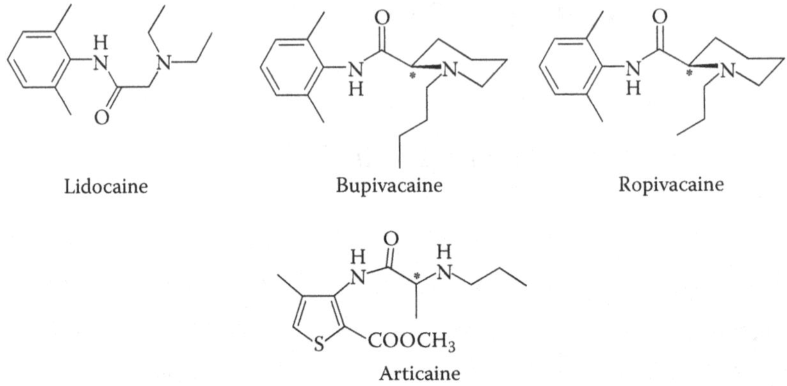

Lidocaine   Bupivacaine   Ropivacaine

Articaine

FIGURE 9.2   Important local anesthetics (type basic anilides).

In 1943, Niels Löfgren presented with lidocaine the group of basic anilides he had developed. Lidocaine is primarily a conduction and infiltration anesthetic but frequently causes allergies upon topical application. From this group, the chiral bupivacaine later assumed a special role in anesthesiology as a long-term anesthetic. Ropivacaine is the first important local anesthetic from the group of basic anilides, which is available as the S-enantiomer. Its clinical profile is better than that of bupivacaine. Above all, the cardiac toxicity of the S-enantiomer is lower than that of the R-enantiomer. In dental medicine, the chiral substance articaine is highly rated because of its rapid penetration of the bone wall (Figure 9.2 and Figure 9.3).

The group of basic ethers comprises only a few representatives, including fomocaine developed by Oelschlaeger in 1967. There are monographs for fomocaine in the German Pharmaceutical Codex (DAC) for both the hydrochloride and the base (Scheme 9.1).

The local anesthetics introduced differ considerably in their i.v. toxicity (Table 9.1).

As lipophilic molecules, all local anesthetics block the ion channels, especially the voltage-dependent sodium channels, by nonspecific membrane expansion (Figure 9.3).

Reduction of membrane permeability means reduced depolarization to complete undepolarization of the nerve fibers. However, the chief process responsible for the local anesthetic action is the interaction of the protonated form with a binding site inside the channels (Figure 9.4).

At higher doses, potassium channels are affected in addition. Up to now, the binding area of the latter has not yet been identified. On the other hand, in the resting

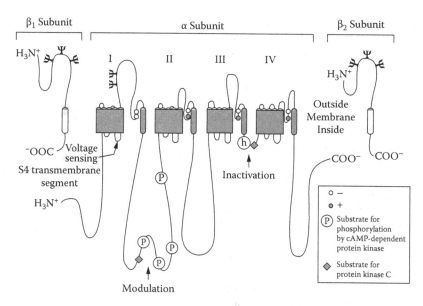

FIGURE 9.3 Structure of the voltage-gated sodium channel.

FOMOCAINE HYDROCHLORIDE

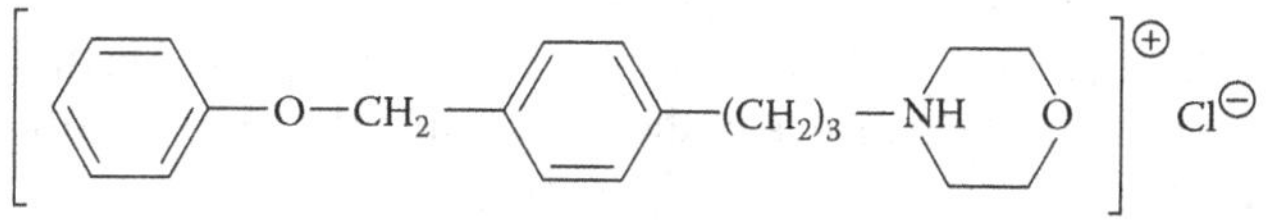

4–[3–(4-Phenoxymethylphenyl)-propyl]morpholine-hydrochloride

SYNONYM: Fomocainum hydrochloricum.

SCHEME 9.1 DAC monograph.

TABLE 9.1
Toxicity of Important Local Anesthetics

	i.v. LD50 in Mouse (mg/kg)	i.p. LD_{50} in Mouse (mg/kg)
Procaine	56	167
Tetracaine	10.5	53
Lidocaine	44	120
Bupivacaine	8	54
Ropivacaine	12	—
Articaine	37	—
Fomocaine	175	230

state sodium ions can penetrate the membrane, which is otherwise impermeable to these ions. This takes place through a hydrophilic pore formed by a protein that spans the entire diameter of the cell wall. The α subunit (M_r = 260 kDa) of this glycoprotein (M_r = 330 kDa) consisting of about 1900 amino acids is responsible for the gate function and the sodium selectivity. The subunits β-1 (M_r = 36 kDa) and β-2 (M_r = 33 kDa) determine the spatial orientation in the membrane. The subunits α-1 and β-1 are linked by low binding, whereas the β-2 subunit is linked covalently to the α-subunit by disulfide bridges. The homologous transmembrane sectors of the protein designated I to IV in the detailed representation in Figure 9.2 are oriented to each other in such a way that they can form a channel under appropriate conditions. The protein segment 6 from substructure IV contains the local anesthetic binding area with the amino acid L-phe

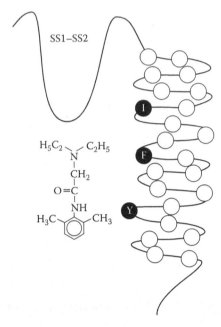

FIGURE 9.4 Binding site for local anesthetics inside the sodium channel.

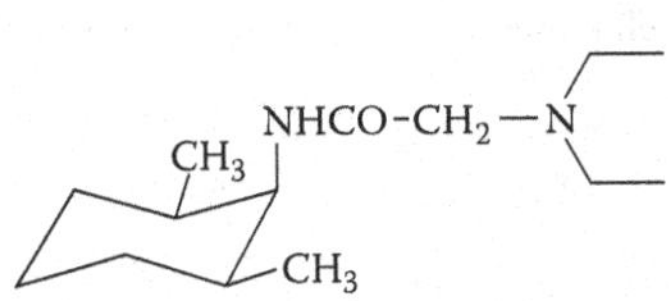

in position 1764 and L-tyr in 1771. Thus, dispersion forces and hydrophobic forces can act between the aromatic rings of the protein and the local anesthetic as well as the alkyl bridge.

cis-cis-hexahydroderivative of lidocaine

SCHEME 9.2 Derivative of lidocaine.

The importance of aromaticity for the local anesthetic effect could already be established by Oelschläger et al.[1] in 1969 by the synthesis and pharmacological testing of the three stereoisomeric hexahydro derivatives of lidocaine. In conduction and topical anesthesia, only the cis-cis isomer was about half as active as lidocaine (Scheme 9.2). The two other stereoisomers were inactive.

SYNTHESES OF FOMOCAINE (INN)

The first technical synthesis of fomocaine (Figure 9.5) entails a Blanc chloromethylation as the key reaction. This was optimized very carefully to keep the percentage of the ortho isomer low.[2]

The second technique (Figure 9.6) also works without regioisomers. Essentially, it entails a Willgerodt–Kindler reaction in which the thioamide is reductively desulfurated with Raney nickel.[3]

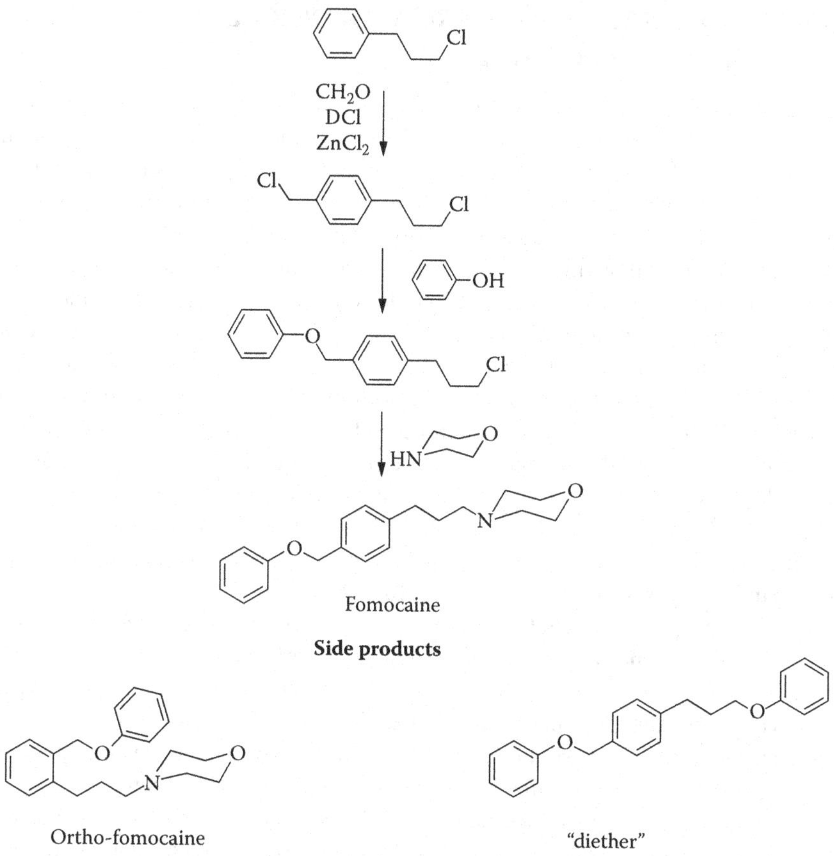

Fomocaine

Side products

Ortho-fomocaine   "diether"

FIGURE 9.5   First technical synthesis of fomocaine.

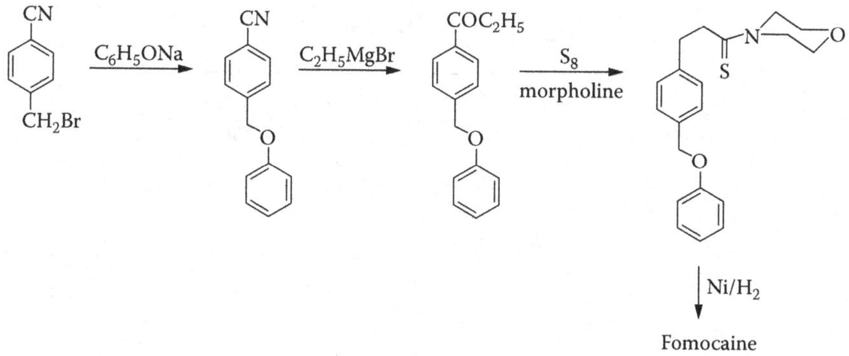

FIGURE 9.6   Second technical synthesis of fomocaine.

PHARMACOKINETIC AND PHARMACOLOGICAL PROPERTIES OF FOMOCAINE

The pharmacokinetic and pharmacological properties of fomocaine have been investigated in detail. Its protein binding is 95%. Fomocaine is mainly bound by serum albumin. The globulins are responsible for about 10% of the protein binding. High-resolution NMR spectroscopy (Jardetzky method) showed that the binding to serum albumin is essentially mediated by the two aromatic rings and the alkene groups, whereas the morpholine ring does not bind. Surprisingly enough, the erythrocytes are not a deep compartment for fomocaine. In topical anesthesia, fomocaine proved to be as effective as tetracaine, whereas it was much more active than procaine and lidocaine in conduction anesthesia. The low toxicity of fomocaine is an advantage. With an LD_{50} of 175 mg/kg in intravenous administration to the mouse, fomocaine is tolerated 17 times better than tetracaine and 4 times better than lidocaine. The low toxicity can be readily explained by the minimal effects on the cardiovascular system. Even at a high dosage, fomocaine brings about only a very brief fall in blood pressure. The low toxicity is due to the morpholine residue.[4] For example, as a topical anesthetic, the piperidine derivative is three times more toxic than fomocaine. More than 20 base variations were tested (Table 9.2).

The spasmolytic effect ($BaCl_2$ spasm of the isolated guinea pig intestine) has the same order of magnitude as papaverine.[5] A sensitizing effect of fomocaine could not be demonstrated in guinea pigs after epicutaneous or subcutaneous application.[6] Experiments with fomocaine administered to test subjects in the repeated patch test for 40 days did not give rise to any skin inflammation. It was established in the Ames test with a certainty of 90% that fomocaine has no carcinogenic and no mutagenic effects.

The biotransformation of fomocaine was also investigated for new indications with two animal species (dog [beagle] and rat) with ^{14}C-labeled fomocaine. The synthesis of ^{14}C-fomocaine starts with 4-bromocinnamic acid[7] (Figure 9.7 and Figure 9.8).

TABLE 9.2
Influence of The Basic Moiety on Toxicity (Selection)

Basic Moiety [R]	i.v. LD_{50} Mouse (mg/kg)
	175
	150
	65
	36
	30

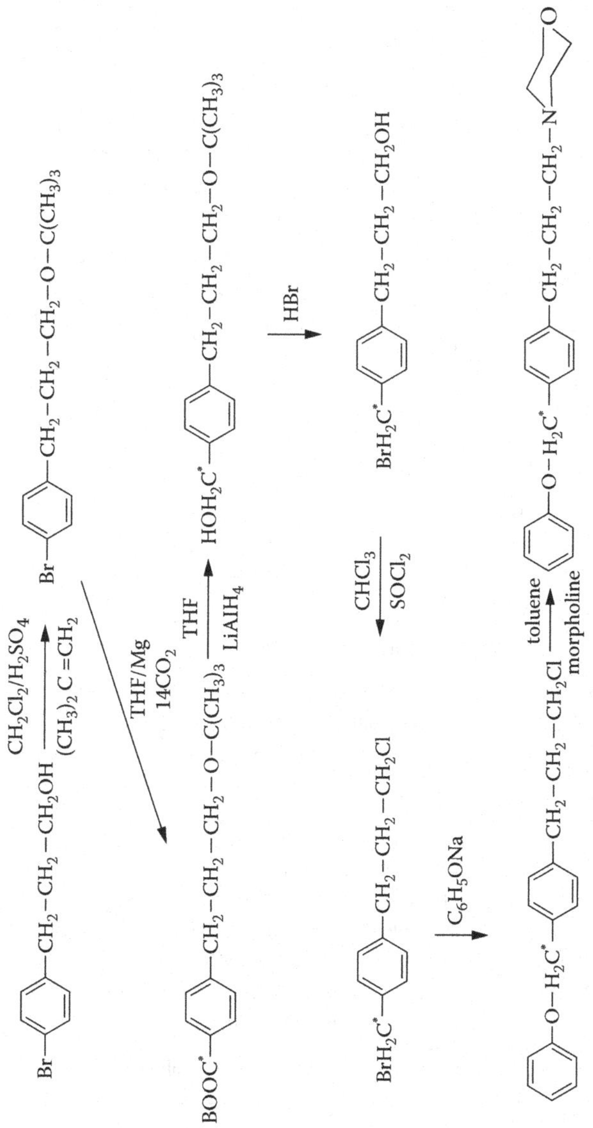

FIGURE 9.7 Synthesis of ^{14}C-fomocaine.

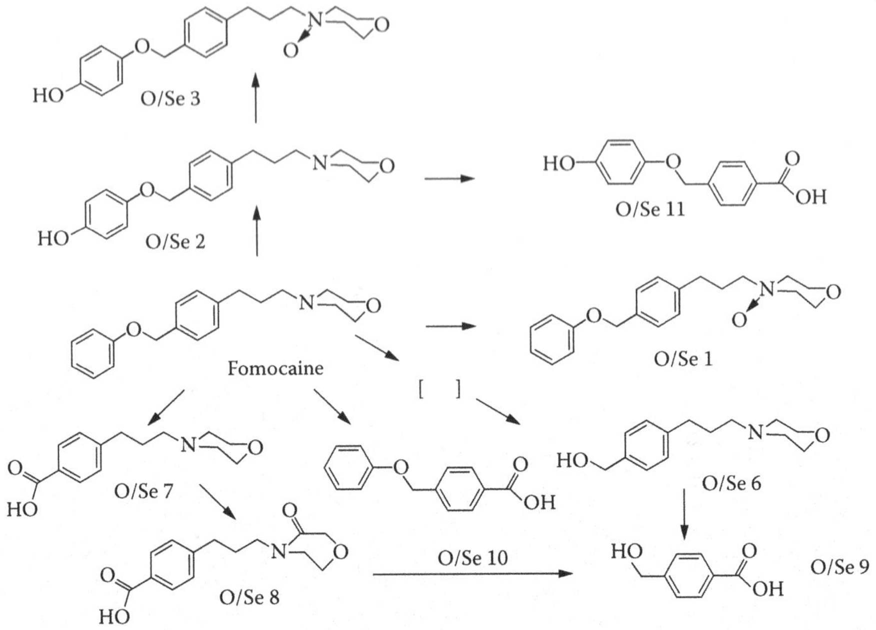

FIGURE 9.8   Metabolites in beagle.

TABLE 9.3
Results with ¹⁴C-fomocaine (Beagle)

After 48 h	38.7% of radioactivity found in urine and feces
After 168 h	94.3% of radioactivity in urine and feces

Identification of Metabolites in Urine

4.0%	Fomocaine
	Fomocaine-N-oxide (traces)
38.3 %	4-Hydroxy-fomocaine
1.7%	4-Hydroxy-fomocaine-N-oxide
9.3%	Morpholino-propyl-benzoic acid
6.1%	Morpholino-propyl-benzyl-alcohol
1.8%	4′-(Phenoxymethyl)-3-phenylpropionic acid
8.0%	4′-(Phenoxymethyl)-benzoic acid

Note: 69.2% of radioactivity found as defined metabolites.

The results with ¹⁴C-fomocaine administered to beagle are shown in Table 9.3. The hydroxylation of the aromatic exo-ring (O/Se 2) with subsequent conjugation predominates in the beagle, whereas cleavage of the ether bond predominates in the rat (O/Se 7). The pattern of the biotransformation in humans largely corresponds to that in the beagle. Altogether, fomocaine forms 14 metabolites (Table 9.4).

TABLE 9.4

Fomocaine Metabolites Demonstrated in Beagle and Rat Up To Now

1. 4-[3-(4-Phenoxymethylphenyl)-propyl]-morpholine-N-oxide (O/Se 1)
2. 4-[3-(4-(4'-Hydroxyphenoxy)-methylphenyl)-propyl]-morpholine (O/Se 2)
3. 4-[3-(4-(4'-Hydroxyphenoxy)-methylphenyl)-propyl]-morpholine-N-oxide (O/Se 3)
4. 4-[3-(4-(2'-Hydroxyphenoxy)-methylphenyl)-propyl]-morpholine (O/Se 4)
5. 4-[3-(4-(2'-Hydroxyphenoxy)-methylphenyl)-propyl]-morpholine-N-oxide (O/Se 5)
6. 4-[3-(4-Hydroxymethylphenyl)-propyl]-morpholine (O/Se 6)
7. 4-[3-(Morpholine-4-yl)-propyl]-benzoic acid (O/Se 7)
8. 4-[3-(Tetrahydro-1,4-oxazine-3-on-4-yl)-propyl]-benzoic acid (O/Se 8)
9. 4-Hydroxymethyl-benzoic acid (O/Se 9)
10. 4-Phenoxymethyl-benzoic acid (O/Se 10)
11. 4-(4'-Hydroxyphenoxy)-methyl-benzoic acid (O/Se 11)
12. 3-(4-Phenoxymethyl)-phenyl-propionic acid (O/Se 12)
13. Morpholine
14. Phenol

Their structure has been clarified and they have been synthesized. In addition, their physical–chemical properties, their pharmacological action, and their toxicity were investigated in detail.[8] With the exception of one compound (compound 4), they do not show any pharmacological action and are, moreover, atoxic. N-oxidation is of subordinate importance. Thiofomocaines are attacked preferentially at the sulfur atom.

STRUCTURE–ACTIVITY RELATIONSHIPS

The structure–activity relationships in the fomocaine group were clarified comprehensively in numerous studies comprising more than 100 substances.[9] A shortening or increasing of the alkane side chain leads to a reduction of the topical anesthetic action. The activity optimum is associated with the propane chain. Exchanging the morpholine residue for other base residues generally led to an increase in the toxicity. The effect depends on the aromaticity of the exo-ring. The corresponding cyclohexyl derivative of fomocaine has only a very weak activity. The introduction of fluorine and bromine into the p-position of the phenolic moiety enhances the effect but causes irritation of the conjunctiva of the test animal at the same time. The introduction of polar groups, e.g., a p-nitro, p-amino, or p-acetamino group into the exo-ring greatly reduces the activity. The corresponding β-naphthyl derivative is inactive. Bioisosteric substitution did not result in any advantage.

A field that is largely neglected in local anesthetic research is the question of the influence of chirality on the local anesthetic action. Scanty information is available merely for the group of basic anilides. S-enantiomer, the recently introduced ropivacaine is longer acting and shows better cardiac tolerance than the R-enantiomer. After we had found that fomocaine acts on sodium, potassium, and calcium channels with specific fomocaine-binding sites, whether enantiomeric fomocaines show a different effect and toxicity became a pressing question. A total of 14 fomocaines were therefore synthesized achirally and first of all separated via their diastereomeric

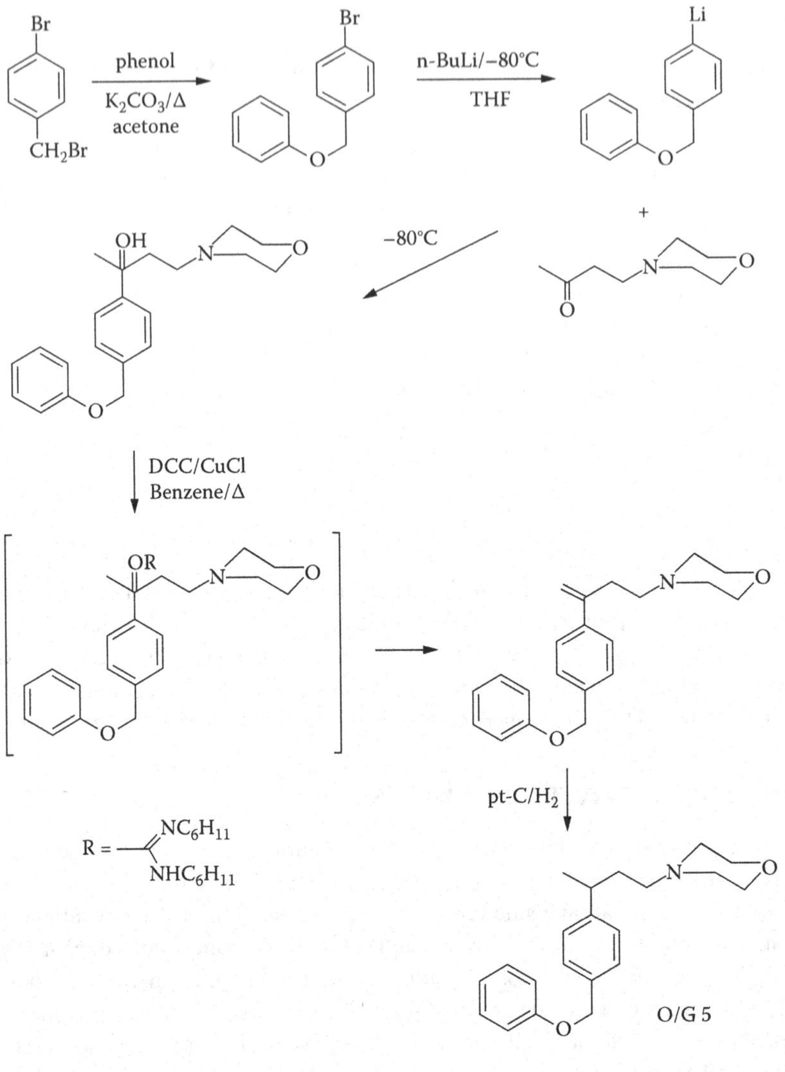

FIGURE 9.9 Synthesis of the chiral fomocaine derivative O/G 5.

salts with R(-) and S(+) camphor sulfonic acid. We recently also have achieved separation on chiral Daicel OD columns on a preparative scale. The achiral synthesis of fomocaine derivatives entailed multifarious difficulties. The synthesis of O/G 5 (Figure 9.9) is taken as an example.[10]

The result of pharmacological testing was surprising.[11] There were no noteworthy differences between the effect of the enantiomers of O/G 3 and O/G 5. In contrast to the racemate of O/G 3, the racemate of O/G 5 has a good topical anesthetic action, but this was weaker than that of fomocaine. We interpret the findings so far as an indication that a specific receptor for topical anesthetics that was postulated long ago does not exist. We also examined the enzymatic degradation of the enantiomers of O/G 3 and O/G 5 with the 10,000 × g supernatant of fresh porcine liver homog-

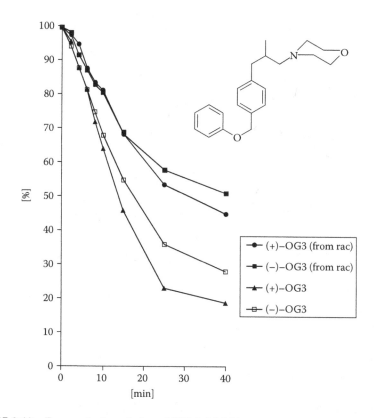

FIGURE 9.10 Enzymatic degradation of O/G 3 (10,000 × g supernatant).

enate (Figure 9.10 and Figure 9.11). It was shown that the (-) enantiomer of O/G 5 is degraded much more rapidly than its optical antipode. The same observation applies to the optical enantiomers of O/G 3.

In numerous studies, regioisomers of fomocaine were also investigated with regard to their local anesthetic action. The synthesis of these regioisomers required new approaches (Figure 9.12). The ortho-isomer proved to be just as active as fomocaine,[12] whereas in the meta-series the compound of O/C 17 with a shortened alkene side chain was much stronger as a topical anesthetic but had a weaker effect in conduction anesthesia than fomocaine.

A derivative in the ortho series, namely O/Mor 2 (Scheme 9.3), proved to be an effective oral anti-arrhythmic in clinical studies.

Investigations with "inverse" fomocaine (Figure 9.13) showed that the tissue tolerance depends on the position of the basic group. For example, as with fomocaine, the compound O/VH 1 acted as a topical anesthetic but gave rise to necroses in infiltration anesthesia.

The cumulative total of all pharmacological and toxicological investigations in fomocaines (especially fomocaine) is becoming crucially important for ongoing developments on new indications, including a new drug for treating migraines. The starting point was the investigation of Maizels et al.,[13] who tested the intranasal

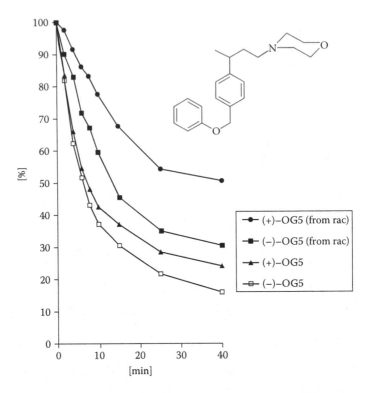

FIGURE 9.11   Enzymatic degradation of O/G 5 (10,000 × g supernatant).

administration of lidocaine solutions for treatment of migraines. In a prospectively randomized placebo-controlled double-blind study, it was found in 67 women and 14 men that there was an appreciable reduction of headaches in 50% of the patients at least 15 min after administration of 4% lidocaine solution, whereas headaches were reduced in only 6 patients in the control group that had received placebo. Vomiting and photophobia were significantly reduced. The authors explain the effect in terms of paralysis of the sphenopalatine ganglion.

OUTLOOK

With regard to the development of an antimigraine drug, we have also tested the bis-[2-hydroxy-ethyl]amine derivative besides fomocaine (DAC) (Scheme 9.4) as well as further compounds.

   As a base, the former shows a much better water solubility, and its local anesthetic action exceeds that of fomocaine with roughly the same toxicity. Both fomocaines should be administered as solids with an inhaler, i.e., not as solutions (Table 9.5).

   If local anesthesia becomes established for treatment of migraines, an appreciable risk is entailed in the long-term regular use of lidocaine. As shown by experiments with rats, 2,6-xylidine, which has a carcinogenic action in high doses, is formed during body passage.[14] On the other hand, the use of fomocaine does not entail a

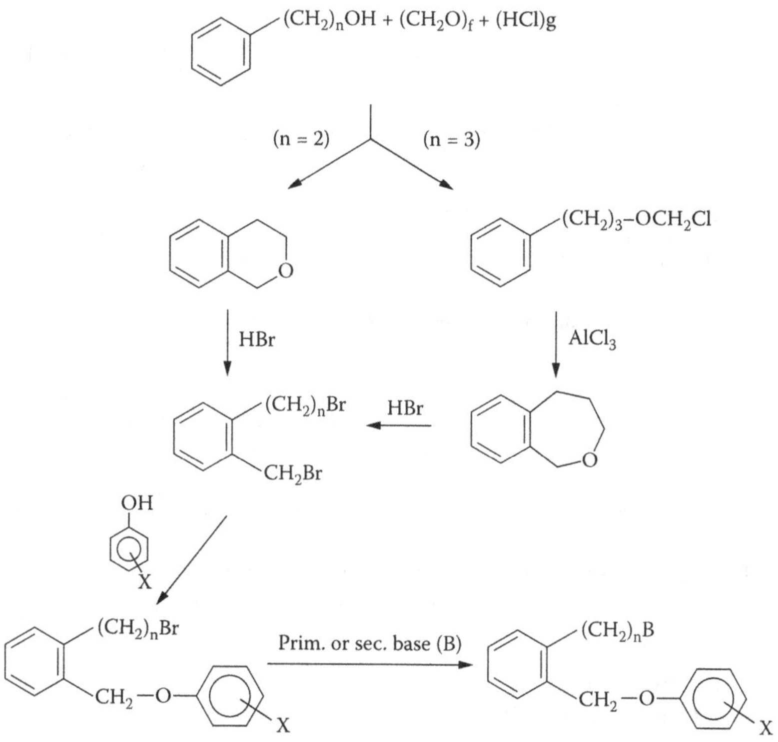

FIGURE 9.12 Synthesis of ortho-fomocaines.

risk because the biotransformation leads to atoxic metabolites and both derivatives under testing are tolerated much better than lidocaine. In addition, the two fomocaines will be tested for their suitability as neurotherapeutics.

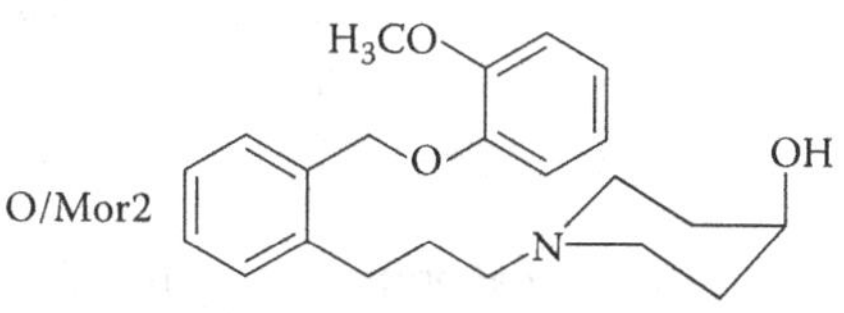

SCHEME 9.3 Anti-arrhythmic O/Mor 2.

CONCLUSION

The fomocaines are a large group of more than 300 base-substituted aryl(alkyl)-benzyl ethers, which are characterized by remarkable physical and chemical stability. As ion-channel blockers, they act as local anesthetics and have an anti-arrhythmic action on heart muscle (class I B), besides having a musculotropic spasmolytic action. Their toxicity largely depends on the basic components. The 4-[3-(4-phenoxymethylphenyl)-propyl]-morpholine (fomocaine INN) for which monographs have been published both for the hydrochloride and the base in the German Pharmaceutical Codex (DAC) has been used for purposes of treatment since 1967. Its pharmacological

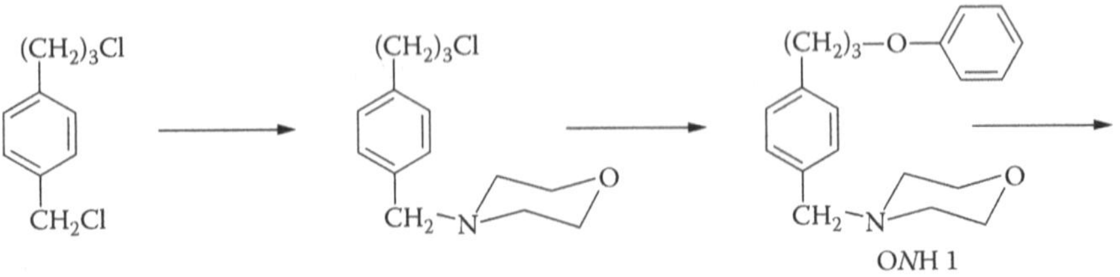

Topical Anaesthesia: similar to Tetracaine
DL_{50} (mouse i.v.): 117 mg/kg
all inverse fomocaines cause necrosis

FIGURE 9.13 Synthesis of inverse fomocaines.

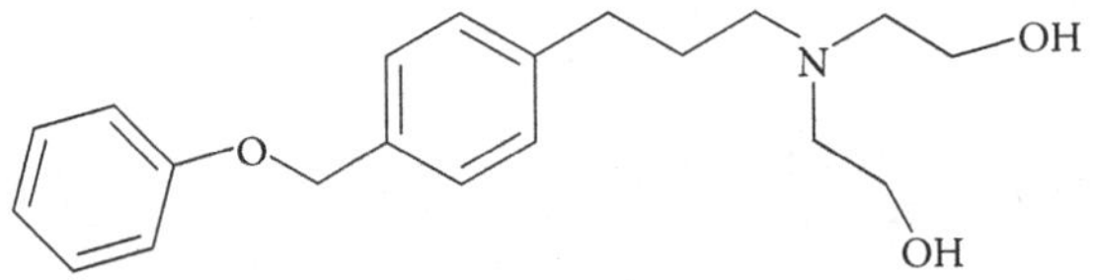

SCHEME 9.4 Structure of the bis-[2-hydroxy-ethyl]amine derivative.

TABLE 9.5
Comparison of Physical Data

Fomocaine (A) versus bis-[2-hydroxy-ethyl]-amino-derivative (B)

		A	B
pK_a		7.1	6.1
D	(HPLC)	3.7	2.7
P	(HPLC)	3.9	2.7
Solubility	(HaO)	6.5×10^{-5} m/L	1.6×10^{-4} m/L
Clouding	pH	5.7	7.1

and toxicological properties were elucidated in comprehensive studies; in particular, the pharmacology and toxicology of its 12 most important metabolites and their physicochemical properties. Fomocaine is accordingly a highly effective topical conduction anesthetic of low toxicity without sensitizing properties. Derivatives are undergoing clinical testing for further indications. Structure–activity relationships based on numerous compounds that have been meticulously tested pharmacologically reveal the importance of the alkyl bridge, the aromaticity of the exo-ring, the regioselectivity, and the chirality.

ACKNOWLEDGMENT

This research was supported by Deutsche Forschungsgemeinschaft, Bonn-Bad Godesberg, and Fonds der Chemischen Industrie, Frankfurt a. M.

REFERENCES

1. Oelschläger, H. et al., Synthese und Wirkung der Ringhydrierungsprodukte des Lidocains, *Arch. Pharm.* (Weinheim, Ger.), 308, 603, 1975.
2. Oelschläger, H. et al., Neue Umweitfreundliche Synthese des 3-(4-Chlormethyl-phenyi)-propylchlorids, eines Zwischenprodukts der technischen Fomocain-Produktion, *Pharm. Ztg. Wiss.*, 4/136, 159, 1991.
3. Oelschläger, H. et al., Über eine neue Fomocain-Synthese (10. Mitt. über Synthesen neuer Verbindungen mit lokal-anästhetischer Wirkung), *Arzneim.-Forsch./Drug Res.*, 27, 1625, 1977.
4. Oelschlager, H., Darstellung des lokalanasthesierenden Aminoethers N-[γ-(4-Phenoxy-methyl-phenyl) propyl]-morpholin und Beziehungen zwischen Toxizität und basischer Komponente bei N-[γ-(4-Phenoxymethyl-phenyl)-propyl]-aminen (3. Mitt. über Synthesen neuer Verbindungen mit lokalanästhetischer Wirkung), *Arzneim.-Forsch./Drug Res.*, 9, 313, 1959.
5. Nieschulz, O., Hoffmann, L., and Popendiker, K., Zur Pharmakologie des N-[4-(Phe-noxy-methyl)-y-phenyl-propyl]-morpholin, eines Aminoethers mit lokalanästhetischer Wirkung, *Arzneim.-Forsch./Drug Res.*, 8, 539, 1958.
6. Oelschläger, H. et al., N-[y-(4-Phenoxymethyl-phenyl)-propyl]-morpholin, ein neues Oberflachenanasthetikum von bemerkenswerter Vermiglichkeit (6. Mitt. über Synthesen neuer Verbindungen mit lokalanästhetischer Wirkung), *Arzneim.-Forsch./Drug Res.*, 18, 729, 1968.
7. Oelschläger, H. and Ewald, H.W., Synthese von ^{14}C-Fomocain (10. Mitt. über Synthesen neuer Verbindungen mit lokalanästhetischer Wirkung), *Arch. Pharm. (Weinheim, Ger.)*, 319, 113, 1986.
8. Seeling, A., Leuschner, F., and Oelschläger, H., Die Metabolite des Fomocain (DAC) aus physikalisch-chemischer, pharmakologischer und toxikologischer Sicht, *Arzneim.-Forsch/Drug Res.*, 51, 7, 2001.
9. Fleck, C. et al., Effects and toxicity of new fomocaine derivatives and of 2-hydroxy-propyl-beta-cyclodextrin inclusion compounds in rats, *Arzneim.-Forsch/Drug Res.*, 54, 265, 2004.
10. Oelschläger, H. et al., Synthese und pharmakoiogische Wirkung chiraler Fomocaine ({4-[2-Methyl-3-(morpholin-4-yl)propyl]benzyl}-phenyl-ether und {4-[l-Methyl-3-(morpholin-4-yl)propyl]benzyl}-phenyl-ether (13. Mitteilung zur Synthese neuer Verbindungen mit lokalanästhetischer Wirkung), *Pharmazie*, 56, 620, 2001.
11. Fleck, C. et al., Local anaesthetic effectivity and toxicity of fomocaine, five N-free fomocaine metabolites and two chiralic fomocaine derivatives in rats compared with procaine, *Arzneim.-Forsch./Drug Res.*, 51, 451, 2001.
12. Oelschläger, H., Dünges, W., and Götze, G., Darstellung lokalanästhesierender ortho-aryloxy-methylierter (ω-Arnino-alkylbenzole (8. Mitt, über Synthesen neuer Verbindungen mit lokalanästhetischer Wirkung), *Arzneim.-Forsch./Drug Res.*, 22. 1013, 1972.
13. Maizels, M. et al., lntranasal lidocaine for treatment of migraine: a randomized, double-blind, controlled trial, *J. Am. Med. Soc.*, 276, 319, 1996.
14. U.S. National Toxicology Program: Technical Report Series No. 278; U.S. Dept. of Health and Human Services, NTH, 1990.

10 Ligands for the GABA Recognition Site at the GABA_A Receptor: *Structure–Activity Studies*

Bente Frølund, Uffe Kristiansen,
Povl Krogsgaard-Larsen, and Tommy Liljefors

CONTENTS

ABSTRACT

Besides playing an important role in central neurotransmission processes, $GABA_A$ receptors have been implied in a number of neurological and psychiatric disorders. These observations have focused interest on the molecular mechanisms of receptor activation and the structural basis of receptor–ligand interaction. To study the structural requirements for receptor–ligand interaction, different areas in the ligand-binding site of the $GABA_A$ receptor have been evaluated by development and application of a 3-D pharmacophore model based on a number of known $GABA_A$ agonists. During this study, a series of highly potent and selective $GABA_A$ antagonists, based on the structure of the partial $GABA_A$ agonist 5-(4-piperidyl)-3-isoxazolol (4-PIOL), has been developed. These results demonstrate the presence of a receptor pocket of considerable dimensions containing specific sites for receptor interactions in the vicinity of the 4-position of the 3-isoxazolol ring in 4-PIOL.

INTRODUCTION

The major inhibitory neurotransmitter in the mammalian central nervous system, 4-aminobutyric acid (GABA) mediates its actions through the ionotropic $GABA_A$

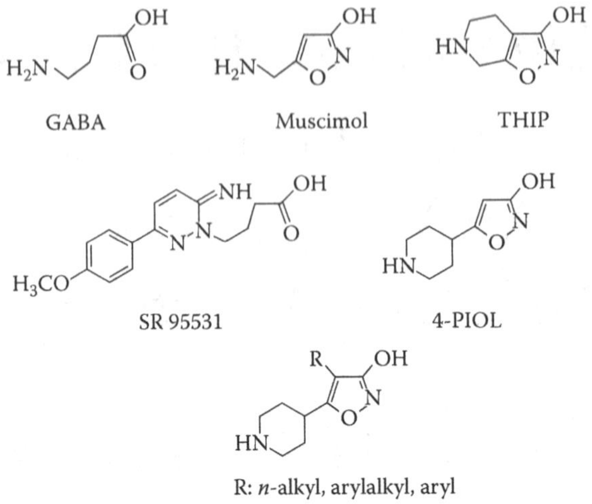

FIGURE 10.1   Structures of GABA, the GABA$_A$ agonists muscimol and THIP, the GABA$_A$ antagonist SR 95531, the low-efficacy partial GABA$_A$ agonist 4-PIOL, and a general structure of a series of 4-substituted 4-PIOL analogues.

and GABA$_C$ receptors and the metabotropic GABA$_B$ receptors.[1,2] Besides playing a role in central transmission processes, these receptors, especially the GABA$_A$ receptors, seem to be involved in certain neurological and psychiatric disorders.[3,4]

GABA$_A$ receptors are built as pentameric assemblies of receptor subunits grouped in different families. Although a large number of subunits have been identified, only a limited number of physiological GABA$_A$ receptors are believed to exist.[5] In addition to the recognition site for GABA, the GABA$_A$ receptor complex comprises modulatory sites for a number of therapeutic agents, including benzodiazepines, barbiturates, neurosteroids, and volatile anesthetics.

There is currently no experimentally determined 3-D structure for the GABA$_A$ receptor complex, but the recently published x-ray structure of a soluble acetylcholine-binding protein[6] has been used for homology building of the extracellular ligand-binding domain of these receptors.[7] However, due to the low degree of sequence identity, it is not straightforward to use such models as templates for the prediction of GABA$_A$ ligand affinity or SAR studies.

The design of selective ligands for the GABA recognition site at the GABA$_A$ receptor is of key importance for elucidation of the structure and function of different subtypes of GABA$_A$ receptors. Within the limited series of compounds showing agonist activity at the GABA$_A$ receptor site, most ligands are structurally derived from the GABA$_A$ receptor agonists muscimol[8] and 4,5,6,7-tetrahydroisoxazolol[5,4-c]pyridin-3-ol (THIP)[8,9] (Figure 10.1), which were developed at the initial stage of the project. A group of structurally different ligands has been identified as selective competitive antagonists for the GABA$_A$ receptor recognition site. The standard GABA$_A$ antagonist, SR 95531 an arylaminopyridazine derivative of GABA, is a representative of this group.[10]

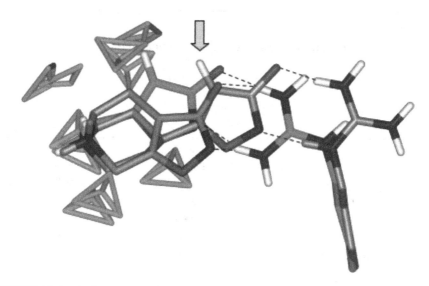

FIGURE 10.2   A pharmacophore model of GABA_A agonists displaying the proposed binding modes of muscimol, THIP, and 4-PIOL. Dashed lines indicate hydrogen bond interactions. The tetrahedrons indicate positions of methyl groups in GABA_A agonists causing strong steric repulsions with the receptor. The arrow points to the 4-isoxazolyl position in 4-PIOL.

STRUCTURE–ACTIVITY STUDIES

Based on pharmacological data for a number of GABA_A ligands, including muscimol, THIP, and 5-(4-piperidyl)-3-isoxazolol (4-PIOL)[11] and a series of analogues, a 3-D pharmacophore for the GABA-binding site has been developed.[12,13] Although muscimol, THIP, and 4-PIOL are assumed to bind to the same binding pocket, the 3-isoxazolol rings of these three compounds do not seem to superimpose and, in contrast to muscimol and THIP, a receptor cavity in the vicinity of the 4-positions of the 3-isoxazolol ring of 4-PIOL seems to exist (Figure 10.2).[12,14]

To investigate this cavity further and extending the pharmacophore model, a number of analogues of 4-PIOL with substituents in the 4-position of the 3-isoxazolol ring have been synthesized.[14,15] Substituents of different size and conformational flexibility such as alkyl, phenylalkyl, diphenylalkyl, and naphthylalkyl have been explored. The compounds were pharmacologically characterized in receptor-binding studies using rat brain membrane preparations and electrophysiologically using whole-cell patch-clamp recordings from cultured cerebral cortical neurons and two-electrode voltage-clamp recordings from human $\alpha_1\beta_2\gamma_{2S}$ GABA_A receptors expressed in *Xenopus* oocytes.[16]

As with 4-PIOL, all of the compounds were shown to interact selectively with the GABA_A-receptor-binding site. The pharmacological data for a representative group of analogues are listed in Table 10.1.

Unbranched alkyl groups as substituents in the 4-position of the 3-isoxazolol ring of 4-PIOL as well as a benzyl group are tolerated. Compounds **1**, **2**, and **3** show affinities for the GABA_A receptor sites comparable to or slightly higher than that of 4-PIOL. The optimal length of the alkyl chain connecting the phenyl group and

TABLE 10.1
Receptor Binding and *In Vitro* Electrophysiological Data

	R	[³H]]muscimol Bindingª K_i (nM)	Electrophysiology[b] IC_{50} (nM)
4-PIOL	H-	9100	110000
1	CH_3-	10000	26000
2	$CH_3(CH_2)_2$-	6600	4600
3	CH_2– (benzyl)	3800	4000
4	$CH_2(CH_2)_2$– (phenyl)	1100	530
5	$(CH_2)_2$– (diphenylmethyl)	68	20
6	CH_2– (2-naphthylmethyl)	49	370
7	CH_2– (1-naphthylmethyl)	100	480
8	CH_2– (biphenylmethyl)	400	710
9	(cyclohexenyl)	220	159
SR 95531	–	74	240

ª Standard receptor binding on rat brain synaptic membranes, $n = 3$.

[b] Whole-cell patch-clamp recordings from cerebral cortical neurons cultured for 7–9 days, $n = 6$–17.

Source: Frølund, B. et al., A novel class of potent 4-arylalkyl substituted 3-isoxazolol $GABA_A$ antagonists: Synthesis, pharmacology and molecular modeling, *J. Med. Chem.*, 45, 2454, 2002.

the 3-isoxazolol moiety was found to be
three carbon atoms (compound **4**).

The addition of another phenyl group
to the terminal carbon atom of the tri-
methylene group of compound **4** gave the
3,3-diphenylpropyl compound, **5**, which
displayed a 16-fold increase in affinity
relative to the 3-phenylpropyl analog, **4**,
indicating that the added phenyl group
in the 3,3-diphenylpropyl compound
is occupying a position in the receptor
cavity favorable for binding. A further
increase in affinity was achieved by
introducing the 2-naphthylmethyl group
into the 4-position of the 3-isoxazolol
ring of 4-PIOL to give compound **6**. This
led to a 78-fold increase in affinity as
compared to that of the benzyl analog, **3**.
Introduction of the isomeric 1-naphthyl-
methyl group proved to be less favorable,
compound **7** showing a decrease in binding affinity compared to the 2-naphthyl-
methyl analog, **6**. The marked effect on
affinity of an additional phenyl group in
the high-affinity compounds **5** and **6** can-

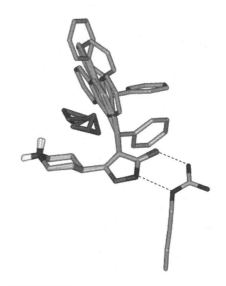

FIGURE 10.3   A superimposition of a series
of 4-substituted 4-PIOL compounds previously
studied[14] illustrating the large space spanned
by the 4-substituents. The tetrahedrons indicate positions of methyl groups in GABA_A
agonists causing strong steric repulsions with
the receptor.

not be rationalized by an increase in lipophilicity or differences in conformational
energy. Therefore, strong interaction between the binding cavity of the GABA_A
receptor and the unsubstituted aromatic rings may exist.[14] A superimposition of the
bioactive forms of the two compounds (Figure 10.3) shows that one of the phenyl
rings in compound **5** overlaps closely with the distal ring of the 2-naphthylmethyl
analog, **6**, indicating that this position is highly favorable for GABA_A receptor binding. By enlarging the aromatic system by attaching a phenyl group to the 4-position
of the phenyl group of **3** to give the 4-biphenylmethyl analog **8**, the receptor affinity
was increased tenfold relative to that of the benzyl analog **3**. These results reflect the
large space spanned by the high-affinity compounds in this series. The length of the
4-biphenylmethyl substituent in **8** is 9.7 Å, measured from the methyl carbon to the
4-hydrogen atom of the distal ring, and one of the phenyl rings in **5**, extends 6.4 Å in
a direction almost perpendicular to the rings in **6** and **8**.

To explore the space surrounding the 2-naphthyl ring system of the high-affinity
compound **6** in the receptor cavity, a bromosubstituent was introduced in different positions. In this study, the 1-position of the ring system (compound **10**) was
shown to be the most favorable position in terms of increasing the affinity for the
GABA_A receptor (Table 10.2).[15] The influence of different substituents in the 1-posi-
tion of the naphthyl ring system on binding to the GABA_A receptor was further
investigated. Introduction of a chloro (**11**), fluoro (**12**), cyano (**13**), methylthio (**14**),
or even a phenyl substituent (**15**) in the 1-position of the 2-naphthyl ring system had

TABLE 10.2

Receptor Binding and *In Vitro* Electrophysiological Data

	R	[³H]muscimol Binding[a] K_i (nM)	Electrophysiology[b] IC_{50} (nM)
6	H	49	370
10	Br	10	42
11	Cl	16	113
12	F	19	242
13	CN	28	140
14	CH₃S	28	61
15	C₆H₅	21	198

[a] Standard receptor binding on rat brain synaptic membranes, $n = 3$.

[b] Whole-cell patch-clamp recordings from cerebral cortical neurons cultured for 7–9 days, $n = 6$–17.

Source: Frølund, B. et al. Potent 4-aryl- or 4-arylalkyl-substituted 3-isoxazolol GABA(A) antagonists: Synthesis, pharmacology, and molecular modeling, *J. Med. Chem.*, 48, 427, 2005.

limited effects on the affinity compared to **6** (Table 10.2). This strongly suggests that these substituents are located in a water-exposed receptor area and are only partially desolvated in the receptor cavity. On the basis of molecular modeling studies, these substituents seem to occupy a previously unexplored receptor region in the pharmacophore model (Figure 10.4).[15]

As a result of attaching a phenyl group directly to the 4-position of the 3-isoxazolol ring in 4-PIOL, compound **9** shows a 40-fold higher affinity for the GABA_A receptor relative to that of the parent compound, whereas the affinity of the 4-benzyl-4-PIOL compound is only about 3 times higher than that of the 4-methyl-substituted compound. Thus, the interaction between a phenyl group

FIGURE 10.4 Regions in the vicinity of the 4-position of 4-PIOL in the direction of the C4-H bond of the 3-isoxazolol ring previously explored by aromatic substituents[14] and the new regions explored using compounds **9** and **15**.

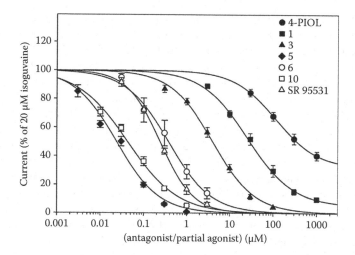

FIGURE 10.5 Effect of the antagonists on the response to 20 μM of the full GABA$_A$ agonist isoguvacine using whole-cell patch-clamp recordings from cultured cerebral cortical neurons. A total of 20 μM isoguvacine and varying concentrations of antagonists were applied simultaneously to the cells. The response of 20 μM isoguvacine alone has been set as 100%, and the other responses are expressed as a fraction of this. The response to isoguvacine is progressively reduced with increasing concentrations of the antagonist. The number of cells (*n*) tested for each compound varied from *n* = 6 to 17.

directly attached to the 3-isoxazolol ring and the receptor is significantly stronger than that displayed by a phenyl group linked to the 4-PIOL system via a methylene group. As for the substituents in compounds **10–15**, the phenyl in compound **9** represents a highly favorable position for an aromatic ring in a region, which has not been studied previously (Figure 10.4).[15]

In the whole-cell patch-clamp studies on cortical neurons, 4-PIOL and selected analogues were characterized as competitive GABA$_A$ receptor antagonists by their ability to inhibit the current induced by the GABA$_A$ agonist isoguvacine in a dose-dependent manner (Figure 10.5). The determined IC$_{50}$ values, shown in Table 10.1, correspond well with recent data from voltage clamp studies on recombinant human $\alpha_1\beta_2\gamma_{2S}$ GABA$_A$ receptors expressed in oocytes and show, with a few exceptions, a fairly good correlation with the receptor-binding data.[16]

4-PIOL has previously been characterized as a partial agonist in whole-cell patch-clamp studies on cerebral cortical neurons and cultured hippocampal neurons.[17,18] Only the methyl and ethyl analogues in the present series retained detectable ability to induce agonist effects as shown using cultured cerebral cortical neurons and recombinant receptors.[14,16] In the study using recombinant GABA$_A$ receptors, the weak responses of the methyl and ethyl analogues were potentiated by simultaneously administered lorazepam and inhibited by the competitive antagonist SR 95531.[16]

As seen from the data in Table 10.1 and Table 10.2, the major consequence of replacing the hydrogen in the 4-position of the 3-isoxazolol ring of 4-PIOL by the lipophilic and bulky aryl or arylalkyl groups is a change of the pharmacology of the compounds from moderately potent, low-efficacy partial GABA$_A$ receptor agonist activity to potent antagonist effects. The 2-naphthylmethyl (**6**), 3,3-diphenylpropyl (**5**), and the 1-bromo (**10**) analogues showed the highest $GABA_A$ receptor affinities of the present series of compounds. The affinities were comparable with or markedly higher than that of the standard $GABA_A$ receptor antagonist SR 95531.

As the GABA-binding site in the $GABA_A$ receptor is assumed to be located at the interface between α and β subunits,[19] it may be speculated that the large cavity accommodating the 4-substituent is located in the space between these subunits. It has been proposed that the mechanism for ligand-induced channel opening in nicotinic acetylcholine receptors involves rotations of the subunits surrounding the ligand-binding domain.[20,21] It is likely that the $GABA_A$ and the nicotinic acetylcholine receptors, belonging to the same super family of ligand-gated ion channels, are using the same mechanism for channel opening. Based on this hypothesis, the large substituents in the 4-substituted 4-PIOL analogues may interfere with the conformational channel-opening transition of the $GABA_A$ receptor, resulting in antagonistic effects of the compounds.

CONCLUSION

In the absence of detailed structural information on the $GABA_A$ receptor complex, the design of ligands for this receptor has thus far been based on systematic structural, stereochemical, and bioisosteric approaches. As demonstrated in the present study, combined pharmacological and molecular modeling studies have been useful for predicting the optimal arrangement of the functional groups of ligands for the $GABA_A$ receptor. With this approach, some apparent differences in structure–activity relationships of a series of known $GABA_A$ agonists have been clarified, and new information has been provided regarding the size and dimensions of the receptor-binding site.

REFERENCES

1. Enna, S.J. and Bowery, N.G., *The GABA Receptors*, The Humana Press, Totowa, NJ, 1997.
2. Chebib, M. and Johnston, G.A.R., GABA-activated ligand gated ion channels: medicinal chemistry and molecular biology, *J. Med. Chem.*, 43, 1427, 2000.
3. Thomsen, C. and Ebert, B., Modulators of the GABA receptor. Novel therapeutic prospects, in *Glutamate and GABA Receptors and Transporters: Structure, Function and Pharmacology*, Taylor & Francis, New York, 2002, pp. 407–427.
4. Mehta, A.K. and Ticku, M.K., An update on $GABA_A$ receptors, *Brain Res. Rev.*, 29, 196, 1999.
5. McKernan, R.M. and Whiting, P.J., Which $GABA_A$-receptor subtypes really occur in the brain? *Trends Neurosci.*, 19, 139, 1996.
6. Brejc, K. et al. Crystal structure of an ACh-binding protein reveals the ligand-binding domain of nicotinic receptors, *Nature*, 411, 269, 2001.
7. Ernst, M. et al. Comparative modeling of GABA(A) receptors: limits, insights, future developments, *Neuroscience*, 119, 933, 2003.
8. Krogsgaard-Larsen, P. et al. Structure and biological activity of a series of conformationally restricted analogs of GABA, *J. Neurochem.*, 25, 803, 1975.
9. Krogsgaard-Larsen, P. et al. A new class of GABA agonist, *Nature*, 268, 53, 1977.

10. Wermuth, C.G. et al. Synthesis and structure-activity relationships of a series of amino-pyridazine derivatives of γ-aminobutyric acid acting as selective GABA-A antagonists, *J. Med. Chem.*, 30, 239, 1987.

11. Byberg, J.R. et al. Synthesis and biological activity of a GABA-A agonist which has no effect on benzodiazepine binding and structurally related glycine antagonists, *Drug Des. Delivery*, 1, 261, 1987.

12. Frølund, B. et al. A novel class of potent 3-isoxazolol GABA$_A$ antagonists: Design, synthesis, and pharmacology, *J. Med. Chem.*, 43, 4930, 2000.

13. Frølund, B. et al. Design and synthesis of a new series of 4-alkylated 3-isoxazolol GABA A antagonists, *Eur. J. Med. Chem.*, 38, 447, 2003.

14. Frølund, B. et al. A novel class of potent 4-arylalkyl substituted 3-isoxazolol GABA$_A$ antagonists: Synthesis, pharmacology and molecular modeling, *J. Med. Chem.*, 45, 2454, 2002.

15. Frølund, B. et al. Potent 4-aryl- or 4-arylalkyl-substituted 3-isoxazolol GABA(A) antagonists: synthesis, pharmacology, and molecular modeling, *J. Med. Chem.*, 48, 427, 2005.

16. Mortensen, M. et al. Activity of novel 4-PIOL analogs at human a1b2g2S GABA$_A$ receptors—correlation with hydrophobicity, *Eur. J. Pharmacol.*, 451, 125, 2002.

17. Kristiansen, U. et al. Electrophysiological studies of the GABA$_A$ receptor ligand, 4-PIOL, on cultured hippocampal neurons, *Br. J. Pharmacol.*, 104, 85, 1991.

18. Frølund, B. et al. Partial GABA$_A$ receptor agonists. Synthesis and *in vitro* pharmacology of a series of nonannulated analogs of 4,5,6,7-tetrahydroisoxazolo[4,5-*c*]pyridin-3-ol (THIP), *J. Med. Chem.*, 38, 3287, 1995.

19. Smith, G.B. and Olsen, R.W., Functional domains of GABA$_A$ receptors, *Trends Pharmacol. Sci.*, 16, 162, 1995.

20. Unwin, N. et al. Activation of the nicotinic acetylcholine receptor involves a switch in conformation of the alpha subunits, *J. Mol. Biol.*, 319, 1165, 2002.

21. Unwin, N., Acetylcholine receptor channel imaged in the open state, *Nature*, 373, 37, 1995.

11 Strategies for Development of New Lead Structures for Inhibition of Acetylcholinesterase

*Petra Kapkova, Vildan Alptuzun,
Eberhard Heller, Eva Kugelmann, Gerd
Folkers, and Ulrike Holzgrabe*

CONTENTS

INTRODUCTION

One of the most prominent features of Alzheimer's disease (AD) is a significant deficit in cholinergic transmission in certain brain areas.[1,2] The concentrations of acetylcholine (ACh) decrease by nearly 90% in patients with AD. Therefore, one main focus of AD treatment is the use of agents that increase the availability of intrinsic ACh by inhibiting the acetylcholinesterase (AChE) enzyme. This may restore the cholinergic function in the brain and significantly reduce the severity of dementia. Another key feature in brain pathology of patients with AD is the progressive deposition of the amyloid β (Aβ) peptide in fibrillar form.[3]

The polymerization of normally soluble proteins or peptides in β-pleated sheet form yielding insoluble amyloid deposits is a key step in development of a variety of so-called amyloid diseases such as AD and prion diseases.[4] In AD, amyloid deposits in diffuse form or as plaques are composed of Aβ, a 39- to 43-amino acid peptide derived from a larger parent known as amyloid precursor protein (APP).[5] In vitro studies revealed the prevalence of the higher molecular weight species Aβ 1-42 to extensive amyloid formation.[6] As with other amyloidogenic proteins, the in vitro polymerization of Aβ into fibrils occurs spontaneously; presumably the protein is present at a certain critical concentration.[7]

Several substances of peptidic and non-peptidic origin with anti-amyloid potencies were reported. Among the latter are endogenous compounds such as melatonin, rifampicin, benzofuran, and naphthylazo derivatives,[8–10] classical antibiotics such as tetracycline and doxycycline,[11] and cyclodextrins.[12] Other researchers have patented several isoindoline[13] and piperidine[14] derivatives that inhibit amyloid protein aggregation and deposition.

Other studies suggest a role for AChE not only in the hydrolysis of the neurotransmitter ACh, but also in acceleration of the aggregation of Aβ into amyloid fibrils. Responsible for this activity should be the peripheral anionic site (PAS) of AChE gorge.[15] It is expected that even a peripheral site blocker would prevent the Aβ peptide from interacting with AChE, thus inhibiting a potential contribution of AChE to the fibril formation process. Studies regarding development of such AChE inhibitors were already reported for coumarin derivatives.[16,17]

Consequently, inhibition of AChE (possibly PAS inhibition of AChE), reduction of Aβ production by blocking the APP cleavage enzymes, and inhibition of Aβ fibril formation are strategies for the development of therapeutic interventions in AD. Most AD clinicians and researchers concede that no single treatment or therapeutic approach will alleviate this condition, and that a multiple-functional approach or multipotent drugs are required.[18]

In this context, we report on AChE inhibition by means of a structurally diverse series of pyridinium-type derivatives and the ability of some representatives to block Aβ fibril formation. X-ray structures of the enzyme co-crystallized with various ligands[19–23] provided insights into the essential structural elements and motifs central to the catalytic mechanism and mode of ACh processing. One of the striking structural features of AChE revealed from x-ray analysis is the presence of a narrow, long, and hydrophobic gorge that is approximately 20Å deep. Aromatic residues lining the gorge constitute a series of binding sites for quaternary ammonium moieties.[24] The interaction between highly potent inhibitors such as tacrine and donepezil and the enzyme is characterized by cation–π interactions of protonated nitrogens and tryptophan (Trp84) and phenylalanine (Phe330) residues. Moreover, π–π stacking between the aromatic moieties of the inhibitors and the conserved aromatic amino acids (Trp84 and Phe330), as well as ion–ion-interactions between the protonated nitrogens of the inhibitors and anionic asparaginic acid (Asp72) play a crucial role.

DUO COMPOUNDS AS DITOPIC INHIBITORS

The first aim of this study was to synthesize and evaluate the inhibition of the AChE activity of bisbenzyl ethers of the bispyridinium-type compound TMB-4

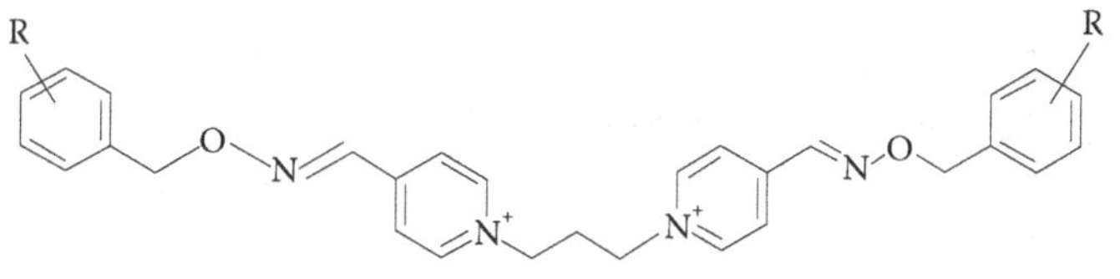

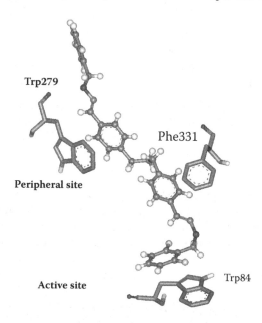

FIGURE 11.1   Skeletons of DUO compounds.

(Figure 11.1), which is analogous to the obidoxime reactivator. The compounds are ditopic ligands that were designed to allow binding to both the active and peripheral sites characterized by two tryptophan residues, Trp84 at the active site and Trp279 at the mouth of the gorge. Among many other inhibitors, ditopic bis-galanthamin,[25] bis-tacrine,[26,27] a huperzine A–tacrine hybrid,[28] and 1,n-bispyridiniumalkanes[29] showed better inhibitory potencies than their respective monomers.

Since substituents attached to the aromatic rings of the benzyl moieties gradually change the electronic and lipophilic properties of these rings, they are supposed to influence the interaction of the entire molecule with the enzyme. Therefore, an attempt was made to achieve a wide variation in lipophilic and electronic properties according to the Craig plot (σ versus π).[30] leading to introduction of halogen atoms, methyl, nitro, methoxy, and cyano groups in ortho, meta, and para positions of the aromatic ring.

The 4,4′-bis-[(benzyloxyimino)-methyl)-1,1′-propan-1,3-diyl-bispyridinium dibromides **1–25** (Table 11.1) were synthesized in two steps. The oxime ethers were

FIGURE 11.2   General binding modes of DUO compounds to active and peripheral sites of AChE.

TABLE 11.1

$IC_{50 \pm}$ SEM Data for Compounds 1 through 26

Compound	Substitution Pattern (R)	$IC_{50} \pm$ SEM [μM]
1	H	3.51 ± 0.70
2	2-Cl	0.58 ± 0.18
3	3-Cl	Not available
4	4-Cl	4.46 ± 1.04
5	2,6-Cl*	0.34 ± 0.05
6	3,5-Cl	0.81 ± 0.09
7	2-CH$_3$	1.70 ± 0.24
8	3-CH$_3$	2.07 ± 0.31
9	4-CH$_3$	7.96 ± 1.83
11	2-CN	1.32 ± 0.21
12	3-CN	1.42 ± 0.34
13	4-CN	9.00 ± 0.44
14	2-NO$_2$	1.37 ± 0.17
15	3-NO$_2$	0.92 ± 0.08
16	4-NO$_2$	0.72 ± 0.18
17	2-OCH$_3$	2.81 ± 0.39
18	3-OCH$_3$	1.95 ± 0.25
19	4-OCH$_3$	20.66 ± 1.69
20	2,6-OCH$_3$	11.37 ± 0.73
21	2,5-OCH$_3$	11.72 ± 2.25
22	2-Br	1.15 ± 0.16
23	2,6-F	2.68 ± 0.36
24	1-naphthyl	0.88 ± 0.09
25	2-naphthyl	1.36 ± 0.06
26 (tacrine)		0.044 ± 0.004

Note: Pharmacological data = mean ± SEM. Measurements were performed as
    described in Reference 32.
* DUO3.

formed via a phase transfer reaction between the pyridine-4-carbaldehyde oxime and
the corresponding benzyl halogenide in the presence of a phase-transfer catalyst tet-
ramethylammonium chloride, aqueous sodium hydroxide solution and dichloromethane, or in presence of sodium methylate/methanol. The bispyridinium compounds
can be obtained by conversion of two moles of the oxime ethers with 1,3-dibrom-
propane in acetonitrile. The yields of both reactions are satisfactory, especially after
microwave application.

The inhibitory potency against AChE was evaluated by means of Ellman's test.[31]
The IC_{50} values are shown in Table 11.1. For comparison purposes, the activity of the
potent tacrine inhibitor was also determined and found to be in the same as reported
in the literature. A qualitative inspection of the IC_{50} values of the DUO compounds

revealed the halogenated compounds to be the most active and the chlorine substitution to be superior to fluorine and bromine substitution (2/22 and 5/23). In contrast, compounds bearing one or two methoxy groups were less active, especially in the case of para substitutions. With the exception of nitro substituents, substituents in para positions seemed to be disadvantageous.[32]

To explain the differences in activities of the compounds tested, docking studies investigating the possible binding modes of the compounds in the binding pocket were conducted. Most of the compounds docked display a general binding mode (Figure 11.1). However, not all compounds are likely to interact in the same manner with the protein. In particular, the less active compounds 20 and 21 exhibited a different binding mode in the docking studies[32] (not shown).

The interactions found after docking include π–π stacking and cation-π contacts with amino acid residues of the anionic substrate binding site (Trp84, Phe331 and Tyr334) and the peripheral anionic binding site (Trp279). Additionally, hydrogen bonds to amino acid residues at the bottom of the gorge were observed. These types of interaction are also known for other crystallized inhibitors of AChE (e.g. donepezil, galanthamine). Moreover, all compounds were potentially able to bind inside the active side gorge, although the entire molecule was not able to interact with amino acid residues of the enzyme. This "size problem" of the ligands may be one reason for their reduced activity as compared to other AChE inhibitors. McCammon et al. addressed this problem with extensive molecular dynamics studies.[33] Larger ligands, however, may escape by diffusion before fluctuations open the bottleneck wide enough to allow binding. In order to prove this hypothesis, shorter molecules were synthesized and tested for their inhibitory activity.

OPTIMIZED DITOPIC COMPOUNDS OF SMALLER SIZE

The new series of shortened compounds 27–32 a–c consisted of 2-chlorobenzyl, and dichlorobenzyl and phthalimidomethyl oxime ether, respectively, at one hand with the pyridinium nitrogen in position 3 and pyridiniumpropyl, phenylpropyl, or propane moieties at the opposite end of the molecule.[34] The inhibitory potency of these compounds is displayed in Table 11.2. The compounds 29 characterized by phenylpropyl moieties at one end were the most active inhibitors (IC$_{50}$ of 29c = 0.073 ± 0.02 μM) within the entire series with IC$_{50}$ values in the same range as tacrine. Without describing details,[34] these results confirmed the aforementioned hypothesis. The phenylpropyl substitution at the end of the pyridinium moieties of 29a–c favorably interacted with aromatic residues localized within the AChE gorge. The other parts take the same place as the symmetrical DUO compounds.

Since the compounds have high inhibitory potency and occupy both the active and peripheral sites of AChE, three representative compounds, the monopyridinium derivatives 29b and 29c and bis-pyridinium DUO3 were tested for their ability to block the Aβ fibril formation using a thioflavin T fluorescence assay.[35]

Amyloid formation is a nucleation-dependent process characterized by two phases: a slow nucleation (the protein undergoes a series of association steps to form an ordered oligomeric nucleus) and a faster growth phase (the nucleus rapidly grows

TABLE 11.2

$IC_{50 \pm}$ SEM Data for Compounds 27 through 32, a through c

		IC$_{50}$ ± SEM [µM]		
No.		**a**	**b**	**c**
	R / R`			
	ditopic compounds	0.58 ± 0.18*	0.34 ± 0.22* (DUO3)	4.5 ± 0.19 (WDUO)
27		7.16 ± 0.73	1.44 ± 0.09	13.66 + 2.0
28		1.53 ± 0.12	0.27 ± 0.044	1.72 ± 0.36
29		0.48 ± 0.006	0.18 ± 0.007	0.073 ± 0.02
30	propanol	6.19 ± 0.66	2.26 ± 0.007	11.09 ± 1.9
31	propyl	6.70 ± 1.83	2.39 ± 0.36	3.78 ± 0.68
32	ethyl	13.10 ± 2.08	4.85 ± 0.41	6.02 ± 0.61

to form larger polymers). In the steady state phase, the ordered aggregate and the monomer appear to be at equilibrium.[36,37] Based on the course of Aβ fibril formation curves of our compounds (not shown), it can be stated that DUO3 retards the nucleation phase of amyloidogenesis, whereas **29b** seems to be active in inhibition of fibril elongation and growth.[34] These results make these compounds interesting, particularly their dual functions (inhibition of AChE and Aß fibrillogenesis) and they may lead to development of anti-AD drugs.

TERTIARY COMPOUNDS OF SERIES

Although the results of studies of the shortened compounds **29** are promising, one must remember that the compounds must penetrate the blood–brain barrier (BBB) to block amyloid formation in the brain. Since quaternary compounds cannot pass the BBB, two series of corresponding tertiary compounds characterized by piperidine rings instead of pyridinium moieties were synthesized and evaluated for their AChE inhibitory potency.

One series was derived from the most potent **29c** compound characterized by a phthalimido moiety. Interestingly, all compounds were found to inhibit AChE in the micromolar range of concentration (Figure 11.3). However, compound **33** which is structurally closest to **29c** was synthetically not available. Therefore, phthalimido analogs of donepezil that are also related to our series were synthesized and also found to be less active,[38] indicating a different and less favorable mode of binding to AChE.

A second series involved synthesis of compounds derived from **29b** which is characterized by a dichlorobenzylether (Figure 11.4). The results were similar; again the compounds were active in the micromolar range of concentration even though the pKa values (about 7) indicate the possibility of being charged at the enzyme and a sufficient high portion of the molecules will be uncharged for the passage through the BBB. This is surprisingly also true for compound **36** which is structurally very similar to **29b**. Preliminary docking studies revealed a different binding

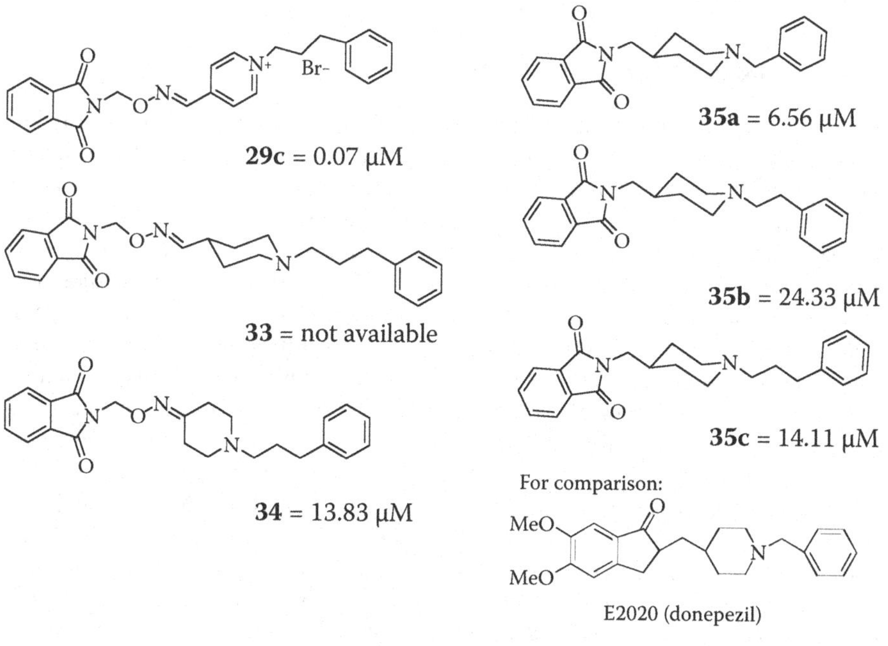

29c = 0.07 μM

33 = not available

34 = 13.83 μM

35a = 6.56 μM

35b = 24.33 μM

35c = 14.11 μM

For comparison:

E2020 (donepezil)

FIGURE 11.3   Short tertiary compounds derived from WDUO and 29c, respectively.

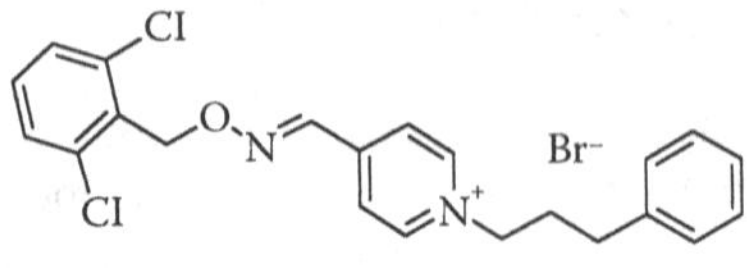

29b = 0.18 μM

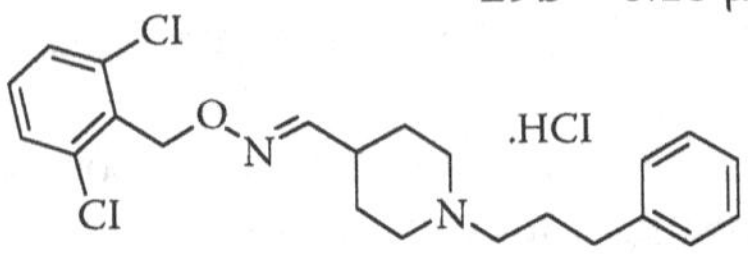

36 = 19.13 μM

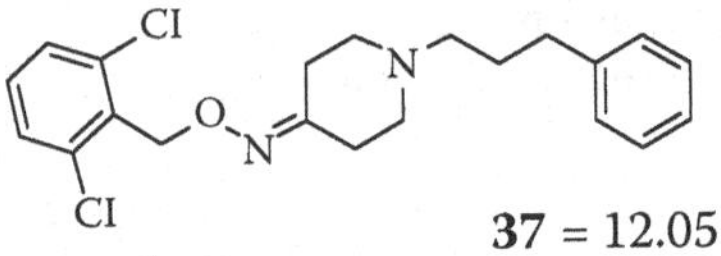

37 = 12.05 μM

FIGURE 11.4   Short tertiary compounds derived from DUO3 and 29b, respectively.

mode caused by a "wrong" positioning of the more voluminous piperidine rings in comparison to the pyridinium rings of compounds **29.** This may explain the weak inhibitory activity.

RECOMBINATORIAL CHEMISTRY

Drug discovery remains a challenging effort. In this regard, we report on a random strategy for generation of small compound libraries.[39,40] The idea is based on the use of γ irradiation as an initiator of random free radical recombinations in aqueous or alcohol solutions of educts. By exposure of a sample to γ irradiation, primary and secondary products of solvent radiolysis are generated and they can react with the dissolved compound and built products.[41,42] The starting compound should be a druggable substance consisting of pharmacophoric elements that we expect to be rearranged in a new way through γ irradiation. Resulting new compounds may represent new candidates for clinical trials and ultimately new therapies. "Random" indicates only that the origin and properties of products obtained by means of γ irradiation cannot be foreseen. Otherwise, as this approach relies on strong deterministic principles of physics, the results should be reproducible.

The samples of potent AChE inhibitors, tacrine and DUO3 (**5**), were subjected to γ rays from a ^{60}Co source. The solvents were selected based on solubility of substances and with intent to obtain all possible polar hits. Tacrine hydrochloride was dissolved in methanol, ethanol, ethylene glycol, and propanol and in a mixture of ethylene glycol and methanol. DUO3 was dissolved in DMSO and a suspension of DUO3 in H_2O was also irradiated.

Via γ irradiation, a small mixture-based compound library was generated. The most promising library was obtained from irradiation of tacrine in water and metha-

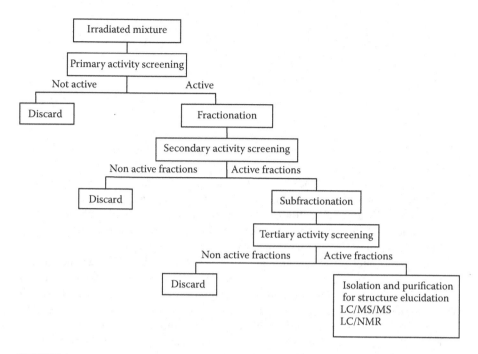

FIGURE 11.5 Workflow of recombinatorial chemistry.

nol. Irradiation of tacrine in the other solvents revealed only a small number of new compounds. The DUO library was not further considered because only a small number of new compounds were generated due to the poor solubility of DUO.

Research efforts focused on bioassay-guided fractionation of both tacrine mixtures and isolation and characterization of biologically active principles. Figure 11.5 depicts the workflow.

Mass spectrometry showed that the radical recombination led to new compounds in the range of 180 to 500 mass units. This meets the mass criterion of Lipinski's "role of five" for drug-like molecules.[43] Furthermore, the method showed reproducible results related to the quality and quantity of members.[44]

Moreover, a previously unknown AChE inhibitor was found, i.e., the hydrophilic aminoacridine derivative, ATAM (9-amino- 5,6,7,8-tetrahydro-acridin-4yl)-methanol) that we expect to be metabolized fast (possibly by glucuronidation or sulfatation) and produce a more favorable clinical profile with fewer side effects than tacrine. ATAM inhibited AChE by half maximal concentration. $IC_{50} = 125 \pm 5$ nM, cf. tacrine, $IC_{50} = 44 \pm 4$ nM (Figure 11.6). The purity of ATAM was about 91%. In addition, hydroxytacrine was isolated from the library produced by irradiation in water.

Most of the new compounds formed by γ irradiation were products of radical chemistry. Free solvent radicals formed by γ irradiation seemed to be the main players in the formation of new compounds. Thus, reagents initiating the formation of radicals should be applicable in recombinatorial chemistry. Since γ irradiation is a highly sophisticated and expensive method, we aimed for proof of principle in the next step. We checked whether Fenton's reagent, one of the best known, cheapest and

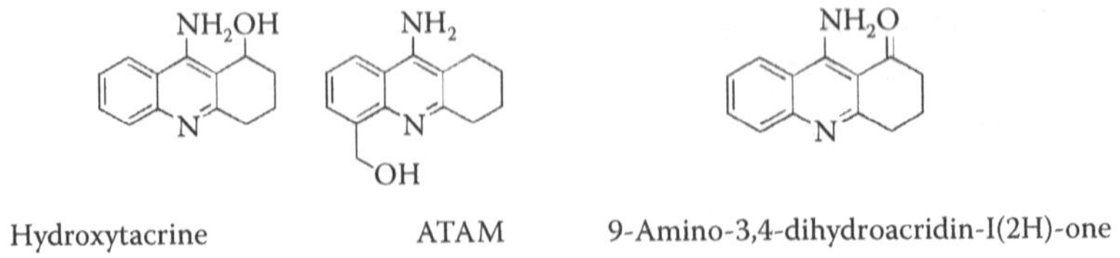

Hydroxytacrine   ATAM   9-Amino-3,4-dihydroacridin-I(2H)-one

FIGURE 11.6 Compounds derived from recombinatorial chemistry.

easiest to handle among all the electron transfer reagents, can serve as an initiator of the production of a small compound library of active substances. The great advantage of the application of Fenton's reagent is the ability to vary the solvent used as well as other reaction conditions in order to influence product composition.

Thus, to a solution of 10^{-4} mol ferrous sulfate in sulfuric acid and 10^{-4} mol tacrine in a methanol–water (15:85) mixture an aqueous H_2O_2 solution (35%) was added and stirred for 2 h at room temperature. After evaporation of the solvent, the residue was purified by column chromatography on aluminum oxide and analyzed by HPLC/MS. The methanol–water mixture was chosen for two reasons: (1) the high solubility of tacrine hydrochloride in water, methanol, and ethanol and the poor solubility of Fenton's reagent in organic solvents and (2) because the results should be comparable to those obtained by γ irradiation in water and methanol systems. Interestingly these preliminary studies revealed the generation of a small compound library and in fact a product composition similar to those obtained by γ irradiation in methanol and water (Figure 11.7). Both 1-hydroxytacrine (a) and ATAM (b) could be identified.

To determine limits and possibilities, conditions were varied and led to differences in product compositions. First the concentration of hydrogen peroxide was varied and found to have great impact. At hydrogen peroxide concentrations less than 100-fold excess in comparison to tacrine, no reactions could be observed. A 200-fold excess of hydrogen peroxide, however, led to a notable conversion and distinct yields of ATAM and 1-hydroxytacrine; increasing the excess to 300 had no further effect. Thus, a 200-fold excess of Fenton's reagent was found to be necessary to provide a satisfying conversion of tacrine. The increase of reaction time from 2 to 24 h provided no advantages. It can be concluded that under these conditions, reaction time is shorter than 2 h.

Trials to increase the methanol concentration to more than 15% failed, as ferric sulfate precipitated and the yields decreased. In order to exclude a possible influence of atmospheric oxygen on the reaction, one experiment was carried out under nitrogen atmosphere. However, no changes of the composition of the developed compound library appeared. The addition of EDTA had no effect on any yields although yield increases are described in the literature in similar cases. Only the combination of EDTA addition and increased reaction time led to a slight increase of yields of 1-hydroxytacrine.

A decreasing pH value resulted in an increasing conversion of tacrine. Decreasing the pH value from 5 to 4 and 2 increased the yields of ATAM and an unknown product X, whereas at pH 5 the highest yields of 1-hydroxytacrine were obtained.

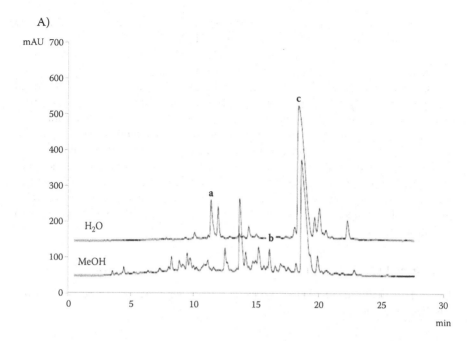

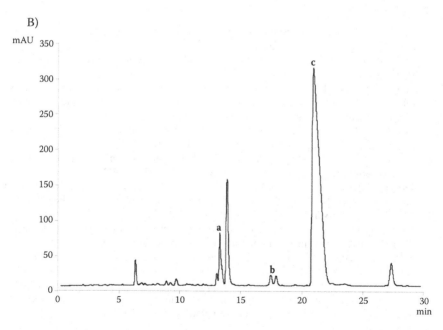

FIGURE 11.7   A. HPLC separation of irradiated tacrine in water and methanol: (a) 1-hydroxytacrine, (b) ATAM, (c) tacrine. HPLC conditions: SYNERGY™ MAX-RP, 150 × 4.6 mm, 4 μm; mobile phase: A: 10 mM ammonium acetate, 0.08% formic acid, pH 3.5; B: methanol; elution rate: 10 to 40% B in 30 min; flow rate: 1 ml/min; detection 254 nm. b. HPLC separation with Fenton's reagent converted tacrine in a water and methanol mixture.

Isolation and spectroscopic investigations revealed the unknown compound X to be 9-amino-3,4-dihydroacridin-l(2H)-one (Figure 11.6).

Taken together, the findings based on the application of Fenton's reagent in recombinatorial chemistry confuted the assumption of Ploszynska et al.[45] that reactive species generated in Fenton's system are not the same as the species generated during water radiolysis, and revealed the method to be a pilot experiment for γ irradiation and thus a powerful tool for the generation of small-compound libraries in drug discovery. Additionally, the method may serve to screen metabolic profiles of new biologically active compounds because the underlying chemistries of both reactions, Fenton's one-electron oxidation and cytochrome P450 oxidations, are similar.

Recombinatorial chemistry has turned out to be a special form of radical chemistry. The produced libraries are different from those created traditionally by "ionic" chemistry. Thus, recombinatorial chemistry opens doors to compound libraries with new structural spaces.

ACKNOWLEDGMENTS

The authors thank the DAAD for financial support for V.A. and the Fonds der Chemischen Industrie.

REFERENCES

1. Holzgrabe, U. et al., Targeting acetylcholinesterase to treat neurodegeneration, *Expert Opin. Ther. Targets*, 11, 161, 2007
2. Ibach, B. and Haen, E., Acetylcholinesterase inhibition in Alzheimer's disease, *Curr. Pharm. Design*, 10, 231, 2004.
3. Francis, P.T., Nordberg, A., and Arnold, S.E., A preclinical view of cholinesterase inhibitors in neuroprotection: do they provide more than symptomatic benefits in Alzheimer's disease? *Trends Pharmacol. Sci.*, 26, 104, 2005.
4. Koo, E.H., Lansbury, P.T., Jr., and Kelly, J.W., Amyloid diseases: abnormal protein aggregation in neurodegeneration, *Proc. Natl. Acad. Sci. USA*, 96, 9989, 1999.
5. Hardy, J. and Selkoe, D.J., The amyloid hypothesis of Alzheimer's disease: progress and problems on the road to therapeutics, *Science*, 297, 353, 2002.
6. Jarrett, J., Berger, E.P., and Lansbury, P.T., The carboxy terminus of the beta amyloid protein is critical for the seeding of amyloid formation: implications for the pathogenesis of Alzheimer's disease, *Biochemistry*, 32, 4693, 1993.
7. Goldsbury, C.S. et al., Studies on the *in vitro* assembly of a beta 1-40: implications for the search for beta fibril formation inhibitors, *J. Struct. Biol.*, 130, 217, 2000.
8. Talaga, P., Beta-amyloid aggregation inhibitors for the treatment of Alzheimer's disease: dream or reality? *Mini Rev. Med. Chem.*, 1, 175, 2001.
9. Allsop, D. et al., 3-p-Toluoyl-2-[4'-(3-diethylaminopropoxy)-phenyl]-benzofuran and 2-[4'-(3-diethylaminopropoxy)-phenyl]-benzofuran do not act as surfactants or micelles when inhibiting the aggregation of beta-amyloid peptide, *Bioorg. Med. Chem. Lett.*, 11, 255, 2001.
10. Parker, M.H. et al., Synthesis of (-)-5,8-dihydroxy-3R-methyl-2R-(dipropylamino)-1,2,3,4-tetrahydronaphthalene: an inhibitor of beta-amyloid(1-42) aggregation, *Bioorg. Med. Chem.*, 10, 3565, 2002.
11. Forloni, G. et al., Anti-amyloidogenic activity of tetracyclines: studies *in vitro*, *FEBS Lett.*, 487, 404, 2001.

12. Danielsson, J. et al., Two-site binding of beta-cyclodextrin to the Alzheimer A beta(1-40) peptide measured with combined PFG-NMR diffusion and induced chemical shifts, *Biochemistry*, 43, 6261, 2004.
13. Lai, Y. et al., PCT Int. Appl., WO 007696, 2000.
14. Ho, J. K. et al., PCT Int. Appl., WO 033489, 2003.
15. De Ferrari, G.V. et al., A structural motif of acetylcholinesterase that promotes amyloid beta-peptide fibril formation, *Biochemistry*, 40, 10447, 2001.
16. Bartolini, M. et al., beta-Amyloid aggregation induced by human acetylcholinesterase: inhibition studies, *Biochem. Pharmacol.*, 65, 407, 2003.
17. Piazzi, L. et al., 3-(4-[[Benzyl(methyl)amino]methyl]phenyl)-6,7-dimethoxy-2H-2-chromenone (AP2238) inhibits both acetylcholinesterase and acetylcholinesterase-induced beta-amyloid aggregation: a dual function lead for Alzheimer's disease therapy, *J. Med. Chem.*, 46, 2279, 2003.
18. Francotte, P. et al., New trends in the design of drugs against Alzheimer's disease, *Curr. Med. Chem.*, 11, 1757, 2004.
19. Harel, M. et al., Quaternary ligand binding to aromatic residues in the active-site gorge of acetylcholinesterase, *Proc. Natl. Acad. Sci. USA*, 90, 9031, 1993.
20. Silman, I. et al., Three-dimensional structures of acetylcholinesterase and of its complexes with anticholinesterase agents, *Biochem. Soc. Trans.*, 22, 745, 1994.
21. Harel, M. et al., The X-ray structure of a transition state analog complex reveals the molecular origins of the catalytic power and substrate specificity of acetylcholinesterase, *J. Am. Chem. Soc.*, 118, 2340, 1996.
22. Greenblatt, H.M. et al., Structure of acetylcholinesterase complexed with (-)-galanthamine at 2.3 A resolution, *FEBS Lett.*, 463, 321, 1999.
23. Kryger, G.I., Silman, J.L. and Sussman, J.L., Three-dimensional structure of a complex of E2020 with acetylcholinesterase from *Torpedo californica*, *J. Physiol. Paris*, 92, 191, 1998.
24. Zacharias, N. and Dougherty, D.A., Cation-pi interactions in ligand recognition and catalysis, *Trends Pharmacol. Sci.*, 23, 281, 2002.
25. Guillou, C. et al., Potent acetylcholinesterase inhibitors: design, synthesis and structure-activity relationships of alkylene linked bis-galanthamine and galanthamine-galanthaminium salts, *Bioorg. Med. Chem. Lett.*, 10, 637, 2000.
26. Carlier, P.R. et al., Evaluation of short-tether bis-THA AChE inhibitors: a further test of the dual binding site hypothesis, *Bioorg. Med. Chem.*, 7, 351, 1999.
27. Hu, M.-L. et al., Homodimeric tacrine congeners as acetylcholinesterase inhibitors, *J. Med. Chem.*, 45, 2277, 2002.
28. Carlier, P.R. et al., Potent, easily synthesized huperzine A-tacrine hybrid acetylcholinesterase inhibitors, *Bioorg. Med. Chem. Lett.*, 9, 2335, 1999.
29. Bunyapaiboonsri, T. et al., Dynamic deconvolution of a pre-equilibrated dynamic combinatorial library of acetylcholinesterase inhibitors, *ChemBioChem*, 2, 438, 2001.
30. Craig, P.N., Interdependence between physical parameters and selection of substituent groups for correlation studies, *J. Med. Chem.*, 14, 680, 1971.
31. Ellman, G.L. et al., A new and rapid colorimetric determination of acetylcholinesterase activity, *Biochem. Pharmacol.*, 7, 88, 1961.
32. Kapkova, P. et al., Synthesis, biological activity, and docking studies of new acetylcholinesterase inhibitors of the bispyridinium type, *Arch. Pharm.*, 336, 523, 2003.
33. Shen, T. et al., Molecular dynamics of acetylcholinesterase, *Acc. Chem. Res.*, 35, 332, 2002.
34. Alptüzün, V. et al., Synthesis and biological activity of pyridinium-type acetylcholinesterase inhibitors, *J. Pharm. Pharmacol.*, 55, 1398, 2003.

35. Kapková, P. et al., Search for dual function inhibitors for Alzheimer's disease: synthesis and biological activity of acetylcholinesterase inhibitors of pyridinium-type and their Abeta fibril formation inhibition capacity, *Bioorg. Med. Chem.*, 14, 472, 2006.

36. Harper, J.D. and Lansbury, P.T., Models of amyloid seeding in Alzheimer's disease and scrapie: mechanistic truths and physiological consequences of the time-dependent solubility of amyloid proteins, *Annu. Rev. Biochem.*, 66, 385, 1997.

37. Lansbury, P.T., A reductionist vew of Alzheimer's disease, *Acc. Chem. Res.*, 29, 317, 1996.

38. Kapkova, P. et al., manuscript in preparation, 2007.

39. Bürgisser, E. and Folkers, G., PCT 1994, WO 94/29719.

40. Folkers, G., Kessler, U., and Bürgisser, E., PCT 2000, WO 00/73281.

41. Folkers, G. and Kessler, U., Random chemistry: look for the unexpected, *Curr. Drug Discov.*, 1, 1, 2003.

42. Kessler, U., Dissertation, ETHZ, No.13592 2000.

43. Lipinski, C.A. et al., Experimental and computational approaches to estimate solubility and permeability in drug discovery and development settings, *Adv. Drug Deliv. Rev.*, 46, 3, 2001.

44. Kapkova, P. et al., Random chemistry as a new tool for the generation of small compound libraries: development of a new acetylcholinesterase inhibitor, *J. Med. Chem.*, 48, 7496, 2005.

45. Ploszynska, J., Kowalski, J. and Sobkowiak, A., The reactivity of monosubstituted benzenes toward reactive species formed in the Fe(II)/HOOH aqueous system, *Pol. J. Chem.*, 72, 2514, 1998.

Part III

Drug Design, Chemical
and Molecular Aspects
of Drug Action

12 Discovery of Potent and Selective Inhibitors of Human Aldosterone Synthase (CYP11B2):

A New Target for the Treatment of Congestive Heart Failure and Myocardial Fibrosis—A Review

*R. W. Hartmann, U. Müller-Vieira,
S. Ulmschneider, and M. Voets*

CONTENTS

ABSTRACT

Pathophysiological and clinical studies have demonstrated that aldosterone plays a detrimental role in congestive heart failure (CHF) and myocardial fibrosis. We have identified aldosterone synthase (CYP11B2) as a novel target for the treatment of these diseases. Using our focused compound library—more than 500 compounds that earlier had been synthesized as inhibitors of selected CYP enzymes—we identified hits, which were further optimized structurally. Thus, potent and selective inhibitors of CYP11B2 were discovered. Further structural optimization led to a new class of inhibitors: *E*- and *Z*-configurated heterocycle-methylene-substituted indanes and tetralins. Here, we present our screening system for the evaluation for CYP11B2 inhibitors and summarize extensive structure–activity studies in the classes of compounds just mentioned. Not only were very potent selective CYP11B2 inhibitors discovered but selective CYP11B1 inhibitors and inhibitors of both enzymes were discovered as well. A protein model of CYP11B2 and a pharmacophore model of CYP11B2 inhibitors are presented as well as the experimental validation of the pharmacophore model. The finding that first compounds also showed an inhibition of the aldosterone plasma concentration *in vivo* is very important.

INTRODUCTION

Six P450 enzymes are involved in steroid biosynthesis, two of which are presently used as pharmacological targets. Inhibitors of aromatase (estrogen synthase, CYP19) are first-line therapeutics for the treatment of breast cancer, and inhibitors of 17α-hydroxylase-C17,20-lyase (CYP 17) are in clinical evaluation for the treatment of prostate cancer. Recently, we propagated a third one to be used pharmacologically—aldosterone synthase (CYP11B2).

CYP11B2 is the key enzyme of mineralocorticoid biosynthesis. It catalyzes the conversion of 11-deoxycorticosterone to the most potent mineralocorticoid, aldosterone.[1] The adrenal aldosterone synthesis is regulated by several physiological parameters, such as the renin–angiotensin–aldosterone system (RAAS), and the plasma potassium concentration. Elevations in plasma aldosterone levels increase blood pressure and play an important role in the pathophysiology of myocardial fibrosis and CHF.[2] In two recent clinical studies (RALES and EPHESUS), the aldosterone receptor antagonists spironolactone and eplerenone were found to reduce mortality in patients with chronic CHF and in patients after myocardial infarction, respectively.[3,4]

Spironolactone, however, showed progestational and anti-androgenic side effects.[4,5] Moreover, a correlation between the use of aldosterone antagonists and hyperkalemia-associated mortality was observed.[6] After having proposed aldosterone synthase as a novel pharmacological target as early as 1994,[7] we propagated more recently the blockade of aldosterone formation by inhibition of CYP11B2 for the treatment of hyperaldosteronism, CHF, and myocardial fibrosis to be a better therapeutic strategy than the use of antihormones.[8,9] Nonsteroidal inhibitors are to be preferred, for we expect these to have less side effects on the endocrine system than steroidal compounds. From our work in the field of CYP19 and CYP17, we know that the concept of heme iron complexing compounds is appropriate to discover highly potent and selective inhibitors.[10,11] These compounds interact with the substrate-binding site in the apoprotein moiety. Additionally, they complex the heme iron that is located in the active site as well. This complexation mechanism does not only increase binding affinity of the inhibitors but also prevents oxygen activation of the heme, which is required for the catalytic process. A crucial point in the development of any CYP inhibitor is selectivity. This is especially true of CYP11B2: the inhibitors must not affect 11β-hydroxylase (CYP11B1), which is the key enzyme of glucocorticoid biosynthesis.[8,9] Selectivity toward aldosterone synthase is not easy to reach because CYP11B1 and CYP11B2 have a sequence homology of 93%.[12] Until now, only a few compounds were known to suppress aldosterone formation. Fadrozole, an aromatase (CYP19) inhibitor, which is in use for the treatment of breast cancer, reduced aldosterone and cortisol levels *in vitro*[13] and *in vivo*.[14] Ketoconazole, an antimycotic and unspecific CYP inhibitor, displayed moderate[15] inhibitory activity toward CYP11B2. However, this compound, as well as others,[15] showed little or no selectivity toward other CYP enzymes, or only marginal activity, and can therefore not be used in the treatment of CHF and myocardial fibrosis.

In the following text, we present a review of our recent work aimed at the discovery of potent, selective, and *in vivo* active CYP11B2 inhibitors.

SCREENING SYSTEM FOR CYP11B2 INHIBITORS

For the discovery and structural optimization of CYP11B2 inhibitors, we recently developed important assays.[8] These were used and combined with other assays existing in our group, and the first screening system was published.[9] The latter was further optimized, leading to the comprehensive screening system for the evaluation of CYP11B2 inhibitors shown in Figure 12.1.

Bovine adrenal mitochondria consisting of 18-hydroxylase (CYP18, CYP11B)[16] are used for preselection. Because this enzyme (which catalyzes the synthesis of gluco- and mineralocorticoids) shows a greater evolutionary distance to CYP11B2 than CYP11B1 to CYP11B2, human CYP11B2 expressed in fission yeast (*Schizosaccharomyces pombe*)[17] is used in the next screening step for the determination of activity.[8] Before this or in parallel, the hit compounds of the first assay are evaluated for their oral absorption properties using artificial membrane assays such as PAMPA and PAMPORE and, eventually, CaCo2-cell assays.[9] Highly active compounds, which can be expected to show a good oral absorption, are processed to the next screening step. To ascertain the selectivity of potent compounds, an assay was developed[8] with V79 MZh cells (hamster lung fibroblasts) expressing human CYP11B1 or CYP11B2.[18] Subsequently, the compounds are examined for their sex hormone inhibitory activity (estrogen: CYP19; androgen: CYP17) as well as for eventual inhibition of the cholesterol side-chain cleavage enzyme (CYP11A1). In the next level, the compounds are evaluated for their overall effects on the adrenals. This is performed by using the NCI-H295R adrenal cortical cell line.[19] This cell line represents the properties of all three zones of the adrenal cortex. In the last screening step, the compounds are evaluated for *in vivo* activity using ACTH-stimulated rats. After administration of the compounds, the aldosterone plasma concentrations are determined 2 h later.

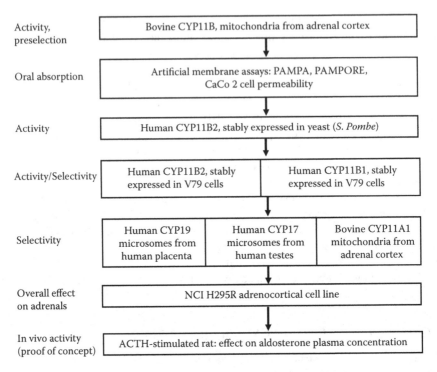

Activity, preselection	Bovine CYP11B, mitochondria from adrenal cortex
Oral absorption	Artificial membrane assays: PAMPA, PAMPORE, CaCo 2 cell permeability
Activity	Human CYP11B2, stably expressed in yeast (*S. Pombe*)
Activity/Selectivity	Human CYP11B2, stably expressed in V79 cells	Human CYP11B1, stably expressed in V79 cells
Selectivity	Human CYP19 microsomes from human placenta	Human CYP17 microsomes from human testes	Bovine CYP11A1 mitochondria from adrenal cortex
Overall effect on adrenals	NCI H295R adrenocortical cell line
In vivo activity (proof of concept)	ACTH-stimulated rat: effect on aldosterone plasma concentration

FIGURE 12.1 Screening system for the evaluation of CYP11B2 inhibitors.

A

n = 1 **1a,b** (*E, Z*)
n = 2 **2a,b** (*E, Z*)

CHART 12.1

HIT-FINDING STRATEGIES

From a screening of our in-house compound library consisting of more than 500 compounds synthesized as potential inhibitors of different CYP enzymes, such as CYP17, CYP19, and CYP5,[10,11] several hits were obtained. Among these were compounds of class A (Chart 12.1). Further structural optimization resulted in the *E* isomers **1a** and **2a**, and *Z* isomers **1b** and **2b**. In this chapter, the results of further structural optimization of compounds **1** and **2**[20,21] are reviewed.

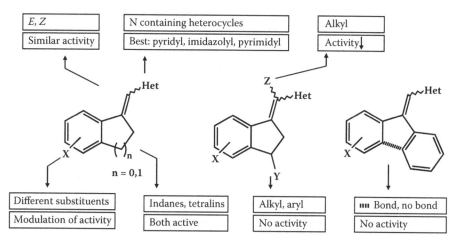

FIGURE 12.2   Structure modification of heteroarylmethylene-substituted indanes.

STRUCTURE–ACTIVITY RELATIONSHIPS

The structure–activity relationships obtained by the evaluation of approximately 100 compounds synthesized are summarized in Figure 12.2.

Interestingly, *E*- and *Z*-configured compounds showed similar potency. This was subsequently explained by molecular modeling, and the corresponding pharmacophore model is shown in the following text. The heterocycles necessary for complexing the heme iron showed different activities. The most potent inhibitors were observed in the classes of pyridyl-, imidazolyl-, and pyrimidyl-substituted compounds. The introduction of a substituent into the exocyclic double bond decreased activity. Positions 2 and 3 of the core molecule were also discovered to be very sensitive toward substitution that led to inactive compounds. Both indane and tetraline moieties are appropriate for potent inhibition. Substitution at the benzene nucleus led to a modulation of activity.[20,21]

Interestingly, compounds showing selectivity toward CYP11B1 and CYP11B2 were obtained. This was not necessarily expected because the homology between the two enzymes is approximately 93%. The selectivity profiles of the *E* and *Z* isomers were quite different. A selection of these enzymes is shown in Figure 12.3.

Fairly selective inhibitors of CYP11B1 were obtained showing selectivity factors between 0.16 and 0.33. On the other hand, compounds that were equally potent inhibitors of both enzymes were discovered, too. This translates to selectivity factors of between 0.8 and 1.2. Most of the compounds, however, turned out to be rather (or very) selective inhibitors of CYP11B2 because structural optimization had been performed toward CYP11B2 inhibition. In terms of selectivity factors, compounds showing values of more than 100 were obtained.

MOLECULAR MODELING STUDIES

With the human CYP2C9 x-ray structure—the first and thus far only human CYP enzyme that could be crystallized[22]—used as a template, protein models of CYP11B2

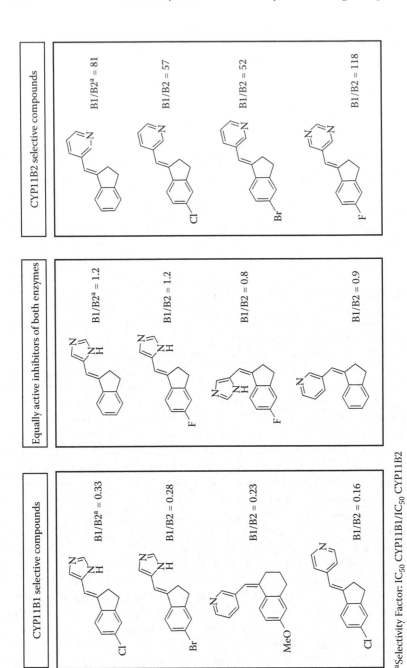

CYP11B2 selective compounds

B1/B2^a = 81

B1/B2 = 57

B1/B2 = 52

B1/B2 = 118

Equally active inhibitors of both enzymes

B1/B2^a = 1.2

B1/B2 = 1.2

B1/B2 = 0.8

B1/B2 = 0.9

CYP11B1 selective compounds

B1/B2^a = 0.33

B1/B2 = 0.28

B1/B2 = 0.23

B1/B2 = 0.16

^{a}Selectivity Factor: IC$_{50}$ CYP11B1/IC$_{50}$ CYP11B2

FIGURE 12.3 Selectivity profiles of selected compounds (CYP11B1/CYP11B2).

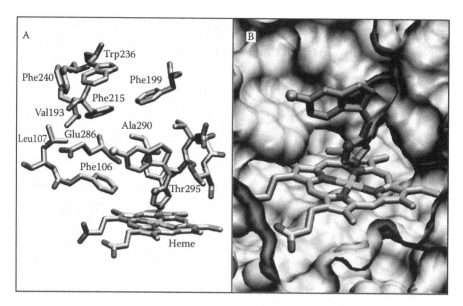

FIGURE 12.4   Structure of the CYP11B2-8b complex.

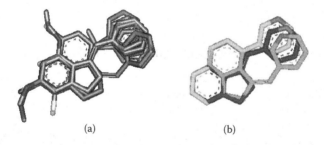

(a)   (b)

FIGURE 12.5   Pharmacophore model for inhibitors of aldosterone synthase (CYP11B2).

and CYP11B1 (for selectivity) were built.[20,21] Figure 12.4 shows details of the active site of CYP11B2 consisting of the heme and a potent inhibitor complexing the heme iron and interacting with select amino acid residues of the active site. Presently, selective compounds for both enzymes are used for the refinement of both protein models.

In addition, the *E*- and *Z*-configurated heterocycle-methylene-substituted tetrahydronaphthalenes and indanes as well as a new class of heterocycle-substituted naphthalenes[23] were used to build up a pharmacophore model for CYP11B2 inhibitors (Figure 12.5).

Superimposition of the compounds (all of them potent CYP11B2 inhibitors), as indicated in Figure 12.5, leads to the pharmacophore model shown. The overlay of the compounds results in nitrogens pointing to a similar direction (to the heme iron). In this superimposition, the rest of the molecules of each compound form a 3-ring core structure.

TABLE 12.1

Biological Activity of Hybrid Compounds A and B

Compound	IC$_{50}$-value (nM)[a]		
	CYP11B1[b]	CYP11B2[c]	Selectivity[d]
A[e]	2452	10	245
B[f]	2896	14	207

a   Mean value of four determinations, standard deviation less than 20%.

b   Hamster fibroblasts (V79 cells) expressing human CYP11B1; substrate deoxycorticosterone, 100 nM.

c   Hamster fibroblasts (V79 cells) expressing human CYP11B2; substrate deoxycorticosterone, 100 nM.

d   IC$_{50}$ CYP11B1/IC$_{50}$ CYP11B2.

e   3-(1,2-Dihydroacenaphthylene-3yl)pyridine.

f   3-(1,2-Dihydroacenaphthylene-5-yl)pyridine.

VALIDATION OF THE PHARMACOPHORE MODEL

For validation of the pharmacophore model, the acenaphthalene compounds A and B were synthesized. Table 12.1 shows that both compounds have a very high activity toward CYP11B2. This result proves the validity of the pharmacophore model. Besides, selectivity of the two compounds is very high, exceeding a value of 200.

Because compounds A and B exhibit a similar potency, additional superimposition studies were performed. From Figure 12.6 it becomes apparent that both compounds can be nicely superimposed, and thus the high potency of compound B can be explained.

FIGURE 12.6   Alignment of hybrid compound A and its isomer B.

IN VIVO ACTIVITY—PROOF OF CONCEPT

After selected compounds had shown very good activity and selectivity profiles *in vitro* and also demonstrated convincing results in the tests for oral absorption

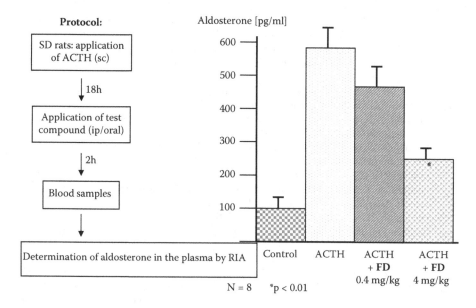

FIGURE 12.7 *In vivo* activity: proof of concept.

and metabolic stability, they were first tested *in vivo*; 18 h after ACTH application to Sprague–Dawley rats to stimulate aldosterone synthesis, the compounds were administered orally. Blood samples were taken 2 h later, and radioimmunoassays for the determination of aldosterone were performed with the plasma. From Figure 12.7 it becomes apparent that ACTH stimulates aldosterone plasma concentration by a factor of approximately 6. Compound FD reduced the ACTH-stimulated aldosterone plasma concentration dose-dependently. At a dose of 4 mg/kg, FD was capable of reducing aldosterone concentration by approximately 60%.

CONCLUSION

It has been demonstrated that it is possible to design very potent and very selective (selectivity factors >200) inhibitors of CYP11B2. These compounds are valuable leads for the development of drugs used for the treatment of CHF and myocardial fibrosis. The compounds shown in this chapter are not only potent and selective *in vitro* but also active *in vivo*. Further optimization is required to obtain a compound that can be used clinically.

REFERENCES

1. Kawamoto, T. et al., Role of steroid 11 beta-hydroxylase and steroid 18-hydroxylase in the biosynthesis of glucocorticoids and mineralocorticoids in humans, *Proc. Natl. Acad. Sci. U.S.A.*, 89, 1458, 1992.
2. Brilla, C.G., Renin-angiotensin-aldosterone system and myocardial fibrosis, *Cardio-vasc. Res.*, 47, 1, 2000.

3. Pitt, B. et al., Eplerenone, a selective aldosterone blocker, in patients with left ventricular dysfunction after myocardial infarction, *N. Eng. J. Med.*, 348, 1309, 2003.
4. Pitt, B. et al., The effect of spironolactone on morbidity and mortality in patients with severe heart failure. Randomized aldactone evaluation study investigators, *N. Engl. J. Med.*, 341, 709, 1999.
5. Khan, N.U.A. and Movahed, A., The role of aldosterone and aldosterone-receptor antagonists in heart failure, *Rev. Cardiovasc. Med.*, 5, 71, 2004.
6. Juurlink, D.N. et al., Rates of hyperkalemia after publication of the randomized aldactone evaluation study, *N. Engl. J. Med.*, 351, 543, 2004.
7. Hartmann, R.W., Selective inhibition of steroidogenic P450 enzymes: current status and future perspectives, *Eur. J. Pharm. Sci.*, 2, 15, 1994.
8. Ehmer, P.B. et al., Development of a test system for inhibitors of human aldosterone synthase (CYP11B2): Screening in fission yeast and evaluation of selectivity in V79 cells, *J. Steroid Biochem. Mol. Biol.*, 81, 173, 2002.
9. Hartmann, R.W., Müller, U., and Ehmer, P.B., Discovery of selective CYP11B2 (aldosterone synthase) inhibitors for the therapy of congestive heart failure and myocardial fibrosis, *Eur. J. Med. Chem.*, 38, 363, 2003.
10. (a) Hartmann, R.W. et al., Pyridyl-substituted tetrahydrocyclopropa[a]naphthalenes: Highly active and selective inhibitors of P450 arom., *J. Med. Chem.*, 38, 2103, 1995. (b) Jacobs, C. et al., 1-Imidazolyl(alkyl)-substituted di- and tetrahydroquinolines and analogues: Syntheses and evaluation of dual inhibitors of thromboxane A_2 synthase and aromatase, *J. Med. Chem.*, 43, 1841, 2000. (c) Recanatini, M. et al., A new class of nonsteroidal aromatase inhibitors: Design and synthesis of chromone and xanthone derivatives and inhibition of the P450 enzymes aromatase and 17α-hydroxylase/C17,20-lyase, *J. Med. Chem.*, 44, 672, 2001. (d) Leonetti, F. et al., Design, synthesis and 3D QSAR of novel potent and selective aromatase inhibitors, *J. Med. Chem.*, 47, 6792, 2004.
11. (a) Wachall, B.G. et al., Imidazole substituted biphenyls: A new class of highly potent and *in vivo* active inhibitors of P450 17 as potential therapeutics for treatment of prostate cancer, *Bioorg. Med. Chem.*, 7, 1913, 1999. (b) Hartmann, R.W. et al., Synthesis and evaluation of 17-aliphatic heterocycle-substituted steroidal inhibitors of 17α-hydroxylase/C17-20-lyase (P450 17), *J. Med. Chem.*, 43, 4437, 2000. (c) Zhuang, Y., Wachall, B.G. and Hartmann, R.W., Novel imidazolyl and triazolyl substituted biphenyl compounds: Synthesis and evaluation as nonsteroidal inhibitors of human 17α-hydroxylase-C17,20-lyase (P450 17), *Bioorg. Med. Chem.*, 8, 1245, 2000. (d) Haidar, S. et al., Effects of novel 17α-hydroxylase/C17,20-lyase (P450 17, CYP 17) inhibitors on androgen biosynthesis *in vitro* and *in vivo*, *J. Steroid Biochem. Mol. Biol.*, 84, 555, 2003.
12. Taymans, S.E. et al., Human CYP11B2 (aldosterone synthase) maps to chromosome 8q24.3, *J. Clin. Endocrinol. Metab.*, 83, 1033, 1998.
13. Häusler, A. et al., Evidence that corticosterone is not an obligatory intermediate in aldosterone biosynthesis in the rat adrenal, *J. Steroid Biochem.*, 34, 567, 1989.
14. Demers, L. M. et al., The effects of CGS 16949A, an aromatase inhibitor on adrenal mineralocorticoid biosynthesis, *J. Clin. Endocrinol. Metab.*, 70, 1162, 1990.
15. Bureik, M. et al., Development of test systems for the discovery of selective human aldosterone synthase (CYP11B2) and 11beta-hydroxylase (CYP11B1) inhibitors. Discovery of a new lead compound for the therapy of congestive heart failure, myocardial fibrosis and hypertension, *Mol. Cell. Endocrinol.*, 217, 249, 2004.
16. Hartmann, R.W. et al., Pyridyl-substituted tetrahydrocyclopropa[a]naphthalenes: highly active and selective inhibitors of P450 arom, *J. Med. Chem.*, 38, 2103, 1995.
17. Bureik, M. et al., Functional expression of human mitochondrial *CYP11B2* in fission yeast and identification of a new internal electrontransfer protein, *etp1*, *Biochemistry*, 41, 2311, 2002.

18. Denner, K. et al., Cloning and stable expression of the human mitochondrial cytochrome P45011B1 cDNA in V79 Chinese hamster cells and their application for testing of potential inhibitors, *Pharmacogenetics*, 5, 89, 1995.
19. Müller-Vieira, U., Angotti, M., and Hartmann, R.W., The adrenocortical tumor cell line NCI-H295R as an *in vitro* screening system for the evaluation of aldosterone synthase (CYP11B2) inhibitors and their potential side effects, *J. Steroid Biochem. Mol. Biol.*, 96, 259, 2005.
20. Ulmschneider, S. et al., Synthesis and evaluation of imidazolylmethylenetetrahydronaphthalenes and imidazolylmethyleneindanes: potent inhibitors of aldosterone synthase, *J. Med. Chem.*, 48, 1796, 2005.
21. Ulmschneider, S. et al., Synthesis and evaluation of (pyridylmethylene)tetrahydronaphthalenes/-indanes and structurally modified derivatives: potent and selective inhibitors of aldosterone synthase, *J. Med. Chem.*, 48, 1563, 2005.
22. Williams, P. et al., Crystal structure of human cytochrome P4502C9 with bound warfarin, *Nature*, 424, 464, 2003.
23. Voets, M. et al., Heteroaryl substituted naphthalenes and structurally modified derivatives: selective inhibitors of CYP11B2 for the treatment of congestive heart failure and myocardial fibrosis, *J. Med. Chem.*, 48, 6632, 2005.

13 Thiocarboxanilides:
A New Class of Nonnucleoside Reverse Transcriptase Inhibitors (NNRTIs) with Great Potential for the Treatment of Human Immunodeficiency Virus Type 1 (HIV-I) Infections

E. De Clercq, A. Karlsson, and J. Balzarini

CONTENTS

SUMMARY

The thiocarboxanilides fulfill the requirements of ideal anti-HIV-1 drug candidates. They are highly potent and selective inhibitors of HIV-1 replication that are targeted at the nonsubstrate binding site of the HIV-1 reverse transcriptase. Like all the other nonnucleoside reverse transcriptase inhibitors (NNRTIs), they lead to the emergence of drug-resistant virus strains. Yet, resistance development in cell culture can be circumvented by (1) switching from one NNRTI (i.e., thiocarboxanilide) to another; (2) combining thiocarboxanilides with other NNRTIs; (3) using the thiocarboxanilides from the start at sufficiently high ("knockout") concentrations; and (4) starting with combinations of thiocarboxanilides and other drugs (whether NNRTIs or not) at knockout concentrations so as to completely suppress virus replication and thus prevent resistance from emerging.

INTRODUCTION

The first compounds ever shown to specifically inhibit HIV-1 (human immunodeficiency virus type 1), but not HIV-2 (human immunodeficiency virus type 2), replication were the 1-(2-hydroxyethoxymethyl)-6-(phenylthio)thymine (HEPT)[1] and tetrahydroimidazo[4,5,1-*jk*][1,4]benzodiazepin-2(1H)-one and –thione (TIBO) derivatives.[2] [The unprecedented specificity of the TIBO derivatives, which were found to inhibit HIV-1 replication at concentrations that were 10,000- to 100,000-fold lower than the concentrations required to affect normal cell viability, was attributed to a specific interaction with the HIV-1 reverse transcriptase (RT).[2]] For the HEPT derivatives it became evident that they also interact specifically with HIV-1 RT after a number of derivatives, such as E-EPU, E-EBU, and E-EBU-dM, had been synthesized that were more active than HEPT itself in inhibiting HIV-1 replication.[3]

Whereas the HEPT and TIBO derivatives were discovered as the result of a systematic evaluation for anti-HIV activity in cell culture, and later proved to achieve their anti-HIV activity through a specific interaction with HIV-1 RT, the following compounds emerged from a screening program for detecting HIV-1 RT inhibitors before they were shown to inhibit HIV-1 replication in cell culture as well: nevirapine (BI-RG-587),[4] pyridinone derivatives (L-696,229 and L-697,661)[5] and bis(heteroaryl)piperazine (BHAP) derivatives (U-88204 and U-90 152).[6]

Following TIBO, HEPT, nevirapine, pyridinone, and BHAP, the 2′,5′ -bis-*O*-(*tert*-butyldimethylsilyl)-3′-spiro-5″-(4″-amino-1″,2″-oxathiole-2″,2″-dioxide)pyrimidine (TSAO) nucleoside analogues (TSAO-T and TSAO-m³T),[7,8] α-anilinophenyl-acetamides (α-APA R89439),[9] and various other compounds have been found to inhibit HIV-1 replication through a specific interaction with HIV-1 RT. These compounds have been collectively referred to as NNRTIs or "nonnucleoside reverse transcriptase inhibitors."[10]

NNRTIs share a number of characteristics that distinguish them from the "conventional" nucleoside-type RT inhibitors such as AZT (3′-azido-2′,3′-dideoxythymidine, zidovudine), DDI (2′,3′-dideoxyinosine, didanosine), DDC (2′,3′-dideoxycytidine, zalcitabine), D4T (2′,3′-didehydro-2′,3′-dideoxythymidine, stavudine), and 3TC ((−)-3′-thia-2′-deoxycytidine, lamivudine). NNRTIs achieve a highly selective suppression of HIV-1 replication in cell culture without affecting the replication of HIV-2 or other retroviruses. They do not require intracellular metabolism, and interact as such with RT in a noncompetitive manner with regard to the natural substrate (dNTP) at an allosteric (nonnucleoside) binding site. However, NNRTIs rapidly lead to the emergence of drug-resistant HIV-1 mutant strains.[11,12] The resistance mutations appeared to be clustered around the putative binding site of the NNRTIs.[13]

As the emergence of virus-drug resistance is the major, if not the only, problem compromising the therapeutic usefulness of the NNRTIs, our efforts should be aimed at strategic approaches to prevent, circumvent, or overcome drug resistance development. Among the different approaches that could be envisaged are the following:[10] (1) switch from one class of NNRTI inhibitors to another (or, even within each class, from one compound to another), because the different compounds do not necessarily lead to cross-resistance; (2) combine different compounds that show differences in their resistance mutation profile, particularly when these mutations appear to antagonize each other; (3) start with sufficiently high drug concentrations so as to "knock out" the virus and prevent breakthrough of resistant virus; and (4), in combining approaches (1) and (3), start with drug combinations that could achieve a total knockout (i.e., clearance of the cells from the virus) at lower drug concentrations than if the compounds were to be used individually.

We will now examine how oxathiin carboxanilide and thiocarboxanilide derivatives qualify as genuine NNRTIs, how their activity compares to that of the other NNRTIs, and how they can be applied in attempts to circumvent resistance development.

INHIBITION OF HIV-1 REPLICATION (CYTOPATHICITY)

Which requirements do the compounds have to fulfill to qualify as NNRTIs? They should specifically interact with HIV-1 RT and inhibit HIV-I, but not HIV-2, replication in cell culture at a concentration that is significantly lower than the concentration required to affect normal cell viability. The latter should be evident from the selectivity index (51), or ratio of the CC_{50} (50% cytotoxic concentration) to the EC_{50} (50% effective concentration) (Table 13.1). Based on these premises, several classes of compounds (Figure 13.1) could be considered as NNRTIs: TIBO derivatives,[2,14] HEPT derivatives,[3,15] nevirapine[4] (as a member of the dipyridodiazepinones), pyridinones,[5] bis(heteroaryl)piperazines,[6] TSAO derivatives,[7,8] α-anilinophenylacetamides,[9] PETT derivatives,[16] quinoxaline 5-2720,[17] thiazolobenzimidazole (N5C 625487),[18] pyrrolobenzodiazepinone, phenyldihydrothiazoloisoindolones (BM 21.1298 and BM + 51.0836),[19] imidazodipyridodiazepine (UK-129,485), chlorophenylsulfonylindole carboxamide (L-737,126), 2-nitrophenyl phenyl sulfone (NPP5), naphthalenone TGG-II-23A, DABO derivatives,[20,21] imidazopyridazinyl phenyl heptanone, pyridylethynyldihydroquinazolinone,[22] benzothiadiazine-l-oxide (NSC 287474),[23] highly substituted pyrrole derivatives, pyrrolobenzothiazepines, and a number of natural products, i.e., calanolides and inophyllums,[24-26] extracted from the tropical rainforest tree *Calophyllum lanigerum* and *Calophyllum inophyllum*, respectively. As is evident from Table 13.1, the most potent and most selective NNRTIs were found to inhibit HIV-1 cytopathicity at nanomolar concentrations with selectivity indices up to 100,000. All the compounds listed in Table 13.1 were found to inhibit HIV-1

TABLE 13.1
Inhibitory Effects of NNRTIs on HIV-1 RT Activity and HIV-1 Cytopathicity

Compound	HIV-1 RT Activity		HIV-1 Cytopathicity	
	IC_{50} (μM)[a]	EC_{50} (μM)[b]	CC_{50} (μM)[c]	SI[d]
TIBO				
R82150	0.139	0.044	552	12545
R82913	0.015	0.033	34	1030
R86183	0.050	0.0046	138	30000
HEPT				
E-EPU	0.27	0.022	146	6636
E-EBU-dM	0.036	0.0022	249	113000
I-EBU (MKC-442)	0.012	0.014	>100	>7000
Nevirapine (BI-RG-587)	0.084	0.048	>50	>1000
Pyridinone				
L-696,229	0.022	≤0.05	>200	>4000
L-697,661	0.019	≤0.012	>60	>4800
BHAP				
U-88204	0.25	0.04	>10	>250
U-90152	0.26	0.01	>100	>10000
TSAO-T	17	0.036	>50	>1400
TSAO-m3T	4.7	0.034	139	4088
α-APA R89439 (Loviride)	0.2	0.013	710	54615
PETT LY297345		0.01	271	27100
Oxathiin carboxanilide UC84 (NSC 615985)		0.22	106	482
Quinoxaline S-2720	0.1	0.003	50	17000
Thiazolobenzimidazole (NSC 625487)	0.5	0.5	50	100

	0.04 (μg/mL)	0.3 (μg/mL)[e]	10 (μg/mL)	
Pyrrolobenzodiazepinone	0.28	0.392		
Phenyldihydrothiazolo-isoindolone (BM 21.1298) and its dimethyl derivative (BM + 51.0836)	0.016	0.01	>50	33
Imidazodipyridodiazepine UK-129,485	0.156	<0.002	>50	>125
Chlorophenylsulfonylindole carboxamide L-737,126	0.003	<0.003	>10	>5000
2-Nitrophenyl phenyl sulfone (NPPS)	0.61	5	125	>5000
Naphthalenone TGG-II-23A	136	26	550	25
DABO (R = cyclohexyl)	1.8	0.8	>335	21
Imidazopyridazinyl phenylheptanone	0.0006	0.01	26	>418
Pyridylethynyl dihydroquinazolinone	0.012	<0.025	>130	2600
Benzothiadiazine (NSC 287474)	1.2	4.04	175	>32
Substituted pyrrole dicarboxylate	0.64	2.3	>362	76
Pyrrolobenzothiazepines	1.5	14	20	>25
Calanolide A	0.07	0.1	55	200
Inophyllum B	0.038	1.4	8.8 (μg/mL)	39
Oxathiin carboxanilide UC84	1.5	0.015 (μg/mL)	16 (μg/mL)	587
Oxathiin thiocarboxanilide UC23		0.007 (μg/mL)	12 (μg/mL)	2286
Thiocarboxanilide UC38		0.009 (μg/mL)		1333

a 50% Inhibitory concentration, or concentration required to inhibit HIV-1 RT activity by 50%.

b 50% Effective concentration, or concentration required to inhibit HIV-1-induced cytopathicity by 50% (in principle, based on cytopathicity of HIV-1 III$_B$ strain for MT-4 cells).

c 50% Cytotoxic concentration, or concentration required to reduce cell (i.e., MT-4 cell) viability by 50%.

d Selectivity index, or ratio of CC$_{50}$ to EC$_{50}$.

e EC$_{90}$.

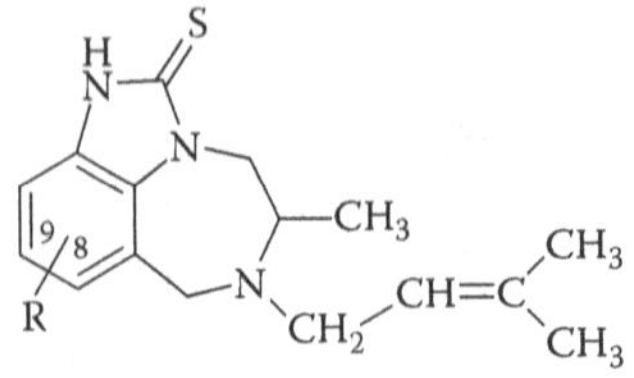

Tetrahydroimidazo(4, 5, 1-<u>jk</u>) (1, 4)-
benzodiazepin-2(1H)-one (<u>TIBO</u>)

R = H   : R82150
R = 9-Cl : R82913
R = 8-Cl : R86183

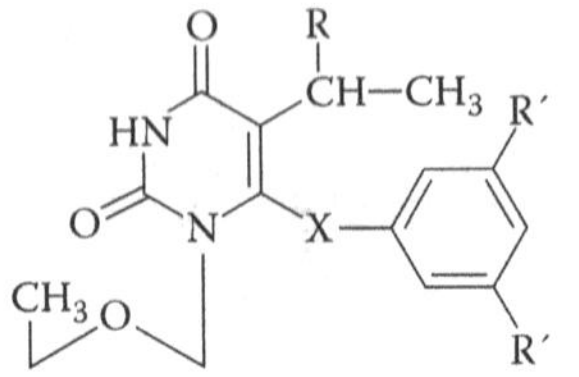

1-(2-Hydroxyethoxymethyl)-
6-(phenylthio)thymine (<u>HEPT</u>)

X = S   R, R′ = H   : E-EPU
X = CH$_2$ R = H R′ = CH$_3$: E-EBU-dM
X = CH$_2$ R = CH$_3$ R′ = H : I-EBU (MKC-442)

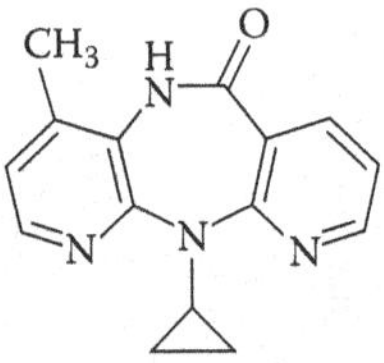

Nevirapine (BI-RG-587)

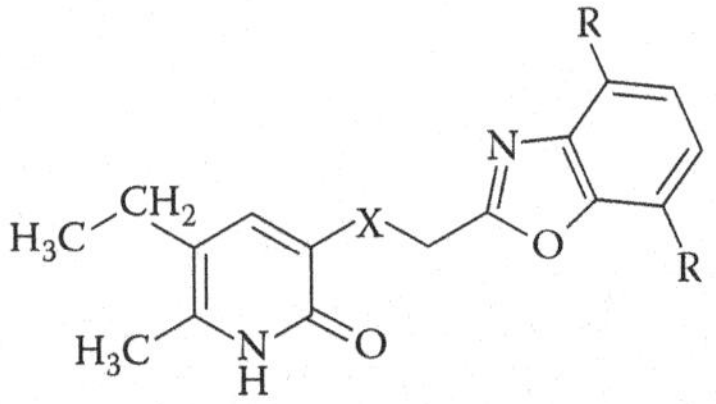

Pyridinone
X = CH$_2$, R = H : L-696, 299
X = NH, R = Cl : L-697, 661

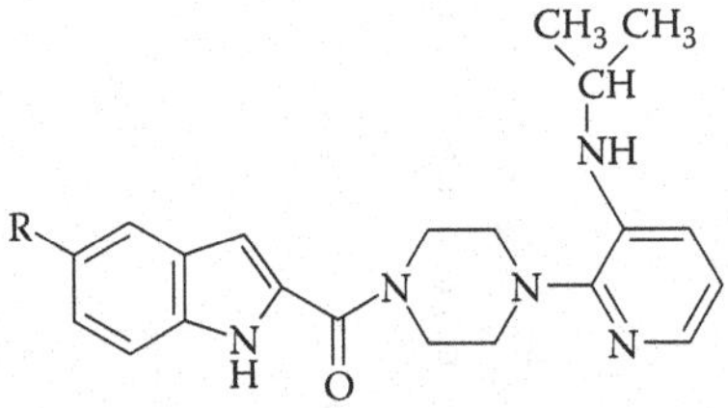

Bis(heteroaryl)piperazine (<u>BHAP</u>)
R = H   : U-88204
R = NHSO$_2$CH$_3$: U-90152

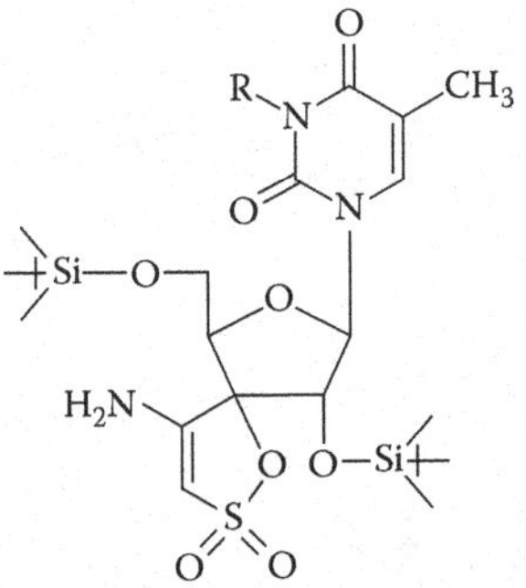

2′, 5′-Bis-O-(tert-butyldimethylsilyl)-
3′-spiro-5′-(4′-amino-1′, 2′-oxathiole-
2′, 2′-dioxide (<u>TSAO</u>) derivative

R = H   : TSAO-T
R = CH$_3$: TSAO-m^{3}T

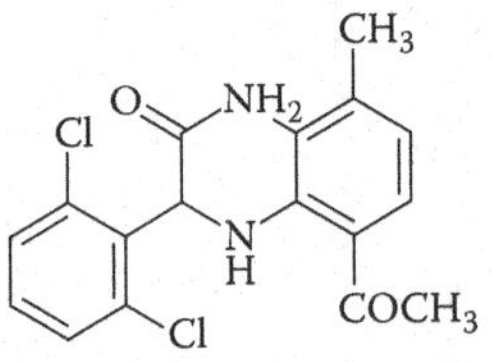

α-Anilinophenylacetamide (<u>α-APA</u>)
R89439 (Loviride)

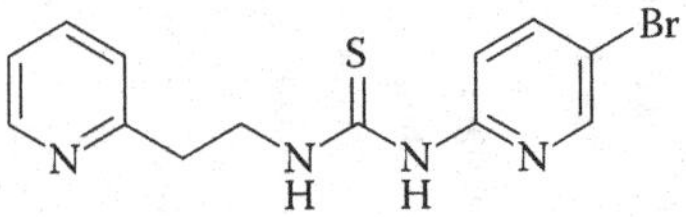

Phenylethylthioureathiazole (<u>PETT</u>)
LY297345

(A)

FIGURE 13.1 (A-D) Structures of NNRTIs.

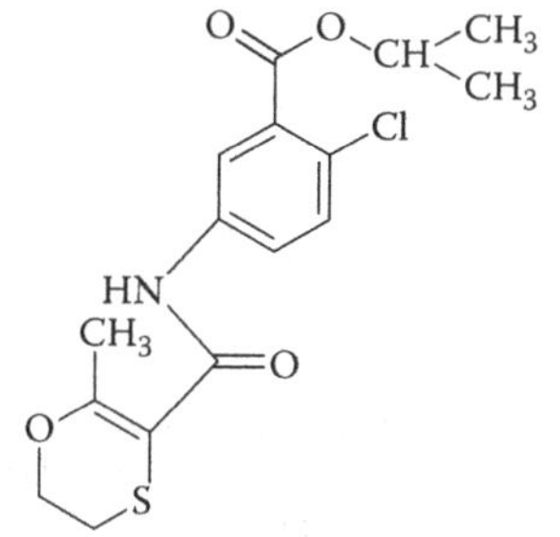

Oxathiin carboxanilide
UC84 (NSC 615985)

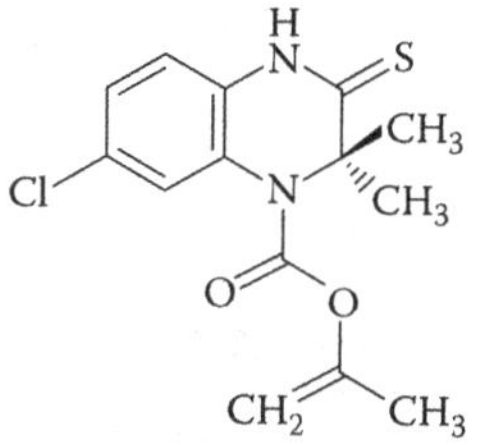

6-Chloro-3, 3-dimethyl-4-(isopropenyl-
oxycarbonyl)-3, 4-dihydroquinoxalin-
2 (1H)-thione (S-2720)

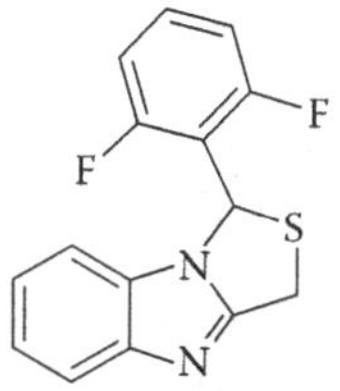

1-(2′, 6′- difluorophenyl)-1H, 3H-thiazolo
[3, 4-a] benzimidazole
(NSC 625487)

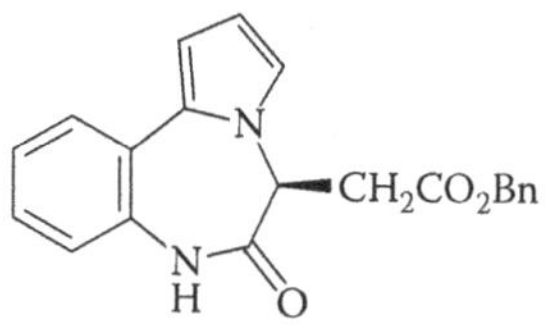

Pyrrolo-[1, 2-d]-(1, 4)-benzodiazepin-6-one

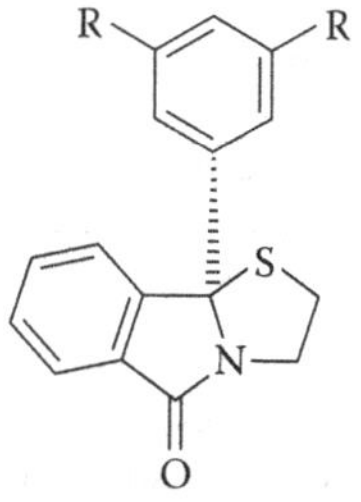

9b-Phenyl-2,3-dihydrothiazolo
[2, 3-a]isoindol-5(9bH) one (R = H)
(BM 21.1298) and its dimethyl derivative
(R = CH₃) (BM + 51.0836)

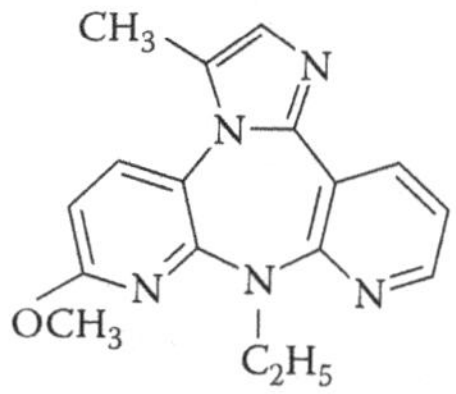

Imidazo[2′, 3′:6, 5]dipyrido
[3, 2-b:2′, 3′-e]-1, 4-diazepine
(UK-129, 485)

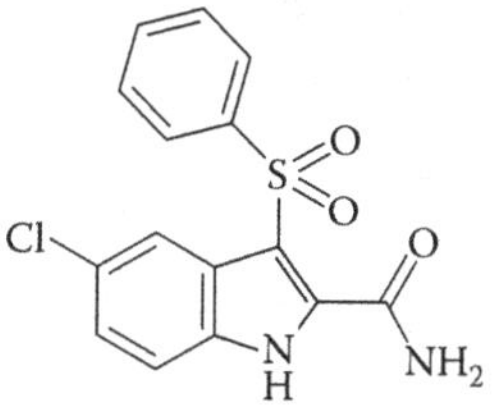

5-Chloro-3-(phenylsulfonyl)-
indole-2-carboxamide
(L-737, 126)

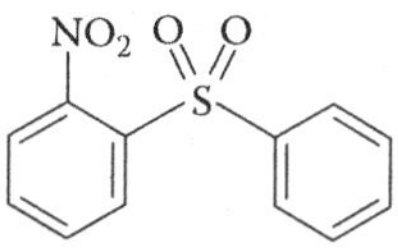

2-Nitrophenyl phenyl sulfone
(NPPS)

(B)

FIGURE 13.1 (CONTINUED)

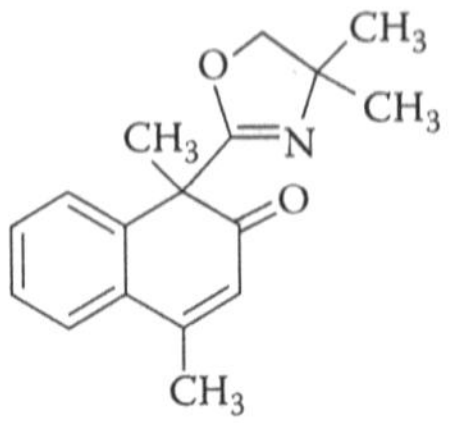

1, 4-Dimethyl-1-(5, 5-dimethyl-
2-oxazolinyl)-naphthalene-2-one
TGG-II-23A

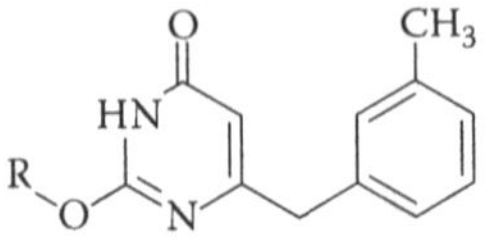

3, 4-Dihydro-2-alkoxy-
6-(3′-methylbenzyl)-4-oxopyrimidine (DABO)
R = cyclohexyl

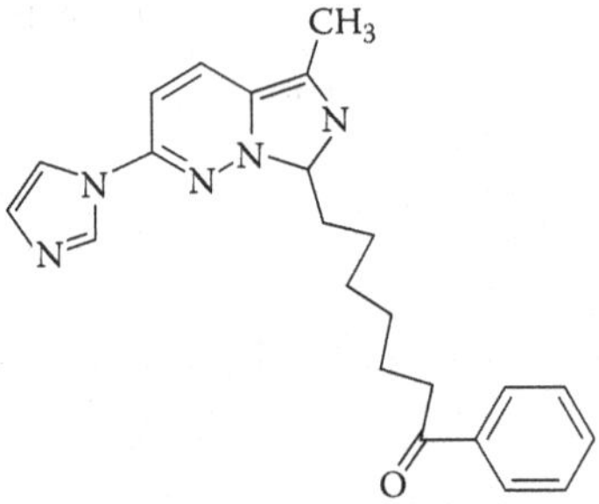

7-[2-(1H-Imidazol-1-yl)-5-methyl-
imidazo[1, 5-b]pyridazin-7-yl]
1-phenyl-1-haptanone

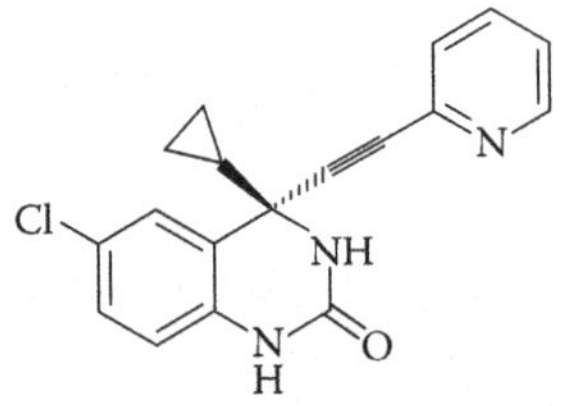

(-)-4-[2-(2-Pyridyl)ethyn-1-yl]-
6-chloro-4-cyclopropyl-3, 4-dihydro-
quinazolin-2 (1H)-one

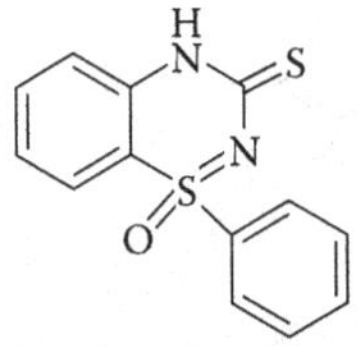

Benzothiadiazine-1-oxide
NSC 287474

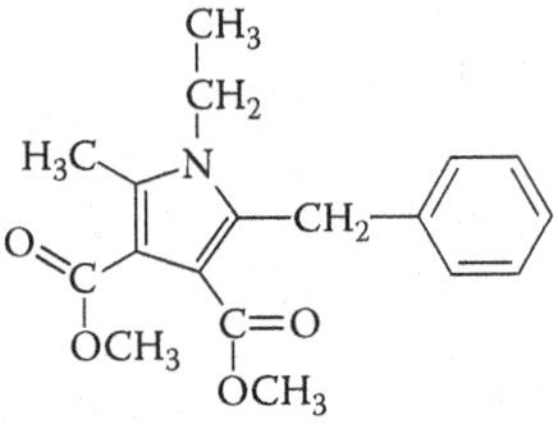

Dimethyl 1-ethyl-2-methyl-5-benzyl
1H pyrrole 3, 4-dicarboxylate

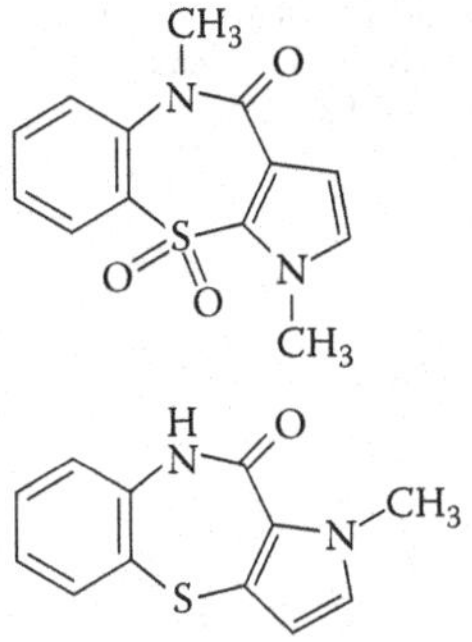

Pyrrolobenzothiazepines:
1, 5-Dimethyl-4-oxo(1H, 5H)-pyrrolo[2, 3-b]
[1, 5]benzothiazepine 10, 10-dioxide &
1-methyl-10-oxo(1H, 9H)-pyrrolo
[3, 2-b] [1, 5] benzothiazepine

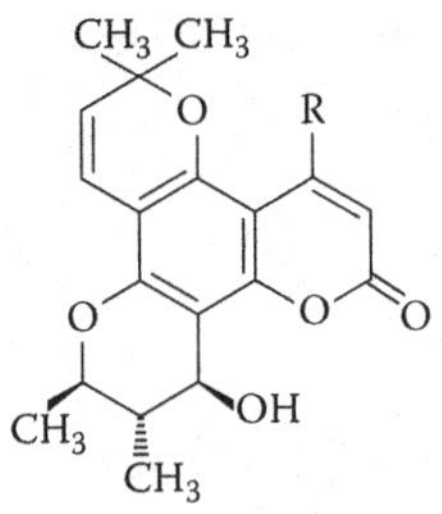

R = propyl   : Calanolide A
R = phenyl   : Inophyllum B

(C)

FIGURE 13.1 (CONTINUED)

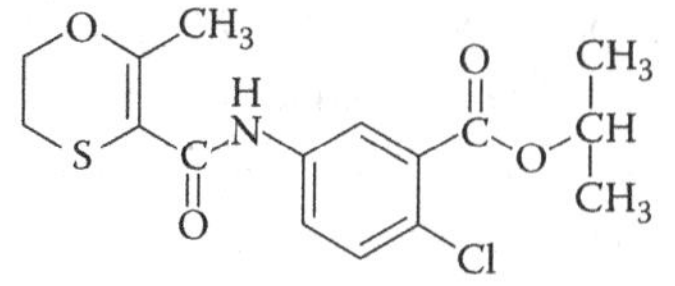

UC84

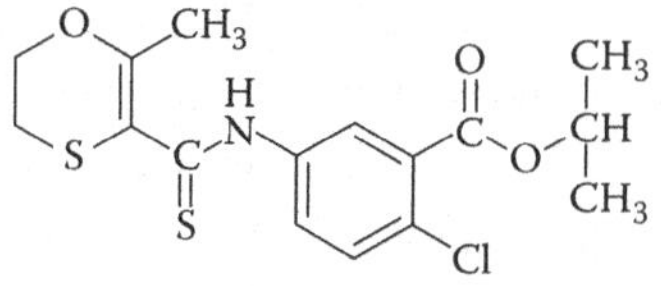

Compound 131
UC23

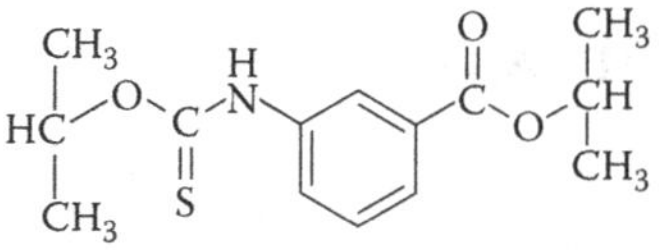

UC38

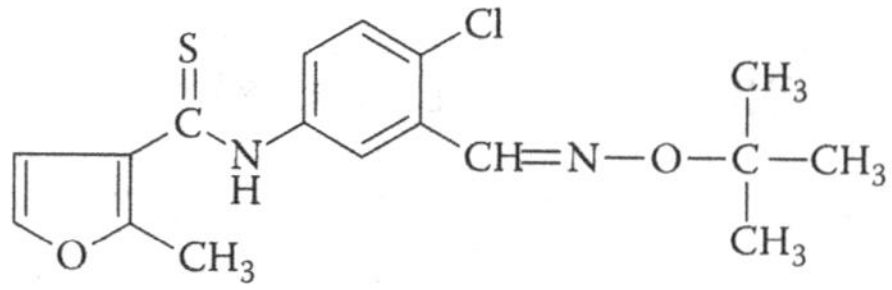

UC10

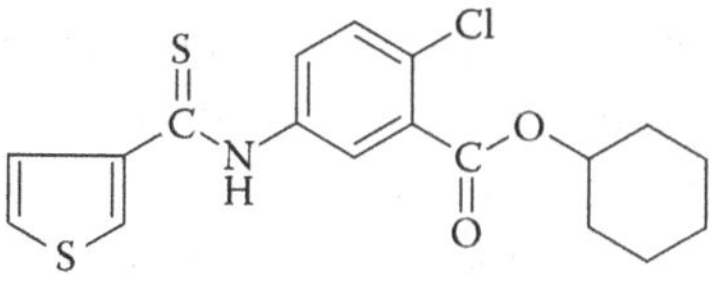

UC16

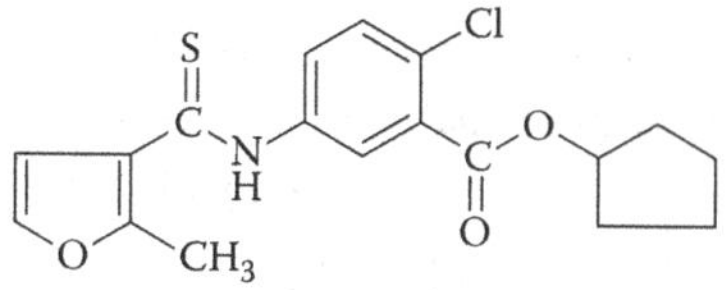

UC42

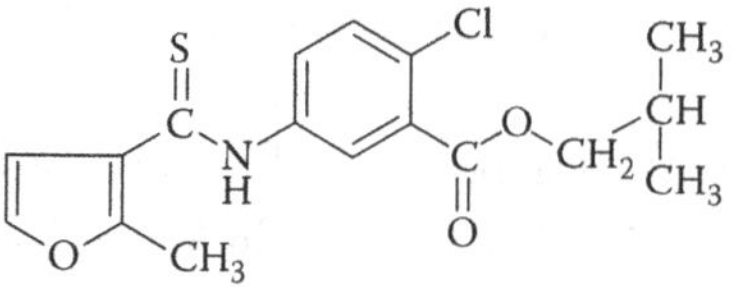

UC68

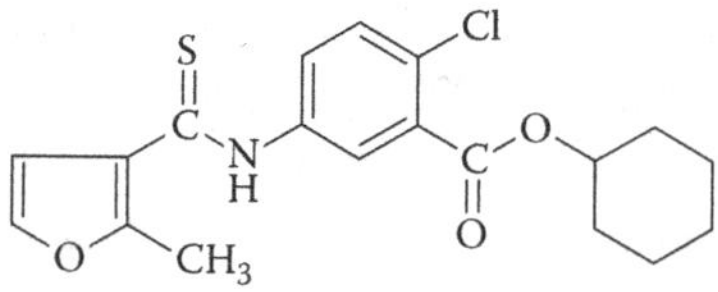

UC81

(D)

FIGURE 13.1 (CONTINUED)

RT activity, albeit at widely varying IC_{50} values, the lowest value (0.65 nM) being recorded for 7-[2-(1H-imidazol-1-yl)-5-methylimidazo-[1,5-b]pyridazin-7-yl]-1-phenyl-1-heptanone. The only compound listed in Table 13.1 for which no inhibitory effect on HIV-1 RT could be witnessed in the original experiments was oxathiin carboxanilide (UC84).

However, in our experiments, oxathiin carboxanilide (UC84) proved to be as good an inhibitor of HIV-1 RT as the other NNRTIs, i.e., TIBO R82913, nevirapine, pyridinone L-697,661, BHAP U88204, and TSAO-m3T;[27,28] in CEM cells, UC84 inhibited HIV-1 replication (cytopathicity) at an EC_{50} of 0.015 pg/mL while being toxic to the host cells at a CC_{50} of 8.8 pg/mL, thus achieving a selectivity index of 587. The anti-HIV-1 potency of UC84 was markedly enhanced upon substituting sulfur for oxygen in the carboxamide moiety of the molecule: the resulting compound 131 (UC23) proved inhibitory to HIV-1 replication at an EC_{50} of 0.007 pg/mL, and as its CC_{50} was 16 pg/mL, compound UC23 achieved a selectivity index of

2286. The thiocarboxanilide UC38 showed a similar potency (EC_{50}: 0.009 pg/mL) and selectivity index (1333). As a rule, all thiocarboxanilide derivatives (i.e., UC10, UC16, UC42, UC68, UC81, and UC84) proved to be potent inhibitors of HIV-1 RT, with IC_{50} values ranging from 0.05 to 0.20 pM, i.e., inhibition values closely to or markedly lower than those noted for the other NNRTIs (TIBO, nevirapine, pyridinone, BHAP, and TSAO).[28]

EMERGENCE OF DRUG-RESISTANT HIV-1 MUTANTS

The NNRTIs rapidly lead to the emergence of drug-resistant HIV-1 mutants, both *in vitro* and *in vivo*.[10,11] The mutations[29] that confer resistance to the NNRTIs appear to be located at positions 98, 100, 101, 103, 106, 108, 179, 181, 188, 190, 230, and 236 of the p66 subunit, and position 138 of the p51 subunit, of the HIV-1 reverse transcriptase (Table 13.2). The most notorious is the 181 Tyr → Cys mutation that leads to resistance to virtually all NNRTIs. Under the continued pressure of the NNRTIs (i.e., BHAP), the 181 Tyr → Cys mutant can further mutate to the 181 Cys → Ile mutant, which is even more resistant to the drugs, except for several HEPT derivatives.[30] Mutations may also appear/disappear in an ordered fashion so as to import a higher level of resistance, as has been noted with the 106 Val → Ala mutation that was replaced by the 190 Gly → Glu mutation under increased pressure of the drug (i.e., quinoxaline S-2720).[31]

The mutations conferring resistance to the NNRTIs appear to be located on RT p66 segments surrounding the binding site for the NNRTIs.[13] The 138 Glu → Lys mutation confers resistance (to TSAO) only when located in the p51 (and not p66) subunit because of the proximity of the p51 subunit Glu-138 residue to the NNRTI binding site at the p66 subunit. The fact that the different classes of NNRTIs give rise to different resistance mutations (Table 13.2) may be interpreted to mean that they interact with different affinities or different amino acid residues within the NNRTI-binding site. This, in turn, would suggest that the different NNRTIs should not necessarily lead to cross-resistance. A case in point is TSAO, which invariably elicits resistance based on the 138 Glu → Lys mutation.[32] This mutation—the only resistance mutation that has thus far been identified at the p51 subunit[33]—leads to resistance to the TSAO derivatives but not to any of the other NNRTIs, except for the oxathiin carboxanilide UC84.[33]

In fact, UC84 was found to select for the 138 Glu → Lys mutant in HIV-1-infected CEM cells.[33] The thiocarboxanilide UC38 selected for mutant HIV-1 strains in which 101 Lys and 190 Gly had mutated to Glu.[27] Other thiocarboxanilide derivatives, such as UC10, UC16, UC42, UC68, and UC81, selected for the 100 Leu → Ile, 101 Lys → Ile, 103 Lys → Asn/Thr, and 141 Gly → Glu mutations in the HIV-1 RT.[28] Thus, the thiocarboxanilide derivatives selected for some mutations, i.e., at positions 101 (Lys → Ile), 103 (Lys → Thr), and 141 (Gly → Glu), which had not been noted previously for the other NNRTIs (Table 13.2).

TABLE 13.2
Mutations in the HIV-1 RT Conferring Resistance to NNRTIs

Amino Acid Number	Mutation		
	Codon	Amino Acid	Compound(s)
98	GCA → GCA	Ala → Gly	Nevirapine, Pyridinone
100	TTA → ATA	Leu → Ile	TIBO, Nevirapine, Pyridinone, BHAP
101	AAA → GAA	Lys → Glu	Pyridinone, BHAP
103	AAA → AAC	Lys → Asn	TIBO, Nevirapine, Pyridinone, BHAP
106	GTA → GCA	Val → Ala	TIBO, HEPT, Nevirapine, BHAP, TSAO
108	GTA → ATA	Val → Ile	TIBO, HEPT, Nevirapine, Pyridinone
138[a]	GAG → AAG	Glu → Lys	TSAO
179	GTT → GAT	Val → Asp	TIBO, Pyridinone
181	TAT → TGT	Tyr → Cys	TIBO, HEPT, Nevirapine, Pyridinone, BHAP, TSAO, α-APA, Quinoxaline
181	TGT → ATT	Cys → Ile	TIBO, Nevirapine, Pyridinone, BHAP, TSAO, Quinoxaline
188	TAT → CAT	Tyr → His	TIBO, HEPT, Pyridinone
188	TAT → TGT	Tyr → Cys	TIBO, Nevirapine, Pyridinone
190	GGA → GAA	Gly → Glu	Quinoxaline, TIBO, Nevirapine, Pyridinone, BHAP
190	GGA → GCA	Gly → Ala	Nevirapine
230	ATG → ATA	Met → Ile	TIBO, Nevirapine, Pyridinone, α-APA
236	CCT → CTT	Pro → Leu	BHAP
138[a]	GAG → AAG	Glu → Lys	UC84
101	AAA → GAA	Lys → Glu	UC38
190	GGA → GAA	Gly → Glu	
100	TTA → ATA	Leu → Ile	UC68
103	AAA → AAC	Lys → Asn	UC10, UC81, UC16
103	AAA → ACA	Lys → Thr	UC42
101	AAA → ATA	Lys → Ile	UC16
141	GGG → GAG	Gly → Glu	

[a] Mutation interfering with the NNRTI is located in the p51 subunit of HIV-1 RT [80,81]; all other mutations interfering with the NNRTIs are located in the p66 subunit.

CROSS-RESISTANCE OF THIOCARBOXANILIDE DERIVATIVES WITH OTHER NNRTIs

Although oxathiin carboxanilide UC84 was virtually inactive against mutant HIV-1 strains containing the TIBO resistance mutation (100 Leu → Ile), or nevirapine resistance mutation (106 Val → Ala), or TSAO resistance mutation (138 Glu → Lys), or pyridinone resistance mutation (181 Tyr → Cys) in their reverse transcriptase (Table 13.3), minor structural changes in the molecule such as replacing the oxygen of the carboxamide moiety by sulfur (as in compound UC23) restored the activity against these virus mutant strains: compound UC23 was only twofold more active than the parent compound UC84 against wild-type HIV-1, but 30- to 100-fold more active against HIV-1 mutant strains that contained the 100-Ile, 106-Ala, 138-Lys, or 181-Cys mutations in their reverse transcriptase.[33]

In contrast with the oxathiin carboxanilide UC84, the thiocarboxanilide derivatives UC10, UC16, UC42, UC68, and UC81, similar to compound UC23, also proved markedly active against the 100-Ile, 106-Ala, 138-Lys, and 181-Cys mutants (Table 13.3). This is most remarkable for the 181-Cys mutant, because this mutant shows resistance and cross-resistance to most other NNRTIs, including TIBO, TSAO, nevirapine, and pyridinone.[28] The 181 Tyr → Cys mutation even arises in the presence of paired combinations of NNRTIs and should, by all means, be avoided or suppressed during anti-HIV chemotherapy. The thiocarboxanilide derivatives fulfill this premise. Not only do they not select for this mutation but they also suppress the replication of the 181-Cys mutant strain.

Depending on the nature of the mutation, the mutant virus strains arising under thiocarboxanilide therapy (i.e., 100 Leu → Ile, 103 Lys → Asn, 103 Lys → Thr, and 101 Lys → Ile/141 Gly → Glu) lost or retained marked sensitivity to one another and other NNRTIs (Table 13.4). For example, the HIV-1/UC16 strain, containing the double (101 Lys→ Ile/141 Gly → Glu) mutation, retained marked sensitivity to virtually all NNRTIs, including the thiocarboxanilides UC68 and UC81. In contrast, the HIV-1/UC68 strain, containing the 100 Leu → Ile mutation, lost its sensitivity to UC16 as well as nevirapine and TIBO. As a rule, all mutant strains selected by the thiocarboxanilide derivatives retained sufficient sensitivity to some, if not most, of the thiocarboxanilides and other NNRTIs (Table 13.4).

SWITCHING FROM ONE COMPOUND TO ANOTHER

If resistance would develop to oxathiin carboxanilide UC84, therapy could be readily switched to one of the thiocarboxanilide derivatives, which retained sensitivity to the resistant (i.e., 100 Leu → Ile, 103 Lys → Asn, 106 Val → Ala, 138 Glu → Lys, or 181 Tyr→ Cys) mutant strains.[28] There are plenty of other examples where switching from one compound to another restores sensitivity. For example, 5-chloro-3-(phenylsulfonyl)indole-2-carboxamide retains activity against those HIV-1 strains that, because of the 103 Lys→ Asn or 181 Tyr → Cys mutation, have acquired resistance to other NNRTIs such as TIBO, nevirapine, pyridinone, and BHAP. The α-APA derivative R89439 (loviride) is very active against the 100 Leu → Ile mutant, which is highly resistant to TIBO R82913 and R86183; and, within the TIBO class, a minor

TABLE 13.3

Activity of Thiocarboxanilide Derivatives Against Mutant HIV-1 Strains that are Resistant to Other NNRTIs

$EC_{50}{}^a$ (μM)

Compound	HIV-1(III$_B$) Wild-Type	HIV-1/ TIBO R82150 100 Leu→Ile	HIV-1/ TIBO R82913 103 Lys→Asn	HIV-1/ Nevirapine 106 Val→Ala	HIV-1/TSAO-m^3 T138 Glu→Lys	HIV-1/Pyridinone L-697661 181 Tyr→Cys
UC84	0.042	≤100	>50	>50	≥50	42
UC10	0.14	0.24	(3	0.37	0.21	0.21
UC16	0.079	2.2	(3	0.19	0.20	0.20
UC42	0.014	0.47	1.7	0.11	0.10	0.12
UC68	0.023	0.54	1.3	0.21	0.17	0.20
UC81	0.011	0.45	3.2	0.15	0.09	0.19
TSAO-m^{3}T	0.05	0.08	0.25	>75	>75	3.6
Nevirapine	0.024	0.34	5.1	7.8	0.10	7.8
Pyridinone L-697661	0.02	0.51	1.4	0.69	0.49	10.5
TIBO R82913	0.05	5.4	13.4	1.6	0.93	6.2

[a] 50% Effective concentration or compound concentration required to inhibit virus-induced cytopathicity in CEM cells.

Source: Balzarini, J. et al., Suppression of the breakthrough of human immunodeficiency virus type 1 (HIV-1) in cell culture by thiocarboxanilide derivatives when used individually or in combination with other HIV-1-specific inhibitors (i.e., TSAO derivatives), *Proc. Natl. Acad. Sci. U.S.A.*, 92, 5470, 1995. With permission.

TABLE 13.4
Activity of NNRTIs Against Mutant HIV-1 Strains that Have Become Resistant to Thiocarboxanilide Derivatives

Compound	HIV-1(IIIB) Wild-Type	HIV-1/UC10 103 Lys→Asn	HIV-1/UC16 101 Lys→Ile 141 Gly(Glu	HIV-1/UC42 103 Lys→Thr	HIV-1/UC68 100 Leu→Ile	HIV-1/UC81 103 Lys→Asn
			EC$_{50}$^{a} (µM)			
UC10	0.14	0.67	0.34	1.7	3.6	0.86
UC16	0.079	4.6	0.79	7.2	≥10	0.92
UC42	0.014	0.55	0.30	1.5	1.1	0.39
UC68	0.020	0.43	0.17	1.1	1.4	0.45
UC81	0.011	0.80	0.13	2.9	2.3	0.42
TSAO-m³T	0.05	0.37	0.10	0.5	0.61	0.33
Nevirapine	0.24	5.7	0.81	9.1	≥15	1.3
Pyridinone L-697661	0.02	0.86	0.23	1.4	1.8	0.086
TIBO R82913	0.062	3.4	0.50	≥15	≥15	0.72
Quinoxaline S-2720	0.013	0.097	0.026	0.39	0.36	0.097
MKC-442	0.002	0.26	0.026	1.1	1.3	0.033

^{a} See footnote to Table 13.3.

Source: Balzarini, J. et al., Suppression of the breakthrough of human immunodeficiency virus type 1 (HIV-1) in cell culture by thiocarboxanilide derivatives when used individually or in combination with other HIV-1-specific inhibitors (i.e., TSAO derivatives), *Proc. Natl. Acad. Sci. U.S.A.*, 92, 5470, 1995. With permission.

chemical modification, i.e., shifting the chlorine from the 9-position (R82913) to the 8-position (R86183) suffices to restore activity against the 181 Tyr → Cys mutant.[14] Similarly, pyridinone L-702,019, which differs from its predecessor L-696,229 only by the addition of two chlorine atoms (in the benzene ring) and substitution of sulfur for oxygen (in the pyridine ring), is markedly inhibitory to HIV-1 mutants containing the 103 Lys → Asn or 181 Tyr → Cys mutation.

In attempts to circumvent the problem of virus-drug resistance, one can switch from one NNRTI to another, and given the multitude of NNRTIs now available, this switching procedure can go on for an indefinite time. Moreover, during this process, one can at any time return to an inhibitor that has been used previously and to which the virus is no longer resistant. Should resistance to the first compound disappear after the second compound has been installed, and, conversely, should resistance to the second compound disappear after the first has been reinstalled, then alternating drug therapy may be envisaged as an attractive procedure to overcome the resistance problem. Some mutations, i.e., the HIV-1 RT 236 Pro → Leu mutation causing resistance to BHAP, may increase the sensitivity to other NNRTIs such as TIBO, nevirapine, and pyridinone,[34] and thus again provides a rationale for the alternating or switching use of these drugs.

DRUG COMBINATIONS

Combinations of different anti-HIV drugs, including NNRTIs, can be expected to lead to a synergistic antiviral action and increased antiviral activity spectrum while diminishing the toxic side effects and reducing the risk of drug resistance development. Synergistic anti-HIV activity has been demonstrated for a large number of combinations, including TIBO R82913 (or TIBO R86183) with AZT,[35] nevirapine (BI-RG-587) with AZT, BHAP (U-90152) with AZT, BHAP (U-87201E) with DDI, and MKC-442 with AZT, and MDL-28, 574 (6-O-butyrylcastanospermine).[36,37] The effectiveness of the drug combination regimen may increase proportionally with the number of drugs that take part in the combination.

It was originally proposed that convergent multiple drug combinations, i.e., combinations based on AZT, DDI, and pyridinone or nevirapine, may prevent the development of multidrug resistance, because the multiply-mutated virus would no longer be viable. This assumption has proved faulty, as following convergent combination therapy multidrug-resistant HIV has been isolated that is still viable.[38]

A rational approach toward drug combination should be based on the mutually antagonistic mutation principle.[11] Indeed, some mutations conferring resistance have been found to counteract each other in that they suppress the emergence of resistance to one another. For example, the AZT-resistance mutation at position 215 (Thr → Phe) is suppressed by the NNRTI-resistance mutation at position 181 (Tyr → Cys), and similarly, the 215 mutation (Thr → Tyr/Phe) is also suppressed by the 3TC-resistance mutation 184 (Met → Val).[39] Substitutions at residues 100 (Leu → Ile) and 181 (Tyr → Cys), which engender resistance to TIBO and other NNRTIs, were found to suppress resistance to AZT when co-expressed with AZT-specific substitutions.[40] These data provide a rationale for the combined use of those agents that lead to a mutually antagonistic resistance profile, particularly triple drug combinations containing AZT, 3TC, and an NNRTI such as loviride.

When 3TC, TSAO-m³T, and the thiocarboxanilide UCI0 were used individually, they rapidly led to the emergence of drug-resistant HIV-1 mutants containing the following mutations in their RT: 138 Glu → Lys for TSAO-m³T, 184 Met → Val for 3TC, and 103 Lys → Thr/Asn for UC10.[41] If 3TC was combined with either TSAO-m³T or UC10, or UC42, emergence of drug-resistant virus was markedly delayed or even suppressed. The concomitant presence of the 138 Glu → Lys and 184 Met → IleVal mutations was noted in the RT of those mutant viruses that emerged under combination therapy of 3TC with either TSAO-m³T or UC10. The 103 Lys → Thr/Asn mutation that emerged under the therapy with UC10 alone was no longer detected (which means that it must have been suppressed) after combination of UC10 with 3TC.[41]

HIV-1 KNOCKOUT CONCENTRATIONS

If the HIV-1-specific RT inhibitors BHAP (U-88204 or U-90152) are used from the start at a sufficiently high concentration (i.e., 1 or 3 pM, respectively), they are able to completely suppress virus replication, so that the virus is "knocked out" and does not have the opportunity to become resistant. Not only BHAP but also other NNRTIs have been shown to knock out HIV-1 in cell culture when used at concentrations of 2.5 pg/mL (TIBO R82913, pyridinone L-697,661) or 10 pg/mL (nevirapine, BHAP U-88204).[42] That the virus was really knocked out, and the cell culture thus cleared from the virus infection, was evident from the absence of such viral parameters as viral cytopathicity, p24 antigen production, and, particularly, proviral DNA [as ascertained by two successive 35-cycle rounds of polymerase chain reaction (PCR) analysis].[42]

With quinoxaline S-2720, a total suppression of HIV-1 infection in CEM cell cultures was achieved at compound concentrations (i.e., 0.35 pM), that were 10- to 25-fold lower than those required for BHAP U-88204 and nevirapine to knock out the virus.[31] Also, cell cultures infected with a TSAO-resistant mutant (138 Glu → Lys) virus could be completely protected from virus-mediated destruction at micromolar concentrations by HIV-1-specific RT inhibitors (i.e., TIBO, nevirapine, and BHAP) that do not show cross-resistance to the TSAO mutant strain.[30]

The thiocarboxanilide derivatives UC10, UC16, UC42, UC68, and UC81 were found to completely suppress HIV-1 replication and thus prevent the emergence of drug-resistant virus when used at a concentration of 1.3 to 6.6 pM, that is 10- to 25-fold lower than the concentration required for nevirapine and BHAP U90152 to do so.[28] In the experiment shown in Table 13.5, the thiocarboxanilides were added to the HIV-infected CEM cell cultures at initial concentrations of 0.1 pg/mL (0.3 pM), 0.5 pg/mL (1.4 pM), and 2.5 pg/mL (7 pM). The drug concentrations were kept constant throughout the whole time period of the experiment (10 subcultivations or 35 d). Then the drugs were removed and the cells were further subcultured for at least five additional passages (up to 53 d). With all thiocarboxanilide derivatives at a concentration of 0.5 or 2.5 pg/mL (except for UC16 at 0.5 pg/mL), virus breakthrough could be completely prevented for at least 53 d. On the 35th day (10th subcultivation), no p24 production and no proviral DNA could be detected, and, after removal of the drugs and further subcultivation of the cells in the absence of the drugs, virus did

TABLE 13.5

Inhibitory Effect of Thiocarboxanilide Derivatives on Breakthrough of HIV-1(III$_B$) in CEM Cell Cultures

Compound	Mean Day of Virus Breakthrough (50% Cytopathicity) at Compound[a] Concentration of		
	0.1 µg/mL	0.5 µg/mL	2.5 µg/mL
UC84		10	
UC10	16	>53	>53
UC16	11	16	>53
UC42	19	>53	>53
UC68	17	>53	>53
UC81	17	>53	>53
Nevirapine	6	11	16

[a] Drug removed from the cell cultures after 35 d (10 subcultivations).

Source: Balzarini, J. et al., Suppression of the breakthrough of human immunodeficiency virus type 1 (HIV-1) in cell culture by thiocarboxanilide derivatives when used individually or in combination with other HIV-1-specific inhibitors (i.e., TSAO derivatives), *Proc. Natl. Acad. Sci. U. S.A.,* 92, 5470, 1995. With permission.

not emerge. Under these conditions, the cells may be considered cleared or "cured" from the HIV infection.

HIV-1 KNOCKOUT CONCENTRATIONS OF DIFFERENT DRUGS COMBINED

If different drugs are combined, the drug concentrations required to knock out the virus in cell culture can be markedly reduced. For example, the concentration of BHAP U-90152 needed to totally prevent the spread of HIV-1 is 3 pM, but, if combined with AZT (0.5 pM), the concentration of U-90152 can be lowered to 0.5 pM so as to totally prevent viral spread. Although 3TC and TSAO-m^{3}T, when used individually at a concentration of 0.02, 0.05, 0.1, or 0.4 pg/mL, could not prevent virus breakthrough in HIV-infected CEM cell cultures for longer than a few days, when combined, these drugs totally prevented virus breakthrough for at least 52 d.[41] Similarly (Table 13.6), 3TC and UC10 were unable to prevent virus breakthrough for more than 13 d (3TC at 0.1 pg/mL) or 21 d (UC10 at 0.2 pg/mL).

However, when combined (i.e., 3TC at a concentration of 0.1 pg/mL with UC10 at a concentration of 0.05 or 0.2 pg/mL), the drugs delayed virus breakthrough for at least 52 d. When the cells were further subcultivated in the absence of the compounds for at least 10 additional passages, they remained virus-free, did not produce viral antigen, and did not reveal any detectable proviral DNA in their genome.

When either BHAP or TSAO-m^{3}T or UC42 were added to HIV-infected CEM cell cultures at 0.11 or 0.28 pM (UC42), 0.09 or 0.22 pM (BHAP), or 1.7 or 4.2 pM

TABLE 13.6

Inhibitory Effect of Combinations of 3TC with UC10 or TSAO-m³T on
Breakthrough of HIV-1(III$_B$) in CEM Cell Cultures

	Mean Day of Virus Breakthrough (50% Cytopathicity) 3TC at a Concentration of			
	0 µg/mL	0.02 µg/mL	0.05 µg/mL	0.1 µg/mL
UC10				
at 0 µg/mL	3	4	6	13
0.02 µg/mL	4	14	26	30
0.05 µg/mL	6	33	>52[a]	>52[b]
0.2 µg/mL	21	42	>52[b]	>52[b]
TSAO-m³T				
at 0 µg/mL	3	4	6	13
0.05 µg/mL	4	6	26	30
0.1 µg/mL	6	19	34	>52[a]
0.4 µg/mL	13	23	>52[a]	>52[b]

[a] After washout of the drugs at day 52 post-infection, virus breakthrough occurred a few days later.

[b] After washout of the drugs at day 52 post-infection, virus breakthrough no longer occurred (i.e., the cell cultures remained virus-free for at least 12 subcultivations in the absence of the test compounds).

Source: Balzarini, J. et al., Marked inhibitory activity of non-nucleoside reverse transcriptase inhibitors against human immunodeficiency virus type 1 when combined with (−)2′,3′-dideoxy-3′-thiacytidine, *Mol. Pharmacol.*, 49, 882, 1996. With permission.

(TSAO-m³T), virus breakthrough could be delayed until day 14, or at the most day 20.[28] If, however, UC42 at 0.28 pM was combined with TSAO-m³T at 1.7 or 4.2 pM, virus breakthrough could be delayed for more than 77 days (Table 13.7). If UC42 at 0.11 pM was combined with TSAO-m³T at 1.7 pM, virus breakthrough was delayed for 25 d, but if to this two-drug combination a third compound was added, namely BHAP U-90152 at 0.09 pM, again virus breakthrough was delayed for more than 77 d. These cell cultures became provirus-free as evidenced by PCR; they did not produce any virus and continued to grow in the absence of the drugs as if they were indistinguishable from the uninfected CEM cell cultures. These cells could thus be considered as being cleared or cured from the infection.

CONCLUSION

Various NNRTIs have been reported to specifically inhibit HIV-1: for example, TIBO, HEPT, nevirapine, pyridinone, BHAP, TSAO nucleoside analogues, α-APA, quinoxaline, and oxathiin carboxanilide derivatives. These compounds interact allosterically (i.e., noncompetitively with respect to the dNTPs) with a specific nonsubstrate binding "pocket" site of the HIV-1 RT. The most potent NNRTIs have been found to inhibit HIV-1 replication at nanomolar concentrations. These compounds

TABLE 13.7
Inhibitory Effect of Combinations of UC42 with TSAO-m³T and BHAP U-90152 on Breakthrough of HIV-1(III$_B$) in CEM Cell Cultures

	Mean Day of Virus Breakthrough (50% Cytopathicity) UC42 at a Concentration of		
Compound	0 µM	0.11 µM	0.28 µM
TSAO-m³T			
at 0 µM	5	16	20
1.7 µM	16	25	>77
4.2 µM	16	35	>77

	BHAP U-90152 at a Concentration of		
	0 µM	0.09 µM	0.22 µM
UC42 + TSAO-m³T			
0 µM 0 µM	5	9	16
0.11 µM 1.7 µM	25	>77	42
0.11 µM 4.2 µM	35	62	49
0.28 µM 1.7 µM	>77	>77	>77
0.28 µM 4.2 µM	>77	>77	>77

Source: Balzarini, J. et al., Suppression of the breakthrough of human immunodeficiency virus type 1 (HIV-1) in cell culture by thiocarboxanilide derivatives when used individually or in combination with other HIV-1-specific inhibitors (i.e., TSAO derivatives), *Proc. Natl. Acad. Sci. U. S.A.,* 92, 5470, 1995. With permission.

therefore offer great potential for the treatment of HIV-1 infections. Yet, the virus may rapidly develop resistance to these drugs.

The mutations conferring resistance have been mapped at the HIV-1 RT positions 98 (Ala → Gly), 100 (Leu → Ile), 101 (Lys → Glu), 103 (Lys → Asn), 106 (Val → Ala), 108 (Val → Ile), 138 (Glu → Lys), 179 (Val → Asp/Glu), 181 (Tyr → Cys → Ile), 188 (Tyr → Cys/His), 190 (Gly → Ala/Glu), 230 (Met → Ile), and 236 (Pro → Leu). Most of these mutations must reside in the p66 subunit of HIV-1 RT to confer resistance to the NNRTIs.

The exception is the 138 (Glu → Lys) mutation that accounts for resistance to the TSAO derivatives when it is located on the p51 subunit. These mutations do not necessarily lead to cross-resistance among the various NNRTIs, and, in some cases, they have even proved to be mutually suppressive. The thiocarboxanilide derivatives lead to mutations at HIV-1 RT positions 100 (Leu → Ile), 101 (Lys → Ile/Glu), 103 (Lys → Asn/Thr), and 141 (Gly → Glu).

Several strategies could be envisaged to circumvent or prevent the development of resistance to the NNRTIs: (1) switching from one class of NNRTIs (to which the virus has developed resistance) to another (to which the virus has not developed resistance), and, even within each particular class of NNRTIs, small chemical modifications suffice to reinstall HIV drug sensitivity; (2) combination of different NNRTIs that do not confer cross-resistance, or, that may, in fact, even counteract development of resistance to one another; this combination strategy also extends to the 2',3'-dideoxynucleoside (ddN) analogues that may be combined with one or more NNRTIs based on the mutually suppressive resistance principle; (3) the use of sufficiently high (knockout) concentrations of the NNRTIs from the start so as to completely suppress virus replication and thus prevent resistance from emerging; (4) combining from the start different NNRTIs (whether or not in combination with ddNs) at knockout concentrations so as to completely suppress virus replication and thus prevent the breakthrough of any virus, whether resistant or not, for a much longer time period and at much lower drug concentrations than could be achieved if the drugs had to be used individually. The oxathiin carboxanilide derivatives, particularly the thiocarboxanilides, fulfill the four premises (1, 2, 3, and 4) that make them ideal candidates to be further explored, in combination with other NNRTIs, for their potential in the treatment of HIV-1 infections.

ACKNOWLEDGMENTS

We thank Christiane Callebaut for fine editorial help, and Ann Absillis and Birgitta Wallstrom for excellent technical assistance with the original experiments. This research was supported in part by the Biomedical Research Programme and the Human Capital and Mobility Programme of the European Commission, the Belgian Nationaal Fonds voor Wetenschappelijk Onderzoek (no. 3.3010.91), the Belgian Geconcerteerde Onderzoeksacties (no. 95/5), the Janssen Research Foundation, the Medical Faculty of the Karolinska Institute, the Swedish Society of Medicine, and the Swedish Physicians against AIDS.

REFERENCES

1. Miyasaka, T. et al., A novel lead for specific anti-HIV-1 agents: 1-[(2-hydroxy-ethoxy)methyl]-6-(phenylthio)thymine, *Med. Chem.*, 32, 2507, 1989.
2. Debyser, Z. et al., An antiviral target on reverse transcriptase of human immunodeficiency virus type 1 revealed by tetrahydroimidazo-[4,5,1-*jk*] [1,4]benzodiazepin-2 (1*H*)-one and -thione derivatives, *Proc. Natl. Acad. Sci. U.S.A.*, 88, 1451, 1991.
3. Baba, M. et al., Highly potent and selective inhibition of human immunodeficiency virus type 1 by a novel series of 6-substituted acyclouridine derivatives, *Mol. Pharmacol.*, 39, 805, 1991.
4. Koup, R.A. et al., Inhibition of human immunodeficiency virus type 1 (HIV-1) replication by the dipyridodiazepinone BI-RG-587, *J. Infect. Dis.*, 163, 966, 1991.
5. Goldman, M.E. et al., L-696,229 specifically inhibits human immunodeficiency virus type 1 reverse transcriptase and possesses antiviral activity *in vitro*, *Antimicrob. Agents Chemother.*, 36, 1019, 1992.
6. Romero, D.L. et al., Bis(heteroaryl)piperazine (BHAP) reverse transcriptase inhibitors: structure-activity relationships of novel substituted indole analogues and the identification of 1-[(5-methanesulfonamido-1*H*-indol-2-yl)-carbonyl]-4-[3- [(1-methylethyl)amino]-pyridinyl]piperazine monomethanesulfonate (U-90152S), a second-generation clinical candidate, *J. Med. Chem.*, 36, 1505, 1993.
7. Balzarini, J. et al., [2',5'-Bis-*O*-(*tert*-butyldimethylsilyl)]-3'-spiro-5"-(4"-amino-1",2"-oxathiole-2",2"-dioxide) (TSAO) derivatives of purine and pyrimidinenucleosides as potent and selective inhibitors of human immunodeficiency virus type 1, *Antimicrob. Agents Chemother.*, 36, 1073, 1992.

8. Balzarini, J. et al., Kinetics of inhibition of human immunodeficiency virus type 1 (HIV-1) reverse transcriptase by the novel HIV-1-specific nucleoside analogue [2′,5′-bis-*O*-(*tert*-butyldimethylsilyl)-beta-*D*-ribofuranosyl]-3′-spiro-5″-(4″-amino-1″,2″-oxa-thiole-2″,2″-dioxide)thymine (TSAO-T), *J. Biol. Chem.*, 267, 11831, 1992.

9. Pauwels, R. et al., Potent and highly selective human immunodeficiency virus type 1 (HIV-1) inhibition by a series of alpha-anilinophenylacetamide derivatives targeted at HIV-1 reverse transcriptase, *Proc. Natl. Acad. Sci. U.S.A.*, 90, 1711, 1993.

10. Balzarini, J. et al., Concomitant combination therapy for HIV infection preferable over sequential therapy with 3TC and non-nucleoside reverse transcriptase inhibitors, *Proc. Natl. Acad. Sci. U.S.A.*, 93, 13152, 1996.

11. De Clercq, E., HIV resistance to reverse transcriptase inhibitors, *Biochem. Pharmacol.*, 47, 155, 1994.

12. Schmitt, J.C. et al., Multiple drug resistance to nucleoside analogues and nonnucleoside reverse transcriptase inhibitors in an efficiently replicating human immunodeficiency virus type 1 patient strain, *J. Infect. Dis.*, 174, 962, 1996.

13. Tantillo, C. et al., Locations of anti-AIDS drug binding sites and resistance mutations in the three-dimensional structure of HIV-1 reverse transcriptase. Implications for mechanisms of drug inhibition and resistance, *Mol. Biol.*, 243, 369, 1994.

14. Pauwels, R. et al., New tetrahydroimidazo[4,5,1-*jk*][1,4]-benzodiazepin-2(1*H*)-one and -thione derivatives are potent inhibitors of human immunodeficiency virus type 1 replication and are synergistic with 2′,3′-dideoxynucleoside analogs, *Antimicrob. Agents Chemother.*, 38, 2863, 1994.

15. Baba, M. et al., Preclinical evaluation of MKC-442, a highly potent and specific inhibitor of human immunodeficiency virus type 1 *in vitro*, *Antimicrob. Agents Chemother.*, 38, 688, 1994.

16. De Clercq, E., New developments in anti-HIV chemotherapy, *Curr. Med. Chem.*, 8, 1543, 2001.

17. Kleim, J.P. et al., Activity of a novel quinoxaline derivative against human immunodeficiency virus type 1 reverse transcriptase and viral replication, *Antimicrob. Agents Chemother.*, 37, 1659, 1993.

18. Buckheit, R.W., Jr. et al., Thiazolobenzimidazole: biological and biochemical anti-retroviral activity of a new nonnucleoside reverse transcriptase inhibitor, *Antiviral Res.*, 21, 247, 1993.

19. Mertens, A. et al., Selective non-nucleoside HIV-1 reverse transcriptase inhibitors. New 2,3-dihydrothiazolo[2,3-*a*]isoindol-5(9*bH*)-ones and related compounds with anti-HIV-1 activity, *J. Med. Chem.*, 36, 2526, 1993.

20. Alam, M. et al., Substituted naphthalenones as a new structural class of HIV-1 reverse transcriptase inhibitors, *Antiviral Res.*, 22, 131, 1993.

21. Artico, M., Selected non-nucleoside reverse transcriptase inhibitors (NNRTIs): the DABOs family, *Drug Fut.* 27, 159, 2002.

22. Tucker, T.J. et al., Synthesis of a series of 4-(arylethynyl)-6-chloro-4-cyclopropyl-3,4-dihydroquinazolin-2(1*H*)-ones as novel non-nucleoside HIV-1 reverse transcriptase inhibitors, *J. Med. Chem.*, 37, 2437, 1994.

23. Buckheit, R.W., Jr. et al., Biological and biochemical anti-HIV activity of the benzothiadiazine class of nonnucleoside reverse transcriptase inhibitors, *Antiviral Res.*, 25, 43, 1994.

24. Antonucci, T. et al., Characterization of the antiviral activity of highly substituted pyrroles: a novel class of non-nucleoside HIV-1 reverse transcriptase inhibitor, *Antiviral Chem. Chemother.*, 6, 98, 1995.

25. Artico, M. et al., Pyrrolobenzothiazepines: a new class of nonnucleoside HIV-1 reverse transcriptase inhibitors, *Med. Chem. Res.*, 4, 283, 1994.

26. Patil, A.D. et al., The inophyllums, novel inhibitors of HIV-1 reverse transcriptase isolated from the Malaysian tree, *Calophyllum inophyllum* Linn, *J. Med. Chem.*, 36, 4131, 1993.

27. Balzarini, J. et al., Activity of various thiocarboxanilide derivatives against wild-type and several mutant human immunodeficiency virus type 1 strains, *Antiviral Res.*, 27, 219, 1995.

28. Balzarini, J. et al., Suppression of the breakthrough of human immunodeficiency virus type 1 (HIV-1) in cell culture by thiocarboxanilide derivatives when used individually or in combination with other HIV-1-specific inhibitors (i.e., TSAO derivatives), *Proc. Natl. Acad. Sci. U.S.A.*, 92, 5470, 1995.

29. Mellors, J.W., Larder, B.A., and Schinazi, R.F., Mutations in HIV-1 reverse transcriptase and protease associated with drug resistance, *Int. Antiviral News*, 3, 8, 1995.

30. Balzarini, J. et al., Human immunodeficiency virus 1 (HIV-1)-specific reverse transcriptase (RT) inhibitors may suppress the replication of specific drug-resistant (E138K)RT HIV-1 mutants or select for highly resistant (Y181C→C181I)RT HIV-1 mutants, *Proc. Natl. Acad. Sci. U.S.A.*, 91, 6599, 1994.

31. Balzarini, J. et al., Resistance pattern of human immunodeficiency virus type 1 reverse transcriptase to quinoxaline S-2720, *J. Virol.*, 68, 7986, 1994.

32. Balzarini, J. et al., Sensitivity of (138 Glu→Lys) mutated human immunodeficiency virus type 1 (HIV-1) reverse transcriptase (RT) to HIV-1-specific RT inhibitors, *Biochem. Biophys. Res. Commun.*, 201, 1305, 1994.

33. Balzarini, J. et al., Oxathiin carboxanilide derivatives: a class of non-nucleoside HIV-1-specific reverse transcriptase inhibitors (NNRTIs) that are active against mutant HIV-1 strains resistant to other NNRTIs, *Antiviral Chem. Chemother.*, 6, 169, 1995.

34. Dueweke, T.J. et al., A mutation in reverse transcriptase of bis(heteroaryl)piperazine-resistant human immunodeficiency virus type 1 that confers increased sensitivity to other nonnucleoside inhibitors, *Proc. Natl. Acad. Sci. U.S.A.*, 90, 4713, 1993.

35. Buckheit, R.W., Jr. et al., Structure activity and cross-resistance evaluations of a series of human immunodeficiency virus type-1 specific compounds related to oxathiin, *Antiviral Chem. Chemother.*, 5, 35, 1994.

36. Chong, K.T., Pagano, P.J., and Hinshaw, R.R., Bisheteroarylpiperazine reverse transcriptase inhibitor in combination with 3'-azido-3'-deoxythymidine or 2',3'-dideoxycytidine synergistically inhibits human immunodeficiency virus type 1 replication *in vitro*, *Antimicrob. Agents Chemother.*, 38, 288, 1994.

37. Brennan, T.M. et al., The inhibition of human immunodeficiency virus type 1 *in vitro* by a non-nucleoside reverse transcriptase inhibitor MKC-442, alone and in combination with other anti-HIV compounds, *Antiviral Res.*, 26, 173, 1995.

38. Larder, B.A., Kellam, P., and Kemp, S.D., Convergent combination therapy can select viable multidrug-resistant HIV-1 *in vitro*, *Nature*, 365, 451, 1993.

39. Tisdale, M. et al., Rapid *in vitro* selection of human immunodeficiency virus type 1 resistant to 3'-thiacytidine inhibitors due to a mutation in the YMDD region of reverse transcriptase, *Proc. Natl. Acad. Sci. U.S.A.*, 90, 5633, 1993.

40. Byrnes, V.W. et al., Susceptibilities of human immunodeficiency virus type 1 enzyme and viral variants expressing multiple resistance-engendering amino acid substitutions to reserve transcriptase inhibitors, *Antimicrob. Agents Chemother.*, 38, 1404, 1994.

41. Balzarini, J. et al., Marked inhibitory activity of non-nucleoside reverse transcriptase inhibitors against human immunodeficiency virus type 1 when combined with (−)2',3'-dideoxy-3'-thiacytidine, *Mol. Pharmacol.*, 49, 882, 1996.

42. Balzarini, J. et al., Knocking-out concentrations of HIV-1-specific inhibitors completely suppress HIV-1 infection and prevent the emergence of drug-resistant virus, *Virology*, 196, 576, 1993.

14 Histamine H$_3$-Receptor Agonists and Antagonists: *Chemical, Pharmacological, and Clinical Aspects*

Holger Stark and Walter Schunack

CONTENTS

SUMMARY

With the detection of the third histamine receptor subtype (H$_3$), the neurotransmitter function of histamine could be established. Histamine H$_3$-receptors located on histaminergic neurons control the synthesis and liberation of histamine by acting as autoreceptors and the release of a number of other neurotransmitters by acting as heteroreceptors. The first potent and selective agonist was (R)-α-methylhistamine. Stereoselectivity of the chain-branched histamine derivatives was high with the monomethylated and was even higher with the (R)-α,(S)-β-dimethylated derivative. These compounds are excellent pharmacological tools for H$_3$-receptor-dependent investigations. Their pharmacodynamic advantages were not obvious when (R)-α-methylhistamine was tested *in vivo* in humans because of some pharmacokinetic disadvantages overcome by prodrug formation.

Novel prodrugs incorporating the potent agonist in a bioreversible azomethine bond with 2-hydroxybenzophenone or with related derivatives show high bioavailability, good membrane penetration, and protection against histamine methyltransferase, the major metabolizing enzyme in humans. The balance between hydrolytic

199

stability and lability, which determines drug liberation, was optimized introducing a hydroxy moiety, and the pharmacodynamic properties of the prodrugs can be shifted to peripheral or central tissues by varying the substitution pattern of the promoiety. Halogenated and five-membered heteroaryl promoieties show striking effects on drug delivery and targeting. One of these prodrugs (FUB 94 = BP2.94) is in phase II of clinical trials showing antinociceptive, anti-inflammatory, and anti-ulcer activity. In the field of antagonists, carbamates and ethers were presented as two potential leads with promising *in vitro* and *in vivo* activity and additional high selectivity. Although the carbamates show moderate to good activity *in vitro*, these compounds are highly effective *in vivo* in the CNS even after p.o. application.

Another highly potent and selective compound is presented with the ether derivative FUB 181, a novel potential candidate for the treatment of dementia, epilepsy, and schizophrenia. The potential indications are based on the cognition enhancement and the anticonvulsant effects of H_3-receptor antagonists. The psychotic effects are supposed to be due to reduced density of histamine receptors in the brain of schizophrenic patients and the H_3-antagonist activity of the atypical neuroleptic clozapine. From these data it can be presumed that H_3-receptor ligands will be introduced into therapy in the near future.

INTRODUCTION

During the past decade the neurotransmitter function of the biogenic amine histamine has been established by identifying a third histamine receptor subtype.[1,2] In contrast to the previous known H_1- and H_2-receptors,[3] H_3-receptors may be located presynaptically on the axon terminals of histaminergic neurons (autoreceptor) and on nonhistaminergic neurons (heteroreceptor).[4] By a negative feedback mechanism unknown up to now, activation of H_3-receptors causes inhibition of histamine synthesis in, and histamine release from, histaminergic neurons (Figure 14.1).[1] H_3-het-

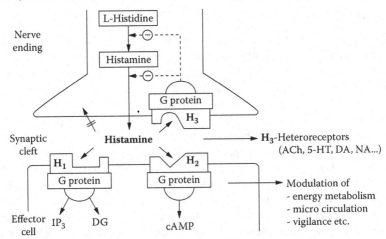

FIGURE 14.1 Schematic illustration of histaminergic neurotransmission. (Abbreviations: Ach, acetylcholine; 5-HT, serotonin; DA, dopamine; NA, noradrenaline; IP_3, inositol triphosphate; DG, diacylglycerol; cAMP, cyclic adenosine monophosphate.)

eroreceptors have been identified on serotonergic, noradrenergic, dopaminergic, and peptidergic neurons in the central nervous system (CNS) and on sympathetic, parasympathetic, and NANC (nonadrenergic noncholinergic) nerve fibers supplying the gastrointestinal, bronchial, and/or cardiovascular system.[4,5]

In contrast to the cloned H_1- and H_2-receptors[6,7] the molecular structure of the H_3-receptor is still unclear. But there is little doubt that it belongs to the superfamily of membrane-bound heptahelical receptors coupled to G proteins like the other histamine receptor subtypes.[8] Further signal transduction remains unclear. Negative coupling to phosphoinositide turnover and involvement of N-type Ca^{2+} channels via a pertussis-insensitive G protein have been shown in different studies.[9–11] In contrast to many other biogenic amines, a reuptake system for histamine has never been found in histaminergic neurons,[12] although some evidence for an equivalent system in other tissues has been described.[13]

Different assays have been established to determine histamine H_3-receptor activity of novel ligands. Competition studies with radioligands,[8–15] H_3-receptor-mediated influence on guinea pig ileum contraction,[16,17] and its effect on noradrenaline release and other neurotransmitters are used for the characterization of new ligands.[18] The main functional *in vitro* test used for the ligands described in this chapter is based on the autoreceptor system measuring the K^+ evoked depolarization-induced release of [³H]histamine from rat synaptosomes.[1] Only very few test systems are described measuring central *in vivo* H_3-receptor activity after p.o. administration.[19] The screening system described in this chapter is based on an increase in the level of the main histamine metabolite N^τ-methylhistamine in brain after p.o. application to Swiss mice.[20]

AGONISTS

The first ligand showing some selectivity and potency to H_3-receptors was the endogenous ligand histamine itself (histamine pD_2-values H_1: 6.7; H_2: 6.0; H_3: 7.2[1]). This observation was the starting point for a number of structural variations of this molecule by different groups.[21] The first really potent and selective compound was the chiral methylated histamine derivative (R)-α-methylhistamine (Table 14.1).[1] This compound, nowadays used as a reference agonist, is extremely useful for H_3-receptor-dependent investigations due to the very poor activity of the (S)-form (distomer), which can be comfortably used for control experiments. Compared with N^α-methylhistamine, another methylated histamine derivative used as H_3-receptor agonist, this is a major advantage.

Although N^α-methylhistamine is three times more potent than histamine at H_3-receptors, it also shows high affinity to H_1- and H_2-receptor subtypes. Chiral dimethylated (R)-α,(S)-β-dimethylhistamine developed later shows slightly higher affinity than (R)-α-methylhistamine, but even more than five log-units higher affinity for H_3-receptors than for H_1- or H_2-receptors.[22] The exceptional selectivity is only found in this stereoisomer of the four possible α,β-dimethylated histamine derivatives. The absolute configuration was confirmed by x-ray structure analysis.[22] It was active *in vivo,* because at a dose of 25 mg/kg p.o. it maximally decreased [³H]histamine formation in rat brain, evaluated after administration of its [³H]precursor, as described by Garbarg et al.[23] Meanwhile, a number of compounds have been found incorporating

TABLE 14.1
H$_3$-Receptor Agonist Activity of Side Chain-Branched Histamine Derivatives

Structure	Name	Relative Activity
	Histamine	100
	(R/S)-β-Methylhistamine	282
	(R/S)-β-Ethylhistamine	0.6
	β,β-Dimenthylhistamine	3.6
	(R/S)-α-Ethylhistamine	0.6
	(R/S)-α-Amino-methylhistamine	Antagonist (pK$_i$ = 5.3)
	α,α-Dimethylhistamine	270
	(R/S)-α-Methylhistamine	3,6
	(R)-α-Methylhistamine	1550
	(S)-α-Methylhistamine	13
	(R)-α,(S)-β-Dimethylhistamine	1800
	(S)-α,(R)-β-Dimethylhistamine	18

the basic moiety in heterocycles as well as in an isothiourea or in related structures.[24–27] Computer-assisted analyses of these ligands have shown that the structures can be superpositioned in one pharmacophore model, supposing interaction with one or more common receptor binding sites in a similar mechanism.[28]

All these compounds have in common that they are highly polar and basic imidazole containing compounds. Their structures are presumed not to cross, or only to a small extent, biological membranes, e.g., the blood-brain barrier,[29] and therefore do not reach the area with the highest density of H₃-receptors. Additionally, many compounds are structurally closely related to histamine and are presumed to act, like the endogenous ligand, as substrates for histamine methyltransferase (E.C. 2.1.1.8), the major inactivating enzyme in humans.[30] The pharmacokinetic disadvantages of (R)-α-methyhistamine, like insufficient oral absorption, poor brain penetration, and rapid metabolism in humans, which stopped clinical investigations, are based on these physicochemical properties and its similarity to histamine.[31] Comparable problems may be presumed with the other imidazole derivatives acting as H₃-receptor agonists. A novel class of partial and full agonists on histamine H₃-receptors structurally very different from the agonists mentioned before have been developed, patented, and evaluated pharmacologically.[32]

PRODRUGS

In order to maintain the excellent pharmacodynamic properties of (R)-α-methylhistamine and to optimize its pharmacokinetic profile, novel prodrugs were designed, prepared by a condensation reaction of (R)-α-methylhistamine and 2-hydroxybenzophenone or related derivatives.[33,34] The resulting azomethine derivatives are highly lipophilic compounds in which the basic primary amine group is masked in a bioreversible manner to a nonbasic imine bond (Figure 14.2). Lipophilic and stability properties of the prodrugs can be influenced by substituents and substitution pattern on

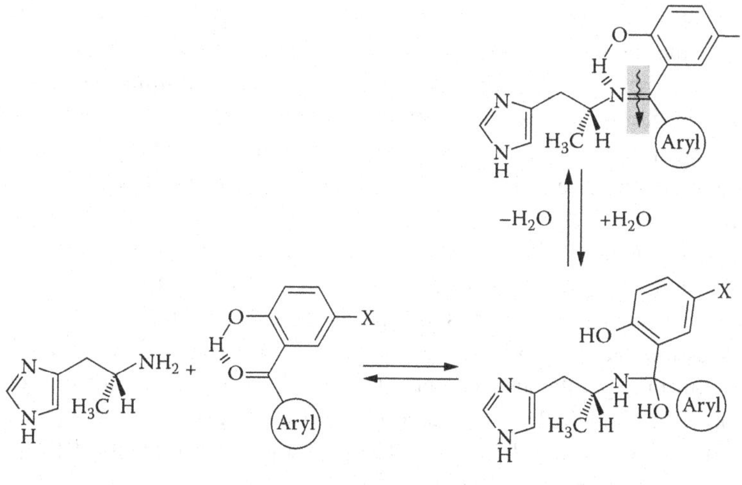

FIGURE 14.2 Formation and hydrolysis of prodrugs via hemiaminal intermediates.

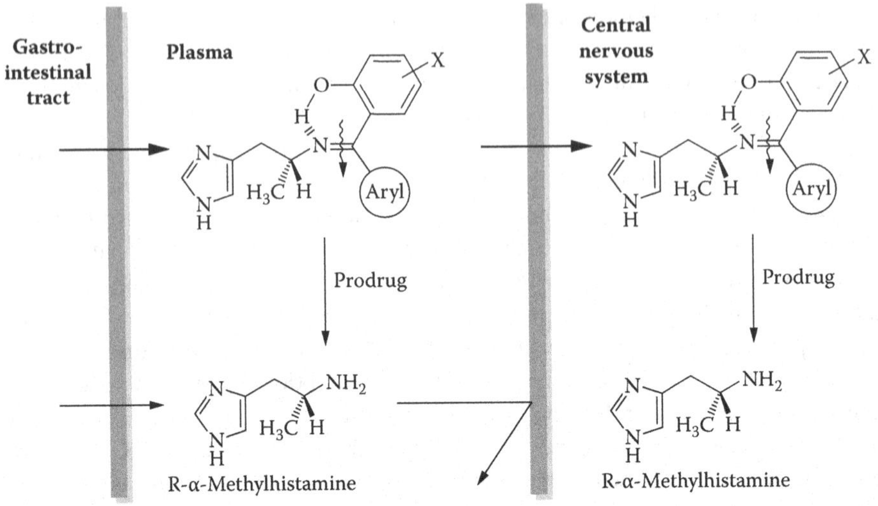

FIGURE 14.3 Penetration of prodrugs through biological membranes. (FUB 94 = BP2.94; X = H, Aryl = phenyl.)

the promoiety to allow drug targeting to central or peripheral tissues.[33,35] Hydrolytic stability of the bioreversible imine bond is mainly increased by an intramolecular hydrogen bond formed by the phenolic hydroxy group and the imine nitrogen (Figure 14.2). X-ray and spectroscopic investigations confirmed this hypothesis.

This prodrug approach was a new conception because, in addition to the change of physicochemical properties, increasing penetration through biological membranes recognition by the major metabolizing enzyme histamine methyltransferase was prevented.[31] Due to the passive diffusion through biological membranes and the chemical hydrolysis, no related enzymes are inhibited, which may have caused interactions and side effects. As a result, it can be observed that some azomethine prodrugs are very effective, after oral administration, regarding the delivery of (R)-α-methylhistamine into the central nervous system and peripheral tissues (Figure 14.3). Furthermore, their pharmacokinetic properties can strongly be influenced depending on the substitution pattern of the aromatic residues. Compounds with halogenated benzophenone promoieties, or especially promoieties containing five-membered heterocycles, appeared to be highly effective for CNS delivery of active (R)-α-methylhistamine (Figure 14.4).[33,35,36]

Astonishing results were obtained by the examination of a pyrrolyl prodrug which exhibited a long-lasting delivery of the active drug into the CNS. Due to this prolonged *in vivo* pharmacokinetics, this prodrug can be regarded as a "retard" prodrug.[35] Investigations for lipophilic parameters of the prodrugs by reversed phase thin layer chromatography and theoretical calculations of logP values showed good correlation with biological *in vivo* data, giving further evidence for a passive diffusion of the prodrugs into the brain.[37] The parent prodrug FUB 94 (BP2.94) is in clinical trials for peripheral diseases, and further prodrugs with more prominent central effects are still in the line.[31,33,45]

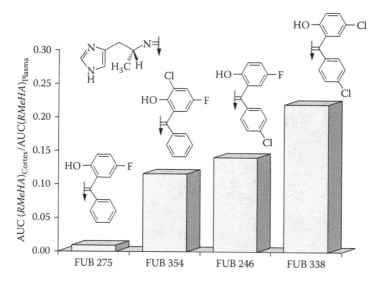

FIGURE 14.4   CNS/Plasma ratio of R-α-methylhistamine (RMeHA) after p.o. administration of prodrugs with high central activity (mouse).

ANTAGONISTS

The first compounds detected as antagonists for the histamine H₃-receptor were taken from the pool of H₂-receptor antagonists. Later on, more effective and more selective antagonists were developed, e.g., thioperamide, clobenpropit (for review see References 4 and 38). Unfortunately, these ligands also showed high affinity for 5-HT₃-receptors[39] and are claimed to possess toxic side effects. A number of different ligands have been described by different groups. With the knowledge of these antagonists and the developments in this field, we found a general construction pattern in which an aromatic heterocycle (in most cases an imidazole ring) is connected by a chain to a polar group which may be connected by another chain to a lipophilic moiety.[21] Whereas the first elements are essential for histamine H₃-receptor binding, the last elements increase affinity.

An extensive research program has been started for the development of novel lead structures starting with structural variations of the endogenous agonist.[40] In a step-by-step modification procedure, carbamate and ether derivatives were found possessing high potency and high selectivity for H₃-receptors.[4,41–44] These compounds have been optimized for central *in vivo* activity after oral application by varying length and structure of the spacer between the carbamate and the lipophilic moiety (Table 14.2).

Carbamate derivatives often do not reach the *in vitro* affinity of compounds like thioperamide or clobenpropit, but selectivity and *in vivo* activity are extraordinary properties of this class of antagonists.[41] Half-life, duration, and metabolism can be targeted with the substitution pattern on the lipophilic moiety. A large variation of substituents and positions is accepted with maintaining receptor affinity. Pharmacokinetic properties differ very much, depending on lead structure and substitution pattern. It

TABLE 14.2

H_3-Receptor Antagonist Activity of Carbamates with Different Chain Length

	in vitro $-\log K_i$	*in vivo* (p.o.) ED_{50} [mg/kg]
	7.0	3.6
	7.9	1.3
	8.0	1.8
	8.0	2.5
	7.6	6.9

can be assumed that, depending on the therapeutic target, it is possible to have a short- or long-lasting compound with drug targeting to central or peripheral tissues.

Most of the benzylic ether derivatives described in Table 14.3 are highly active *in vitro*, but ineffective *in vivo*.[44] The para-iodinated benzyl derivative is used in (^{125}I) labeled form as radioligand (see section "Radioligands").[15,45] The compounds with larger aromatic moieties show moderate to good antagonist *in vivo* activity.[44] One problem with the diphenylmethyl ether derivatives is their relative high H_1-receptor antagonist activity. This pharmacological profile can be explained when taking into account the structural similarity to classical H_1-receptor antagonists, e.g., diphenhydramine.[44] Further optimization of the lead led to 3-phenylpropyl derivatives, which are potent to the same or even a higher degree, showing high H_3-receptor selectivity.[4] One compound, elected for further trials due to its superior chemical and pharmacological properties, is the para-chlorinated 3-phenylpropyl derivative

TABLE 14.3

H₃-Receptor Antagonist Activity of 3-(1H-imidazol-4-yl)propyl Ethers

R	in vitro −log K_i	in vivo ED_{50}	R	in vitro −log K_i	in vivo ED_{50}
(benzyloxy)	7.7	>10 mg/kg	(diphenylmethyloxy)	7.6	2.4 mg/kg
(4-Cl-benzyloxy)	7.7	>10 mg/kg	(bis-fluorophenylmethyloxy)	7.6	1.2 mg/kg
(4-I-benzyloxy)	8.3	>10 mg/kg	(fluorophenyl-phenylmethyloxy)	7.6	1.1 mg/kg
(cyclohexylmethyloxy)	8.0	>10 mg/kg	(4-NO₂-phenyloxy)	8.2	
(biphenylmethyloxy)	7.2	4.0 mg/kg	(4-F-phenylpropyloxy)	8.2	0.76 mg/kg
(naphthylmethyloxy)	7.7	3.2 mg/kg	(4-Cl-phenylpropyloxy)	7.9	1 mg/kg

FUB 181. This compound was tested in functional assays on isolated peripheral organs at histaminergic H_1-, H_2-, and H_3-, serotoninergic 5-HT$_{2A}$- and 5-HT$_3$-, muscarinic M_3-, and adrenergic α_1- and β_1-receptors proving at least a 100 times higher affinity for H_3-receptors than any other of the tested receptors (Figure 14.5).

First pharmacokinetic studies determining the metabolism rate of liberated histamine by the concentration of N^τ-methylhistamine show high efficiency after 1.5 h already and also a duration for more than 4.5 h after p.o. application to mice. FUB 181 increased the turnover rates of noradrenaline and serotonin in some brain regions, but did not affect acetylcholine and choline levels. FUB 181 alone showed a significant ameliorating effect on learning and memory in mice not to be detected

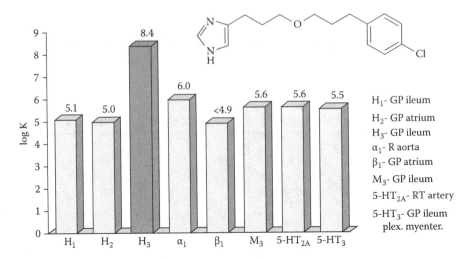

FIGURE 14.5   Selectivity of the H_3-receptor antagonist FUB 181. (Abbreviations: GP, guinea pig; R, rat; RT, rat tail.)

by application of thioperamide or clobenpropit alone, but only when also adding H_2-receptor antagonists.[46] Further studies are necessary to evaluate the therapeutic potential of this compound. Another novel H_3-receptor antagonist, FUB 359, showing even higher activity *in vitro* and *in vivo* as well as higher selectivity compared to FUB 181, is in preclinical trials.[47]

RADIOLIGANDS

Another approach for the evaluation of the therapeutic potential of H_3-receptor ligands is the development of radioligands. Determination of the localization of H_3-receptors in the brain provides valuable information for the potential use of ligands. The differences in the density of H_3-receptors and histamine axons in various brain regions suggest that the majority of the receptors are heteroreceptors. Tritium-labeled (R)-α-methylhistamine and N^α-methylhistamine are extensively used for autoradiographic studies in different tissues and different species.[2,3] Although these ligands are useful tools, they display some disadvantages because agonist binding is much more complex than that of antagonists. Furthermore, [³H]-labeled compounds are less useable for autoradiographic pictures than [¹²⁵I]-labeled compounds.

[¹²⁵I]Iodophenpropit was the first radiolabeled H_3-receptor antagonist.[48,49] Its good pharmacological properties were soon overcome by [¹²⁵I]iodoproxyfan (Figure 14.6),[15,45] which is more selective, more potent, and shows less unspecific binding than [¹²⁵I]iodophenpropit. At the moment [¹²⁵I]iodoproxyfan is the most potent radiolabeled antagonist used in binding studies. The initially described synthesis of the radioligand by a nucleophilic copper(I)-catalysed exchange reaction[45] has been improved by using a [¹²⁵I]iododestannylation procedure under mild conditions.[50] [¹²⁵I]Iodoproxyfan fulfills all criteria of a useful radioligand, e.g., high affinity, selectivity, reversible binding, saturability, and low unspecific binding. These properties

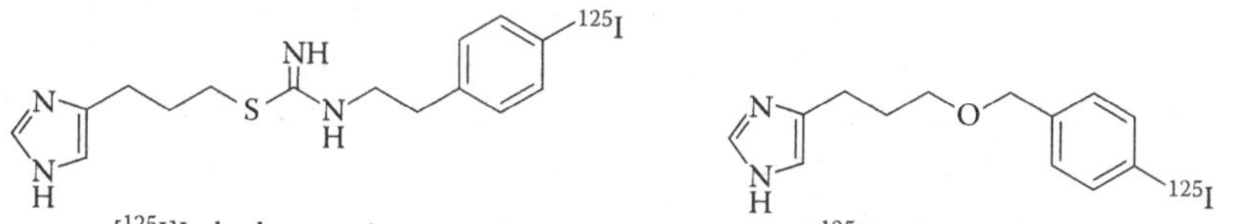

FIGURE 14.6 [^{125}I]-Radiolabeled histamine H_3-receptor antagonists.

account for the well-contrasted autoradiographic pictures taken in short exposure time with [^{125}I]iodoproxyfan, allowing a clearly distinguishable laminated pattern of labeling in the cerebral cortex and hippocampal formation.[15] Other [^{3}H]-labeled compounds of different structures have been described.[4,51]

The first approaches for histamine H_3-receptors ligands in [^{11}C]- or [^{18}F]-abeled form used for positron emission tomography (PET) have also been described, but with limited success up to now.[52–54]

PHARMACOLOGY AND CLINICAL OUTLOOK

In contrast to the widely used H_1- and H_2-receptor antagonists, no H_3-receptor ligands have been marketed thus far. The only exceptions from this statement are that beta-histine, marketed as H_1-receptor agonist for vestibular disturbances like Menière's disease,[55] and clozapine, an atypical neuroleptic, have been shown to possess moderate H_3-receptor antagonist activity.[56,57] It must be stressed that although H_3-receptors have also been identified in the human brain, including their anatomical distribution,[58] most indications are anticipated from *in vitro* and animal studies.

In animal models of migraine, H_3-receptor agonists are inhibiting the plasma protein extravasation within meninges to the same extent as the antimigraine drug sumatriptan, a 5-HT$_1$ agonist.[59,60] Anti-inflammatory effects associated with anti-nociceptive properties were detected in a number of tissues.[61,62] In the dura mater, as in peripheral tissues, sensory nerve fibers and mast cells, actively synthesizing and releasing histamine, form a short, inhibitory feedback loop involving prejunctional H_3-receptors that could regulate the release of pro-inflammatory mediators. It is not clear if the H_3-receptor-mediated inhibition of sensory C-fiber activity is alone responsible for this effect or not. First clinical trials using (R)-α-methylhistamine as an anti-asthmatic drug were stopped due to unsatisfying bioavailability described earlier. These problems were overcome with the prodrug BP2.94, which poorly enters the brain, thus reducing the risk of lateral effects with the therapy of diverse peripheral inflammatory diseases such as arthritis, asthma, inflammatory bowel diseases, etc.[31] Significant decrease of gastric secretion and gastric mucosal protection can be observed by agonist application as well as by prodrug application giving a novel anti-ulcer drug entity.[63] Other indications like hypertension and immunological regulation are also discussed.

For central effects, the sedative and perhaps hypnotic properties of H_3-receptor agonists are predicted due to behavioral changes and sleep modulation in rodents and cats.[64,65] For H_3-receptor antagonists, the opposite effects of the agonists are the most prominent indications. The stimulation of central activity in connection with increasing arousal as well as improvements in learning and memory behavior are effects of antagonists which indicate their use as anti-Alzheimer's drugs.[66,67] This potential indication is forced by the observation that tacrine inhibits histamine degradation more effectively than that of acetylcholine.[68] Cognition enhancement and triggering of central stimulation may be an effect of H_3-receptor modulation in vestibular nuclei and can also be used for the therapy of vertigo or motion-sickness.[69]

Although H_3-receptor antagonists are manifested as stimulating agents, they are able to inhibit electrically evoked convulsive effects.[70,71] This can lead to a new class of anti-epileptic drugs.

The influence of H_3-receptors on the release of a number of neurotransmitters may be leading to an antipsychotic activity of H_3-receptor antagonists. This hypothesis is strengthened by the observation that in schizophrenic patients the level of histamine metabolites is significantly changed, whereas the concentration of metabolites of other neurotransmitters is not changed,[72] and that clozapine, an atypical antipsychotic, displays moderate H_3-receptor antagonist activity.[56,57,73] Because of the lack of reliable test systems for antipsychotics, as well as for drugs for Alzheimers's disease, the first use of H_3-receptor antagonists in patients will give the final outcome of the investigation. Due to the new developments on ligands for H_3-receptors and to the multifactorial influence on neuroregulation, new medical entities based on the modulation of histamine H_3-receptors can be anticipated in the near future.

ACKNOWLEDGMENTS

Our work described herein was supported by the Biomedical and Health Research Programme BIOMED of the European Union and by the Verband der Chemischen Industrie, Fonds der Chemischen Industrie. The valuable scientific contributions to this work in the field of medicinal chemistry by Dr. A. Hüls, Dr. K. Purand, Dr. M. Krause, Dr. R. Lipp, and S. Reidemeister and in the field of pharmacology by Prof. Dr. J.-C. Schwartz, Prof. Dr. E. Schlicker, Dr. J.-M. Arrang, Dr. M. Garbarg, Dr. A. Rouleau, and X. Ligneau are gratefully acknowledged.

REFERENCES

1. Arrang, J.M. et al., Highly potent and selective ligands for histamine H3-receptors, *Nature*, 327, 117, 1987.
2. Schwartz, J.C. et al., Histaminergic transmission in the mammalian brain, *Physiol. Rev.*, 71, 1, 1991.
3. Leurs, R., Smit, M.J., and Timmerman, H., Molecular pharmacological aspects of histamine receptors, *Pharmacol. Ther.*, 66, 413, 1995.
4. Stark, H., Schlicker, E., and Schunack, W., Developments of histamine H_3-receptor antagonists, *Drugs Future*, 21, 507, 1996.
5. Schlicker, E. et al., Modulation of neurotransmitter release via histamine H3 heteroreceptors, *Fundam. Clin. Pharmacol.*, 8, 128, 1994.
6. Yamashita, M. et al., Expression cloning of a cDNA encoding the bovine histamine H1 receptor, *Proc. Natl. Acad. Sci. U.S.A.*, 88, 11515, 1991.
7. Gantz, I. et al., Molecular cloning of a gene encoding the histamine H2 receptor, *Proc. Natl. Acad. Sci. U.S.A.*, 88, 429, 1991.

8. Clark, E.A. and Hill, S.J., Sensitivity of histamine H3 receptor agonist-stimulated [35S]GTP gamma[S] binding to pertussis toxin, *Eur. J. Pharmacol.*, 296, 223, 1996.

9. Cherifi, Y. et al., Purification of a histamine H3 receptor negatively coupled to phosphoinositide turnover in the human gastric cell line HGT1, *J. Biol. Chem.*, 267, 25315, 1992.

10. Endou, M., Poli, E., and Levi, R., Histamine H3-receptor signaling in the heart: possible involvement of Gi/Go proteins and N-type Ca^{++} channels, *J. Pharmacol. Exp. Ther.*, 269, 221, 1994.

11. Oike, M., Kitamura, K., and Kuriyama, H., Histamine H3-receptor activation augments voltage-dependent Ca^{2+} current via GTP hydrolysis in rabbit saphenous artery, *J. Physiol.*, 448, 133, 1992.

12. Neame, K.D., Uptake of histidine, histamine and other imidazole derivatives by brain slices, *Neurochem.*, 11, 655, 1964.

13. Corbel, S. et al., Binding of histamine H3-receptor antagonists to hematopoietic progenitor cells. Evidence for a histamine transporter unrelated to histamine H3 receptors, *FEBS Lett.*, 404, 289, 1997.

14. Jansen, F.P. et al., Characterization of the binding of the first selective radiolabelled histamine H3-receptor antagonist, [125I]-iodophenpropit, to rat brain, *Br. J. Pharmacol.*, 113, 355, 1994.

15. Ligneau, X. et al., [125I]iodoproxyfan, a new antagonist to label and visualize cerebral histamine H3 receptors, *J. Pharmacol. Exp. Ther.*, 271, 452, 1994.

16. Vollinga, R.C. et al., A simple and rapid *in vitro* test system for the screening of histamine H3 ligands, *Meth. Find. Exp. Clin. Pharmacol.*, 14, 747, 1992.

17. Schlicker, E. et al., Novel histamine H3 receptor antagonists: affinities in an H3 receptor binding assay and potencies in two functional H3 receptor models, *Br. J. Pharmacol.*, 112, 1043, 1994, and Erratum, *Br. J. Pharmacol.*, 113, 637, 1994.

18. Clapham, J. and Kilpatrick, G.J., Histamine H3 receptors modulate the release of [3H]-acetylcholine from slices of rat entorhinal cortex: evidence for the possible existence of H3 receptor subtypes, *Br. J. Pharmacol.*, 107, 919, 1992.

19. Tedford, C.E. et al., Pharmacological characterization of GT-2016, a non-thiourea-containing histamine H3 receptor antagonist: *in vitro* and *in vivo* studies, *J. Pharmacol. Exp. Ther.*, 275, 598, 1995.

20. Garbarg, M. et al., S-[2-(4-imidazolyl)ethyl]isothiourea, a highly specific and potent histamine H3 receptor agonist, *J. Pharmacol. Exp. Ther.*, 263, 304, 1992.

21. Lipp, R., Stark, H., and Schunack, W., in *The Histamine Receptor, Ser. Receptor Biochemistry and Methodology*, Vol. 16, Schwartz, J.-C. and Haas, H.L., Eds., Wiley-Liss, New York, 1992, p. 57.

22. Lipp, R. et al., Synthesis, absolute configuration, stereoselectivity, and receptor selectivity of (alpha R, beta S)-alpha,beta-dimethylhistamine, a novel high potent histamine H3 receptor agonist, *J. Med. Chem.*, 35, 4434, 1992.

23. Garbarg, M. et al., *Methods of Biogenic Amine Research*, Parvez, S., Nagatsu, T., Nagatsu, I., and Parvez, H., Eds., Elsevier, Amsterdam, 1983, p. 623.

24. Shih, N.-Y. et al., A novel pyrrolidine analog of histamine as a potent, highly selective histamine H3 receptor agonist, *J. Med. Chem.*, 38, 1593, 1995.

25. Shih, N.-Y. and Green, M.J., *Intern. Pat. Appl. WO* 93,121 07, 1993.

26. Vollinga, R.C. et al., A new potent and selective histamine H3 receptor agonist, 4-(1H-imidazol-4-ylmethyl)piperidine, *J. Med. Chem.*, 37, 332, 1994.

27. Ganellin, C.R. et al., Imetit and N-methyl derivatives. The transition from potent agonist to antagonist at histamine H₃ receptors, *Bioorg. Med. Chem. Lett.*, 2, 1231, 1992.

28. Sippl, W., Stark, H., and Höltje, H.-D., Development of a binding site model for histamine H3-receptor agonists, *Pharmazie*, 53, 433, 1998.

29. Yamazaki, S. et al., The disposition of (R)-alpha-methylhistamine, a histamine H3-receptor agonist, in rats, *J. Pharm. Pharmacol.*, 46, 371, 1994.

30. Hough, L.B., Khandelwal, J.K., and Mittag, T.W., Alpha-methylhistamine methylation by histamine methyltransferase, *Agents Actions*, 11, 425, 1981.
31. Rouleau, A. et al., Bioavailability, antinociceptive and antiinflammatory properties of BP 2-94, a histamine H3 receptor agonist prodrug, *J. Pharmacol. Exp. Ther.*, 281, 1085, 1997.
32. Schwartz, J.C., Arrang, J.M., Garbarg, M., Rouleau, A., Lecomte, J.M., Ligneau, X., Schunack, W., Stark, H., Purand, H., Hüls, A., Reidemeister, S., Athmani, S., Ganellin, C.R., Pelloux-Léon, N., and Tertiuk, W., *Demande de brevet d'invention français FR* 95 03267, 1995.
33. Krause, M. et al., Synthesis, x-ray crystallography, and pharmacokinetics of novel azomethine prodrugs of (R)-alpha-methylhistamine: highly potent and selective histamine H3 receptor agonists, *J. Med. Chem.*, 38, 4070, 1995.
34. Schunack, W. and Stark, H., Design of histamine H3-receptor agonists and antagonists, *Eur. J. Drug Metab. Pharmacokinet.*, 3, 173, 1994.
35. Krause, M. et al., New potent azomethine prodrugs of the histamine H3-receptor agonist (R)-alpha-methylhistamine containing a heteroarylphenyl partial structure, *Arch. Pharm. (Weinheim)*, 329, 209, 1996.
36. Krause, M. et al., Structure-activity relationships of novel azomethine prodrugs of the histamine H3-receptor agonist (R)-alpha-methylhistamine: from alkylaryl to substituted diaryl derivatives, *Pharmazie*, 51, 720, 1996.
37. Krause, M. et al., Relationship between lipophilicity and brain uptake of azomethine prodrugs of a histamine H3-receptor agonist, *Sci. Pharm.*, 64, 503, 1996.
38. Leurs, R., Vollinga, R.C., and Timmerman, H., The medicinal chemistry and therapeutic potentials of ligands of the histamine H3 receptor, *Prog. Drug Res.*, 45, 107, 1995.
39. Leurs, R. et al, Evaluation of the receptor selectivity of the H3 receptor antagonists, iodophenpropit and thioperamide: an interaction with the 5-HT3 receptor revealed, *Br. J. Pharmacol.*, 116, 2315, 1995.
40. Schwartz, J.C., Arrang, J.-M., Garbarg, M., Lecomte, J.-M., Ganellin, C.R., Fkyerat, A., Tertiuk, W., Schunack, W., Lipp, R., Stark, H., and Purand, K., PCT WO93/14070, 1992.
41. Stark, H. et al., Novel carbamates as potent histamine H3 receptor antagonists with high *in vitro* and oral *in vivo* activity, *J. Med. Chem.*, 39, 1157, 1996.
42. Schlicker, E. et al., Potencies of antagonists chemically related to iodoproxyfan at histamine H3 receptors in mouse brain cortex and guinea-pig ileum: evidence for H3 receptor heterogeneity? *Naunyn-Schmiedeberg's Arch. Pharmacol.*, 353, 482, 1996.
43. Ganellin, C.R. et al., A novel series of (phenoxyalkyl)imidazoles as potent H3-receptor histamine antagonists, *J. Med. Chem.*, 39, 3806, 1996.
44. Hüls, A. et al., Diphenylmethyl ethers: synthesis and histamine H_3-receptor antagonist *in vitro* and *in vivo* activity, *Bioorg. Med. Chem. Lett.*, 6, 2013, 1996.
45. Stark, H. et al., [^{125}I]iodoproxyfan and related compounds: a reversible radioligand and novel classes of antagonists with high affinity and selectivity for the histamine H3 receptor, *J. Med. Chem.*, 39, 1220, 1996.
46. Onodera, K. et al., Improvement by FUB 181, a novel histamine H_3-receptor antagonist, of learning and memory in the elevated plus-maze test in mice, *Naunyn-Schmiedeberg's Arch. Pharmacol.*, 357, 508, 1998.
47. Stark, H., Ciproxifan and chemically related compounds are highly potent and selective histamine H_3-receptor antagonists, *Naunyn-Schmiedeberg's Arch. Pharmacol.*, 358, 998, 1998.
48. Menge, W.M.P.B. et al., Synthesis of S-[3-(4(5)-imidazolyl)propyl],N-[2(4-{^{125}I}-iodo phenyl)ethyl]isothiourea sulfate (^{125}I-iodophenpropit), a new probe for histamine H_3 receptor binding sites, *J. Labelled Comp. Radiopharm.*, 31, 781, 1992.

49. Jansen, F.P. et al., The first radiolabeled histamine H3 receptor antagonist, [^{125}I]iodophenpropit: saturable and reversible binding to rat cortex membranes, *Eur. J. Pharmacol.*, 217, 203, 1992.

50. Krause, M., Stark, H., and Schunack, W., Iododestannylation: an improved synthesis of [^{125}I]iodoproxyfan, a specific radioligand of the histamine H3 receptor, *J. Labelled Comp. Radiopharm.*, 39, 601, 1997.

51. Brown, J.D. et al., Characterisation of the specific binding of the histamine H₃-receptor antagonist radioligand [^{3}H]-GR168320, *Eur. J. Pharmacol.* 311, 305, 1996.

52. Yanai, K. et al., Receptor autoradiography with ^{11}C and [^{3}H]-labelled ligands visualized by imaging plates, *Neuroreport* 3, 961, 1992.

53. Windhorst, A.D. et al., Evaluation of [18F]VUF 5000 as a potential PET ligand for brain imaging of the histamine H3 receptor, *Bioorg. Med. Chem.*, 7, 1761, 1999.

54. Airaksinen, A.J. et al., Radiosynthesis and biodistribution of a histamine H3 receptor antagonist 4-[3-(4-piperidin-1-yl-but-1-ynyl)-[11C]benzyl]-morpholine: evaluation of a potential PET ligand, *Nucl. Med. Biol.*, 33, 801, 2006.

55. Tighilet, B., Leonard, J., and Lacour, M., Betahistine dihydrochloride treatment facilitates vestibular compensation in the cat, *J. Vestibul. Res.*, 5, 53, (1995).

56. Kathmann, M., Schlicker, E., and Göthert, M., Intermediate affinity and potency of clozapine and low affinity of other neuroleptics and of antidepressants at H3 receptors, *Psychopharmacology*, 116, 464, 1994.

57. Rodriges, A.A. et al., Interaction of clozapine with the histamine H3 receptor in rat brain, *Br. J. Pharmacol.*, 114, 1523, 1995.

58. Martinez-Mir, M.I. et al., Three histamine receptors (H1, H2 and H3) visualized in the brain of human and non-human primates, *Brain Res.*, 526, 322, 1990.

59. Buzzi, M.G. et al., 5-Hydroxytryptamine receptor agonists for the abortive treatment of vascular headaches block mast cell, endothelial and platelet activation within the rat dura mater after trigeminal stimulation, *Brain Res.*, 583, 137, 1992.

60. Matsubara, T., Moskowitz, M.A., and Huang, Z., UK-14,304, R(-)-alpha-methyl-histamine and SMS 201-995 block plasma protein leakage within dura mater by prejunctional mechanisms, *Eur. J. Pharmacol.*, 224, 145, 1992.

61. Dimitriadou, V. et al., Functional relationships between sensory nerve fibers and mast cells of dura mater in normal and inflammatory conditions, *Neuroscience*, 77, 829, 1997.

62. Imamura, M. et al., Histamine H3-receptor-mediated inhibition of calcitonin gene-related peptide release from cardiac C fibers. A regulatory negative-feedback loop, *Circ. Res.*, 78, 863, 1996.

63. Bertaccini, G. and Coruzzi, G., An update on histamine H3 receptors and gastrointestinal functions, *Dig. Dis. Sci.*, 40, 2052, 1995.

64. Schwartz, J.C. et al., in *Psychopharmacology: The Fourth Generation of Progress*, Bloom, F.E. and Kupfer, D.J., Eds., Raven Press Ltd., New York, 1995, p. 397.

65. Monti, J.M. et al., Effects of selective activation or blockade of the histamine H3 receptor on sleep and wakefulness, *Eur. J. Pharmacol.*, 205, 283, 1991.

66. Miyazaki, S., Imaizumi, M., and Onodera, K., Effects of thioperamide, a histamine H3-receptor antagonist, on a scopolamine-induced learning deficit using an elevated plus-maze test in mice, *Life Sci.*, 57, 2137, 1995.

67. Meguro, K. et al., Effects of thioperamide, a histamine H3 antagonist, on the step-through passive avoidance response and histidine decarboxylase activity in senescence-accelerated mice, *Pharmacol. Biochem. Behav.*, 50, 321, 1995.

68. Morisset, S., Traiffort, E., and Schwartz, J.C., Inhibition of histamine versus acetylcholine metabolism as a mechanism of tacrine activity, *Eur. J. Pharmacol.*, 315, R1, 1996.

69. Yabe, T. et al., Medial vestibular nucleus in the guinea-pig: histaminergic receptors. II. An *in vivo* study, *Exp. Brain Res.*, 93, 249, 1993.

70. Yokoyama, H. et al., Effect of thioperamide, a histamine H3 receptor antagonist, on electrically induced convulsions in mice, *Eur. J. Pharmacol.*, 234, 129, 1993.

71. Yokoyama, K. et al., Clobenpropit (VUF-9153), a new histamine H3 receptor antagonist, inhibits electrically induced convulsions in mice, *Eur. J. Pharmacol.*, 260, 23, 1994.

72. Prell, G.D. et al., Histamine metabolites in cerebrospinal fluid of patients with chronic schizophrenia: their relationships to levels of other aminergic transmitters and ratings of symptoms, *Schizophr. Res.*, 14, 93, 1995.

73. Schlicker, E. and Marr, I., The moderate affinity of clozapine at H3 receptors is not shared by its two major metabolites and by structurally related and unrelated atypical neuroleptics, *Naunyn-Schmiedeberg's Arch. Pharmacol*, 353, 290, 1996.

15 Anti-Inflammatory Actions of Flavonoids and Structural Requirements for New Design

Theoharis C. Theoharides

CONTENTS

ABSTRACT

Flavonoids are low molecular weight compounds that are most concentrated in seeds, citrus fruits, olive oil, tea, and red wine, and that have potent antioxidant, cytoprotective, and anti-inflammatory activities. Flavonoids are composed of a three-ring structure (A, B, and C) with various substitutions; they can be subdivided according to the presence of an oxy group at position 4, a double bond between carbon atoms 2 and 3, or a hydroxyl group in position 3 of the C (middle) ring. Particular hydroxylation patterns of the B ring of the flavones permit them to inhibit histamine, tryptase, interleukin-6, and interleukin-8 release from human umbilical-cord-derived cultured mast cells, as well as from macrophages. The catechol (o-dihydroxy) group in the B ring, as in quercetin, confers potent inhibitory ability, while a pyrogallol (trihydroxy) group, as in myricetin, produces even higher activity. However, addition of one hydroxyl group on position 2' of the B ring, as in the flavonol morin, renders this compound inactive. The C2-C3 double bond of the C ring appears to increase scavenger activity because it confers stability to the phenoxy-radicals produced, while the 4-oxo (keto double bond at position 4 of the C ring) increases free radical scavenger activity by delocalizing electrons from the B ring. The 3-OH group on the C ring appears to be critical for anti-inflammatory activity (Table 15.1). Inhibition of mast cell secretion was shown to be mediated by a 78-kD phosphoprotein which

TABLE 15.1
Characteristics of Flavonoid Structure for Anti-Inflammatory Activity

The catechol (o-dihydroxy) group in ring B.

A pyrogallol (trihydroxy) group in ring B may confer even higher activity.

The C2-C3 double bond of the C ring appears to increase scavenger activity.

The 4-oxo (keto double bond at position 4 of the C ring), especially in association with the C2-C3 double bond, increases free radical scavenger activity.

The 3-OH group on the C ring generates an extremely active scavenger.

The combination of C2-C3 double bond and 4-oxo group appears to be the best combination.

has been cloned and serves as a bridge between the cell surface and the cytoskeleton. Phosphorylation at particular sites in the C-terminus unfolds the three-dimensional structure of this protein, making actin-binding sites accessible; crosslinking with actin in the cytoskeleton prevents secretion of inflammatory mediators. These properties present unique opportunities for the synthesis of new compounds for the treatment of inflammatory and possibly proliferative disorders.

INTRODUCTION

Over 4000 structurally unique flavonoids have been identified in fruits, vegetables, nuts, seeds, herbs, spices, stems, flowers, and citrus fruits, as well as tea and red wine. These low molecular weight substances are phenylbenzo-pyrones (phenylchromones) with a common three-ring nucleus. They are usually subdivided according to their substituents into flavanols, anthocyanidons, flavones, flavanones, and chalcones. This basic structure is comprised of two benzene rings (A and B) linked through a heterocyclic pyran or pyrone ring (C) in the middle. This subdivision is primarily based on the presence (or absence) of a double bond on position 4 of the C (middle ring) (Figure 15.1), the presence (or absence) of a double bond between carbon atoms 2 and 3 of the C ring, and the presence of hydroxyl groups in the B ring. In the flavonoid structure, a phenyl group is usually substituted at the 2-position of the pyrone ring. In isoflavonoids, the substitution is at the 3-position. Flavonoids and tocopherols share a common structure, namely the chromane ring. On average, the daily American diet was estimated to contain approximately 1 g of mixed flavonoids expressed as glycosides. However, the average intake of mixed flavonoids was only 23 mg/d based on data from the 1987–1988 Dutch National Food Consumption Survey. The amount of 23 mg/d was mostly flavonols and flavones measured as aglycones. The corresponding amount of daily aglycones consumed in the United States would be about 650 mg/d. These quantities could provide pharmacologically significant concentrations in body fluids and tissues. Flavonoid dietary intake far exceeds that of vitamin E, a monophenolic anti-oxidant, and of α-carotene on a mg/d basis (For reviews see References 1 and 2.)

Flavonoids have survived in vascular plants throughout evolution for over one billion years, possibly due to the important properties they possess, most notably anti-oxidant activity and inhibition of leukocyte function.[3,4] A suggestion of clinical

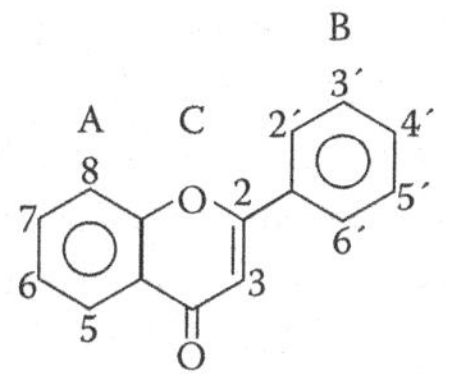

	5	7	2′	3′	4′	5′
Quercetin	OH	OH	–	OH	OH	–
Kaempferol	OH	OH	–	–	OH	–
Myricetin	OH	OH	–	OH	OH	OH
Morin	OH	OH	OH	–	OH	–

FIGURE 15.1 Comparative structure of certain flavonoids, showing the hydroxylation pattern of the B ring. The presence of hydroxyl groups at positions 3′, 4′, and 5′ confer anti-inflammatory activity, while addition of a hydroxyl group at position 2′ nearly eliminates such activity. The oxy group at position 4 and a hydroxyl group at positions 4 and 3, respectfully, of the middle C ring increase anti-inflammatory activity.

benefit in patients with allergic disease treated with rutin was recorded as early as 1952.

The flavonoids are formed in plants and participate in the light-dependent phase of photosynthesis during which they catalyze electron transport. They are synthesized from the aromatic amino acids, phenylalanine and tyrosine, which are converted to cinnamic acid and para-hydroxycinnamic acid, respectively, by the action of phenylalanine and tyrosine ammonia lyases. Cinnamic acid condenses with acetate units to form the cinnamoyl structure of the flavonoids. There is then alkali-catalyzed condensation of an ortho-hydroxyaceto-phenone with a benzaldehyde derivative generating chalcones and flavonones, as well as a similar condensation of an ortho-hydroxyacetophonone with a benzoic acid derivative (acid chloride or anhydride) leading to 2-hydroxyflavanones and flavones. In plants, flavonoids generally occur as glycosylated and also as sulfated derivatives. Biotransformation in the intestine by ring scission under the influence of intestinal microorganisms can release these phenolic acid derivatives. Ingested flavonoids are converted into sulfated and/or glucuronidated metabolites that are excreted in the bile. For instance, plasma quercetin concentrations following ingestion of fried onions containing quercetin glycosides equivalent to 64 mg of quercetin aglycone led to peak plasma levels of 196 μg/mL after 2.9 h, with a half-life of 16.8 h.

Adverse reactions to flavonoids in humans appear to be rare, even though essentially all human beings have daily physical contact with flavonoid-containing foods and plants.

MAST CELL INHIBITORY ACTIVITY

Mast cells derive from distinct bone marrow precursors[5] and enter the tissues as undifferentiated cells where they mature under the influence of "microenvironmental" conditions; these include stem cell factor (CSF or c-kit ligand) and interleukins

3, 4, and 6.[6] Rat connective tissue mast cells (CTMC), found primarily in skin and lungs, contain rat mast cell protease I (RMCP I), whereas mucosal mast cells (MMC) contain RMCP I.[7] All human mast cells (HMC) contain the proteolytic enzyme tryptase, but human CTMC contain yet another protease, chymase.[7] Mast cells are the main source of tissue histamine, which is released from secretory granules when mast cells are triggered with IgE and specific antigen.[6] Such granules also store numerous other vasoactive, pro-inflammatory, and nociceptive molecules,[8,9] some of which may be released differentially without exocytosis.[10–13] In addition to the well-known immunologic stimuli, CTMC also secrete in response to a number of neuropeptides.[14]

Mast cells play a central role in the pathogenesis of diseases such as allergic asthma, rhinoconjunctivitis, asthma, anaphylaxis, urticaria, and systemic mastocytosis; they are now considered to be important players in other chronic inflammatory disorders, such as arthritis and inflammatory bowel disease.[6,14,15] Mast cells may also participate in sterile inflammatory conditions exacerbated by stress such as atopic dermatitis, interstitial cystitis, irritable bowel syndrome, migraines, and multiple sclerosis.[14] Basophils, the circulating "equivalent" of the tissue mast cells, are considered as important cells in the pathogenesis of late phase allergic reactions.[16–19]

Both mast cells and basophils possess high affinity receptors for IgE in their plasma membranes. Cross-linking of these receptors is essential to trigger the secretion of histamine and other preformed, granule-associated mediators and to initiate the generation of newly formed phospholipid-derived mediators. Various flavonoids have been shown in a number of systems to inhibit this secretory process.[20,21] Definitive evidence of flavonoid regulation of secretion was first provided in studies of the secretion of histamine from rat mast cells and basophils stimulated with antigen. Quercetin, kaempferol, and myricetin were found to inhibit the release of rat mast cell histamine.[22,23] Quercetin also induced histamine accumulation, expression of mRNA for RMCP II,[24] and development of secretory granules in rat basophil leukemia (RBL) cells.[24] Flavonoids have also been shown to have antiproliferative effects on a number of transformed cells.[1]

Inhibitory activity of some flavon-3-ols was associated with the following structural features: a C4 keto group, an unsaturated double bond at position C2-C3 in the γ-pyrone ring, and an appropriate pattern of hydroxylation in the B ring. These characteristics were near identical to those identified for other inhibitory activities. The flavonoid glycosides, rutin and naringin, were inactive *in vitro* as were the flavanones (reduced C2-C3 bond) taxifolin and hesperitin. Morin, catechin, and cyanidin were also inactive.[1] It is important to note that, while quercetin, kaempferol, and myricetin were potent inhibitors of histamine release from rat peritoneal mast cells, morin was not. The addition of a single hydroxyl group at position 2′ appears to be sufficient to reduce inhibition of mast cell secretion. This hydroxyl group may be interacting with the oxygen at position 1 forming a cyclic structure that possibly interferes with some key biological event.

The basophil histamine-releasing effect of different secretagogues could be inhibited by some, but not all, of 11 flavonoids representing several chemical classes: (a) anti-IgE or concanavalin A (IgE-dependent histamine releasing agents); (b) the chemoattractant peptide, f-MetLeuPhe or the tumor promoter phorbol ester, TPA (both

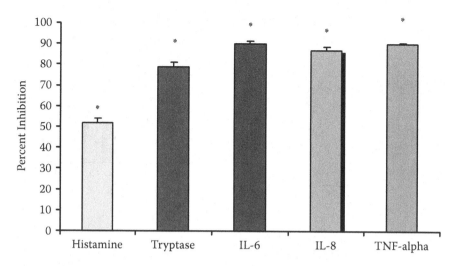

FIGURE 15.2 Inhibition of release of inflammatory molecules from human umbilical cord blood-derived mast cells (hCBMCs) by pretreatment with quercetin (0.5 mM) for 5 min before stimulation with anti-IgE for either 30 min for histamine and tryptase, and 24 hours for IL-6, IL-8, and TNF-α ($n = 6$, *$p<0.05$). Note that inhibition was 80% or higher for all mediators except for histamine.

f-MetLeuPhe and TPA are receptor-dependent, IgE-independent, histamine-releasing agents); and (c) the divalent cation ionophore A23187 (bypasses receptor-dependent processes and carries Ca^{2+} directly into the cytoplasm).[1] The effect of quercetin to uniformly inhibit basophil histamine secretion stimulated by a variety of agonists strongly suggests that there is a final common pathway utilized by each of these agonists which is sensitive to quercetin and other structurally similar flavonoids.

We were the first to show that while cromolyn was a weak inhibitor of histamine release from human umbilical cord-derived cultured mast cells, quercetin was a much better inhibitor with morin having intermediate activity. Quercetin, moreover, was a potent inhibitor of release of both the unique mast cell proteolytic enzyme tryptase, as well as of the cytokines interleukin-6, interleukin-8, and tumor necrosis factor (Figure 15.2). Cromolyn could not inhibit the release of these molecules, whereas morin was considerably weaker than quercetin.[25] These results indicate that findings from rodent mast cells are not necessarily applicable to human cells, and that quercetin is a better mast cell inhibitor than cromolyn. Quercetin could also inhibit selective release of IL-6 in response to IL-1 (Figure 15.3).

Fewtrell and Gomperts[22] and Middleton et al.[1] demonstrated that only activated mast cells or activated basophils were affected by quercetin and other inhibitory flavonoids.[1] In other words, the unstimulated cells could be exposed to the flavonoids, be washed, and subsequently be shown to react normally to a secretagogue with histamine release. They also observed that pretreatment of rat mast cells with disodium cromoglycate (cromolyn), a "mast cell stabilizing" anti-allergic drug, for 30 min completely abolished the inhibition normally observed upon subsequent exposure to quercetin, added together with antigen. This finding suggested that cromolyn and

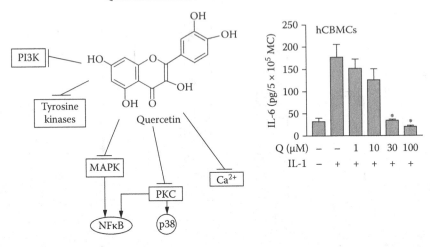

FIGURE 15.3 Structure of quercetin with possible targets, inhibition of which may explain its ability to inhibit mast cell secretion. The insert shows a dose-response of the inhibitory activity of quercetin (5 min pretreatment) on release of IL-6 from hCBMCs stimulated by IL-1 (100 mM) for 24 hours. ($n = 8$, *$p < 0.05$.) MAPK, mitogen activation protein kinase; MC, mast cells; PI3K, phosphotidyl inositol-3 kinase.

quercetin acted at the same or a closely associated molecular site. It was not known, however, just what cellular component in activated mast cells or basophils first interacts with cromolyn or active flavonoids to inhibit the secretory process.

One possible explanation was considered by the similarity in the structure of quercetin with that of cromolyn. It had been shown that secretagogue-induced exocytosis in rat mast cells was associated with stimulation of Ca^{2+}-dependent protein phosphorylation.[26] Purified rat peritoneal mast cells which had been labeled with ^{32}P and then stimulated by addition of compound 48/80 resulted in the phosphorylation of four proteins of apparent molecular weights 78, 68, 59, and 42 kD. Phosphorylation of the proteins with apparent molecular weights of 68, 59, and 42 kD was evident within 10 s after addition of 48/80; phosphorylation of the 78 kD molecular weight protein, however, was not evident until 30–60 s after addition of the secretagogue.[26] These experiments clearly indicated that the exocytosis of the mast cell was associated with phosphorylation of certain proteins, while recovery from secretion was related to phosphorylation of a unique protein.

We then showed that cromolyn promoted the incorporation of radioactive phosphate into a single rat mast cell protein of apparent molecular weight 78 kD.[27] The time course and dose-dependence of phosphorylation of this protein closely paralleled inhibition of mast cell secretion.[27] This finding provided an insight into the mechanism of inhibition by cromolyn of mast cell secretion triggered by an immunologic stimulus, anti-rat IgE. In additional experiments, we showed that other inhibitors of rat mast cell histamine secretion, also increased the incorporation of radioactive phosphate into a single protein band with an apparent molecular weight of 78 kD.[28]

We subsequently showed that the 78 kD mast cell phosphoprotein had high homology to moesin,[29] a member of the ezrin-radixin-moesin family of proteins,[30] which have been shown to regulate signal-transduction by coupling the cell surface to the cytoskeleton.[31] Phosphorylation of this protein was shown to take place by a calcium and phorbol ester-independent PKC isozyme.[32] More recently, this 78 kD phosphoprotein was cloned and was shown to be identical to moesin;[33] it was further shown that its phosphorylation by cromolyn induced some conformational change that permitted covalent binding to actin and resulted in preferential clustering around the mast cell secretory granules, thus possibly preventing them from undergoing exocytosis.[33] Because of its apparent involvement in mast cell inhibition, this protein was also called MAst CEll DegranulatiON Inhibitory Agent (MACE-DONIA).[14] This protein was associated with the plasma membrane, but there were no intramembranous domains. During secretion, the protein could not be recognized by immunocytochemistry and gave the impression it had disappeared; however, Western blot analysis showed that it had not been secreted. Pretreatment with quercetin prevented this apparent "disappearance," suggesting some conformational rearrangements that changed the antigenicity of this protein.

DISCUSSION

The present findings extend previous reports showing that quercetin inhibits stimulated histamine release from rat CTMC[22,34] and MMC,[35,36] from human lung and intestinal mast cells,[37] as well as from activated basophils.[23] Various flavonoids, but not morin, had previously also been shown to inhibit polymorphonuclear leukocyte function[38] and lymphocyte secretion of interleukin-2[39] in concentrations similar to the ones used here. Our results indicate that the structural requirements of flavonoids with respect to the inhibition of both proliferation and accumulation of mediators was the same as it had previously been reported for inhibition of rat mast cell secretion.[1,25] Similarly, morin could not inhibit mitogen-induced lymphocyte proliferation[40] or proliferation of a human lymphoblastoid cell line.[41] The fact that morin was inactive supports the notion that the effect of flavonoids is fairly specific. Morin has one additional hydroxyl group in the $2'$ position of the B ring,[4] not shared by the other flavonoids tested, suggesting that there must exist some steric interference that leads to loss of inhibitory activity. The structural requirements for the inhibitory effect reported here may be related to the ability of quercetin and other flavonoids to inhibit various enzymatic systems in vitro,[1,4] such as protein kinase C,[42, 43] which regulates secretion.[44]

PKC is the ubiquitous, largely Ca^{2+}- and phospholipid-dependent, multifunctional serine- and threonine-phosphorylating enzyme. It is involved in a wide range of cellular activities including tumor-promotion, mitogenesis, secretory processes and inflammatory cell function. Accumulated evidence indicates that quercetin may exert its mast cell inhibitory effect by inducing the phosphorylation of MACEDONIA through activation of the protein kinase C (PKC) calcium and phorbol ester independent isozyme zeta.[45] However, it may also be blocking mast cell activation by inhibiting other PKC isozymes critical in stimulus-response coupling. PKC could be inhibited in vitro by certain flavonoids that appeared to competitively block the

ATP binding site on the catalytic unit of PKC.[1] Quercetin was also shown to inhibit tyrosine kinases (PTK) which are implicated in the regulation of cell transformation and cell growth, gene expression, cell–cell adhesion interactions, cell motility, and shape. Again, it acted as a competitive inhibitor of ATP binding and the pattern of B-ring hydroxylation, C2-C3 unsaturation, and C4 keto groups were recognized as strongly affecting inhibitory activity.[1]

The inhibitory effect of flavonoids on secretory processes is not limited to basophils and mast cells, but they are also capable of inhibiting stimulated rabbit neutrophil lysosomal enzyme release, and mitogen-activated adherent human neutrophils. Anti-IgE-induced H_2O_2 generation and human basophil histamine release was also inhibited by quercetin. Oxygen free radicals and nonradical reactive oxygen intermediates released by neutrophils and other phagocytes have been increasingly implicated in inflammatory/immune disorders. Flavonoids could profoundly impair the production of reactive oxygen intermediates by neutrophils and other phagocytic cells. This may be accomplished by interference with NADPH oxidase, a powerful oxidant-producing enzyme localized on the surface membrane of neutrophils. Flavonoids could also inhibit neutrophil myeloperoxidase (MPO), a source of reactive chlorinated intermediates. Impairment by flavonoids of the production of active oxygen intermediates by neutrophils and other phagocytes might contribute to the anti-inflammatory activity of these compounds.[1] Quercetin inhibited the activation of rabbit peritoneal neutrophils stimulated by f-MetLeuPhe, as determined by measurement of degranulation and superoxide formation; quercetin also inhibited tyrosine phosphorylation, mitogen-activated protein kinase (MAPkinase), and phospholipase D. Neutrophil protein tyrosine phosphorylation stimulated by chemotactic factors was diminished by genistein. Ionophore A23187-induced eosinophil secretion of Charcot-Leyden crystal protein and eosinophil cationic protein was inhibited by quercetin, but not by taxifolin (dihydroquercetin), in a concentration-dependent manner. Thus, the activated eosinophil appears to respond to these flavonoids in the same fashion as basophils and mast cells. Eosinophil degranulation stimulated by IgA- or IgG-coated beads was inhibited by genistein; at the same time, several phosphorylated proteins were decreased in quantity, and PLC activation was inhibited.[1]

We have reported that flavone, quercetin, and kaempferol, but not morin, led to accumulation of secretory granules in RBL cells,[25] which do not normally accumulate secretory granules even at stages expressing increased membrane receptors for IgE. RBL cells are considered similar to MMC[46] and have been used as a model for studying IgE-mediated processes leading to secretion of histamine.[47] Quercetin had previously been reported to induce the accumulation of rat mast cell protease II in RBL cells.[24] Moreover, quercetin was reported to "endow" RBL cells with secretory responsiveness to cationic mast cell secretagogues, such as compound 48/80.[48] In an analogous manner, bone marrow-derived mouse mast cells were shown to acquire responsiveness to SP when cultured in the presence of both SCF and IL-4.[49] Flavonoids had previously been shown to inhibit two human lymphoid tissue-derived cell lines.[39] Moreover, quercetin inhibited proliferation of human acute myeloid and lymphoid leukemic cells without affecting normal hematopoiesies.[50] These findings are in accordance with previous studies showing that quercetin blocks proliferation of HL-60 leukemia cells by inducing an accumulation of the cells in the G_2/M phase of the cell cycle [51] and progression of human gastric cancer cells from G_1 to the S phase of growth.[52] Quercetin has been reported to inhibit the growth of estrogen sensitive cells, such as human breast cancer cells in culture.[53] It is interesting that quercetin binds to estrogen type II receptors [41] and human mast cells were shown to express cytoplasmic estrogen receptors.[54] In fact, the growth inhibitory effect of quercetin on a human lymphoblastoid cell line was shown to be through action on a type-II estrogen-binding site, an effect not shared by hesperidin, which does not bind to such sites.[41]

Taken together, such findings indicate that certain flavonoids can induce differentiation of cancer cells, possibly by mimicking the action of certain cytokines. Select flavonoids may, therefore, be useful for the treatment of mast cell proliferative disorders, such as systemic mastocytosis[55] or interstitial cystitis of the bladder,[56] conditions that have been associated with constitutive release of mediators.[57] Flavonoids have the potential to be used as therapeutic agents,[1] especially because their inhibitory effect on mast cell secretion is additive to that recently discovered for chondroitin sulfate,[58] one of the major proteoglycans of rat and human mast cells.[59,60]

ACKNOWLEDGMENTS

These studies were supported in part by Theta Biomedical Consulting and Development Co., Inc. (Brookline, MA, USA).

REFERENCES

1. Middleton, E., Jr., Kandaswami, C., and Theoharides, T.C., The effects of plant flavonoids on mammalian cells: Implications for inflammation, heart disease and cancer, *Pharmacol. Rev.*, 52, 673, 2000.
2. Wang, H.K., The therapeutic potential of flavonoids, *Exp. Opin. Invest. Drugs.*, 9, 2103, 2000.
3. Middleton, E., Jr., The flavonoids, *Trends Pharm. Sci.*, 5, 335, 1984.
4. Middleton, E., Jr. and Kandaswami, C., Effects of flavonoids on immune and inflammatory cell functions, *Biochem. Pharmacol.*, 43, 1167, 1992.
5. Rodewald, H.R. et al., Identification of a committed precursor for the mast cell lineage, *Science*, 271, 818, 1996.
6. Galli, S.J., New concepts about the mast cell, *N. Engl. J. Med.*, 328, 257, 1993.
7. Schwartz, L.B., Mediators of human mast cells and human mast cell subsets, *Ann. Allergy*, 58, 226, 1987.
8. Serafin, W.E. and Austen, K.F., Mediators of immediate hypersensitivity reactions, *N. Engl. J. Med.*, 317, 30, 1987.
9. Grabbe, J. et al., Comparative cytokine release from human monocytes, monocyte-derived immature mast cells and a human mast cell line (HMC-1), *J. Invest. Dermatol.*, 103, 504, 1994.
10. Theoharides, T.C. et al., Differential release of serotonin and histamine from mast cells, *Nature*, 297, 229, 1982.
11. Theoharides, T.C. and Douglas, W.W., Secretion in mast cells induced by calcium entrapped within phospholipid vesicles, *Science*, 201, 1143, 1978.
12. Benyon, R., Robinson, C., and Church, M.K., Differential release of histamine and eicosanoids from human skin mast cells activated by IgE-dependent and non-immunological stimuli, *Br. J. Pharmacol.*, 97, 898, 1989.

13. Gagari, E. et al., Differential release of mast cell interleukin-6 via c-kit, *Blood*, 89, 2654, 1997.
14. Theoharides, T.C., Mast cell: a neuroimmunoendocrine master player, *Int. J. Tissue React.*, 18, 1, 1996.
15. Mekori, Y.A. and Metcalfe, D.D., Mast cells in innate immunity, *Immunol. Rev.*, 173, 131, 2000.
16. Marone, G. et al., Molecular and cellular biology of mast cells and basophils, *Int. Arch. Allergy Immunol.*, 114, 207, 1997.
17. Wedemeyer, J., Tsai, M., and Galli, S.J., Roles of mast cells and basophils in innate and acquired immunity, *Curr. Opin. Immunol.*, 12, 624, 2000.
18. Kay, A.B., Allergy and allergic diseases: First of two parts, *N. Engl. J. Med.*, 344, 30, 2001.
19. Kay, A.B., Allergy and allergic diseases: Second of two parts, *N. Engl. J. Med.*, 344, 109, 2001.
20. Kimata, M., Inagaki, N., and Nagai, H., Effects of luteolin and other flavonoids on IgE-mediated allergic reactions, *Planta Medica*, 66, 25, 2000.
21. Kimata, M. et al., Effects of luteolin, quercetin and baicalein on immunoglobulin E-mediated mediator release from human cultured mast cells, *Clin. Exp. Allergy*, 30, 501, 1999.
22. Fewtrell, C.M. and Gomperts, B.D., Quercetin: a novel inhibitor of Ca^{++} influx and exocytosis in rat peritoneal mast cells, *Biochim. Biophys. Acta.*, 469, 52, 1977.
23. Middleton, E. and Drzewiecki, G., Flavonoid inhibition of human basophil histamine release stimulated by various agents, *Pharmacology*, 33, 3333, 1984.
24. Trnovsky, J. et al., Quercetin-induced expression of rat mast cell protease II and accumulation of secretory granules in rat basophilic leukemia cells, *Biochem. Pharmacol.*, 46, 2315, 1993.
25. Kempuraj, D., Madhappan, B., Christodoulou, S., Boucher, W., Papadopoulou, N., Cetrulo, C.L., and Theoharides, T.C., Flavonols inhibit pro-inflammatory mediator release, intracellular calcium ion levels and protein kinase C theta phosphorylation in human mast cells. *Br. J. Pharmacol.*, 145, 934–944, 2005.
26. Sieghart, W. et al., Calcium-dependent protein phosphorylation during secretion by exocytosis in the mast cell, *Nature*, 275, 329, 1978.
27. Theoharides, T.C. et al., Antiallergic drug cromolyn may inhibit histamine secretion by regulating phosphorylation of a mast cell protein, *Science*, 207, 80, 1980.
28. Sieghart, W. et al., Phosphorylation of a single mast cell protein in response to drugs that inhibit secretion, *Biochem. Pharmacol.*, 30, 2737, 1981.
29. Correia, I. et al., Characterization of the 78 kDa mast cell protein phosphorylated by the antiallergic drug cromolyn and homology to moesin, *Biochem. Pharmacol.*, 52, 413, 1996.
30. Lankes, W.T. and Furthmayr, H., Moesin: a member of the protein 4.1-talin-ezrin family of proteins, *Proc. Natl. Acad. Sci. U.S.A.*, 88, 8297, 1991.
31. Tsukita, S. and Yonemura, S., ERM (ezrin/radixin/moesin) family: From cytoskeleton to signal transduction, *Curr. Opin. Cell Biol.*, 9, 70, 1997.
32. Wang, L. et al., Ca^{2+} and phorbol ester effect on the mast cell phosphoprotein induced by cromolyn, *Eur. J. Pharmacol.*, 371, 241, 1999.
33. Theoharides, T.C. et al., Cloning and cellular localization of the rat mast cell 78kD protein phosphorylated in response to the mast cell "stabilizer" cromolyn, *J. Pharmacol. Exp. Ther.*, 294, 810, 2000.
34. Fewtrell, C.M. and Gomperts, B.D., Effect of flavone inhibitors of transport ATPases on histamine secretion from rat mast cells, *Nature*, 265, 635, 1977.
35. Pearce, F.L., Befus, A.D., and Bienenstock, J., Effect of quercetin and other flavonoids on antigen-induced histamine secretion from rat intestinal mast cells, *J. Allergy Clin. Immunol.*, 73, 819, 1984.

36. Barrett, K.E. and Metcalfe, D.D., The histologic and functional characterization of enzymatically dispersed intestinal mast cells of nonhuman primates: effects of secretagogues and anti-allergic drugs on histamine secretion, *J. Immunol.*, 135, 2020, 1985.

37. Fox, C.C. et al., Comparison of human lung and intestinal mast cells, *J. Allergy Clin. Immunol.*, 81, 89, 1988.

38. Berton, G., Schneider, C., and Romeo, D., Inhibition by quercetin of activation of polymorphonuclear leucocyte functions, *Biochim. Biophys. Acta.*, 595, 47, 1980.

39. Devi, M.A. and Das, N.P., *In vitro* effects of natural plant polyphenols on the proliferation of normal and abnormal human lymphocytes and their secretions of interleukin-2, *Cancer Lett.*, 69, 191, 1993.

40. Namgoong, S.Y. et al., Effects of naturally occurring flavonoids on mitogen-induced lymphocyte proliferation and mixed lymphocyte cuture, *Life Sci.*, 54, 313, 1993.

41. Scambia, G. et al., Type-II estrogen binding sites in a lymphoblastoid cell line and growth-inhibitory effect of estrogen, anti-estrogen and bioflavonoids, *Int. J. Cancer*, 46, 1112, 1990.

42. Pick, M. et al., Flavonoid modulation of protein kinase C activation, *Life Sci.*, 44, 1563, 1989.

43. Ferriola, P.C., Cody, V., and Middleton, E., Protein kinase inhibition by plant flavonoids, kinetic mechanisms and structure-activity relationships, *Biochem. Pharmacol.*, 38, 1617, 1989.

44. Nishizuka, Y., The molecular heterogeneity of protein kinase C and its implications for cellular regulation, *Nature*, 334, 661, 1988.

45. Kandere-Grzybowska, K., Kempouraj, S., Cao, J., and Theoharides, T.C., Regulation of IL-1-induced selective IL-6 release from human mast cells and inhibition by quercetin. *Br. J. Pharmacol.*, 148, 208–215, 2006.

46. Seldin, D.C. et al., Homology of the rat basophilic leukemia cell and the rat mucosal mast cell, *Proc. Natl. Acad. Sci. U.S.A.*, 82, 3871, 1985.

47. Oliver, J.M. et al., Signal transduction and cellular response in RBL-2H3 mast cells, *Prog. Allergy*, 42, 185, 1988.

48. Senyshyn, J., Baumgartner, R.A., and Beaven, M.A., Quercetin sensitizes RBL-2H3 cells to polybasic mast cell secretagogues through increased expression of G_i GTP-binding proteins linked to a phospholipase C signaling pathway, *J. Immunol.*, 160, 5136, 1998.

49. Karimi, K. et al., Stem cell factor and interleukin-4 increase responsiveness of mast cells to substance P, *Exp. Hematol.*, 28, 626, 2000.

50. Larocca, L.M. et al., Antiproliferative activity of quercetin on normal bone marrow and leukaemic progenitors, *Br. J. Haematol.*, 79, 562, 1991.

51. Kang, T. and Liang, N., Studies on the inhibitory effects of quercetin on the growth of HL-60 leukemia cells, *Biochem. Pharmacol.*, 54, 1013, 1997.

52. Mitsumori, Y. et al., The effect of quercetin on cell cycle progression and growth of human gastric cancer cells, *FEBS Lett.*, 260, 10, 1990.

53. Damianaki, A. et al., Potent inhibitory action of red wine polyphenols on human breast cancer cells, *J. Cell. Biochem.*, 78, 429, 2000.

54. Pang, X. et al., Bladder mast cell expression of high affinity estrogen receptors in patients with interstitial cystitis, *Br. J. Urol.*, 75, 154, 1995.

55. Longley, J., Duffy, T.P., and Kohn, S., The mast cell and mast cell disease, *J. Am. Acad. Dermatol.*, 32, 545, 1995.

56. Theoharides, T.C. et al., Activation of bladder mast cells in interstitial cystitis: a light and electron microscopic study, *J. Urol.*, 153, 629, 1995.

57. Granerus, G., Lönnqvist, B., and Roupe, G., No relationship between histamine release measured as metabolite excretion in the urine, and serum levels of mast cell specific tryptase in mastocytosis, *Agents Actions*, 41, C127, 1994.

58. Theoharides, T.C. et al., Chondroitin sulfate inhibits connective tissue mast cells, *Br. J. Pharmacol.*, 131, 10139, 2000.

59. Krilis, S.A. et al., Continuous release of secretory granule proteoglycans from a strain derived from the bone marrow of a patient with diffuse cutaneous mastocytosis, *Blood*, 79, 144, 1992.

60. Nilsson, G. et al., Stem cell factor-dependent human cord blood derived mast cells express alpha- and beta-tryptase, heparin and chondroitin sulphate, *Immunology*, 88, 308, 1996.

16 Molecular Mechanisms of H₂O₂-Induced DNA Damage:
The Action of Desferrioxamine

M. Tenopoulou, P.-T Doulias, and D. Galaris

CONTENTS

AN INTRODUCTION TO OXIDATIVE STRESS

Life on Earth has evolved on the foundation of two main processes: (a) photosynthesizing organisms (plants) that capture solar energy and use it to promote thermodynamically unfavorable reactions leading to the reduction of carbon substances, and (b) organisms that receive the above produced reduced compounds and oxidize them in thermodynamically favorable reactions. The latter processes release large amounts of energy that are captured in the form of chemical energy and used then as the power that supports the maintenance of life. The electrons used for the reductive processes in photosynthetic plants are released during the oxidation of the H_2O to O_2, whereas in the catabolic processes the opposite direction is followed by reduction of molecular oxygen (O_2) to water (H_2O) (Scheme 16.1). The latter reaction is catalyzed by the last enzyme of the respiratory chain, namely cytochrome oxidase (complex IV), which binds molecular oxygen in its active site and reduces it to H_2O

Photosynthesis

$4e^- + 4H^+$

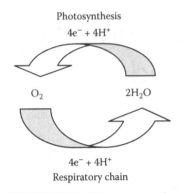

O_2 $2H_2O$

$4e^- + 4H^+$

Respiratory chain

SCHEME 16.1 The oxygen cycle.

by donating four electrons without any release of partially reduced intermediates. However, a small part of the oxygen consumed by mitochondria (usually about 2–4%) is reduced, even under normal conditions, by single electrons leading to the formation of a variety of reactive oxygen intermediates (Scheme 16.2). In this way, aerobic organisms are continuously exposed to endogenously generated reactive oxygen species (ROS) and in spite of the existed anti-oxidant defenses, a steady-state concentration of ROS is always present.[1] In addition, a number of pathological conditions have been shown to be intimately connected to increased steady-state levels of ROS which may be the result of either an increased rate of production or a decreased ability for the removal of these species.[2–5] This temporary imbalance of cellular redox equilibrium is usually defined as "oxidative stress."[6] This imbalance of the cellular redox equilibrium is not, as believed before, an unavoidable side effect of oxygen metabolism but rather it appears to be a carefully regulated process, capable of transferring important signals from cell surface toward the genetic machinery of the cell.[7] Today, redox regulation of gene expression is regarded as a vital mechanism involved in a number of pathophysiological complications in humans and animals.[8,9]

GENERATION OF REACTIVE OXYGEN SPECIES

Oxygen, due to its ubiquitous appearance in biological materials and its ability to accept single electrons from other compounds represents the main source of reactive species in cells. One electron reduction of oxygen leads to the formation of superoxide radicals ($O_2{}^-$) whereas, when it is reduced by two electrons, the product is hydrogen peroxide (H_2O_2). It has to be noted that the latter compound is not a free radical, but it can participate in processes that contribute to formation of free radicals. Ferrous iron, cuprous copper, and several other transition metals are able

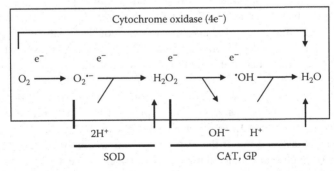

SCHEME 16.2 Controlled generation of ROS during the reduction of oxygen (O_2) to water into the final step of the respiratory chain.

to donate a third electron to H$_2$O$_2$, causing the breakage of the O-O bond. In these cases the one part of the molecule is reduced to the state of water while the other part forms the extremely reactive hydroxyl radical ($\cdot$OH). The latter is one of the most potent oxidants known, reacting indiscriminately with any organic molecule that is going to be in the vicinity of its generation.

Apart from the ROS mentioned above (O$_2$$^{\cdot-}$, H$_2O_2$, and $\cdot$OH), the term is usually expanded to include the electronically excited oxygen (singlet oxygen), peroxynitrite, hypochlorous acid and other activated forms that contain oxygen.

HYDROGEN PEROXIDE

Among all different ROS, hydrogen peroxide (H$_2$O$_2$), located centrally in cellular redox reactions (Scheme 16.2), is likely to be of special importance. Although its chemical reactivity is modest, it exerts toxic effects when cells are exposed to this compound even at relatively low concentrations. Controlled generation of H$_2$O$_2$ appears to be a common phenomenon among different cell types.[10,11] Moreover, it has been shown that it plays important roles in signaling pathways.[12,13] Because H$_2$O$_2$ is able to pass freely across cell membranes, it can transfer information to nearby cells or tissues acting in a paracrine fashion, similar to nitric oxide. Although extensively studied, the exact molecular mechanisms of the generation, as well as the mode of action of H$_2$O$_2$, remain largely elusive. A large number of molecules have been proposed or identified as direct or indirect targets of intracellular H$_2$O$_2$. Protein kinases and phosphatases, proteins containing sulfhydryl groups or iron sulfur clusters, lipids, DNA, and others are among the potential targets.[14–16] Although some indications regarding the involvement of transition metal ions in these processes have been emerged, the exact molecular mechanisms remain mainly hypothetical.[17,18] More studied is the role of transition metal ions (mainly iron and copper) in the process of H$_2$O$_2$-induced damage of cellular components. DNA is regarded as the most sensitive target among all cell components. The prevailing idea is that H$_2$O$_2$ interacts with redox active iron, which is bound in cellular DNA, and the hydroxyl radicals formed by Fenton reaction (reaction 16.1), due to their extreme reactivity, react with DNA in their immediate vicinity.

$$Fe^{2+} + H_2O_2 \longrightarrow Fe^{3+} + \cdot OH + OH^- \qquad (16.1)$$

There are indications, however, that the location of iron may be away from the nucleus, and increases of intracellular levels of Ca^{2+} may be a mediator of DNA damage.[19,20] This presentation will concentrate on some findings from our laboratory that indicate the crucial role of lysosomal iron in the molecular mechanisms underlying H$_2$O$_2$-induced DNA damage.

HYDROGEN PEROXIDE-INDUCED DNA DAMAGE

Cellular DNA is especially sensitive to the action of H$_2$O$_2$ and, as mentioned above, it is widely believed to be mediated by transition metal ions, mainly iron and/or copper, which are able to catalyze the formation of hydroxyl radicals ($\cdot$OH) by

Fenton-type reactions.[21-23] In this regard, the location of redox-active metal ions is of utmost importance for the ultimate effect because ·OH, due to its extreme reactivity, interacts exclusively in the vicinity of the bound metal.[24] Formation of ·OH close to DNA results in its damage, including base modifications, single and double strand breakage, and sister chromatid exchange. In some occasions, single strand break (SSB) formation has been linked to intracellular Ca^{2+} increases, indicating an obligatory intermediary role for Ca^{2+}.[20,25] We have previously shown that H_2O_2-induced DNA damage is Ca^{2+}-dependent at low rates of continuous H_2O_2 production, although it seems likely that a Ca^{2+}-independent mechanism operates at higher rates of H_2O_2 production.[20] When H_2O_2 is added directly as a bolus, as is the case in most studies,[21,25] cells are initially exposed to relatively high concentrations followed by a fast decrease as H_2O_2 is consumed. The rate of removal of H_2O_2 follows first-order kinetics with the rate constant dependent on the kind of cells used, and reciprocally related to the number of cells.[26] However, when the mode of action of H_2O_2 is concentration-dependent,[21] the results may appear inconsistent. In order to avoid this problem, we have used a continuously generating system of H_2O_2 by employing the enzyme glucose oxidase (GO) (reaction 16.2).

$$\text{Glucose} + O_2 \xrightarrow{\text{Glucose oxidase}} H_2O_2 + \text{Gluconate} \qquad (16.2)$$

An appropriate amount of this enzyme was added directly to the growth medium in order to generate constant but relatively low concentrations of H_2O_2.

DETECTION OF SINGLE STRAND BREAKS

Formation of SSBs was detected by the single cell gel electrophoresis or "comet assay" as it is usually called as previously described.[20,27-30] As previously reported with other cells,[20,29,30] exposure of HeLa cells (1×10^5 cells per ml) in growth medium containing 10% fetal calf serum to continuously generated H_2O_2 caused a dose dependent induction of SSBs (Figure 16.1). It is important to mention that in the experiment shown in Figure 16.1 the cells remained viable for at least 60 min of incubation, as determined by propidium iodide exclusion using flow cytometry (results not shown).

PROTECTION BY IRON CHELATORS

A number of iron chelators with different chemical structures were used in our laboratory in order to evaluate the molecular mechanisms of H_2O_2-induced DNA damage. Previous studies have shown that 1,10 phenantroline (but not 1,7–henenthroline) was able to protect the DNA of Jurkat cells in culture after exposure to H_2O_2.[31] It was known that 1,10 phenanthroline is an effective iron chelator[32] whereas its 1,7 counterpart, due to different position of a nitrogen atom in its molecule, was unavailable for iron binding. These results clearly indicate that the iron-binding capacity of 1,10 phenanthroline is responsible for the observed protective effect. In addition to 1,10 phenanthroline, other iron chelators (Figure 16.2) like 2,2′ dipyridyl and TPEN

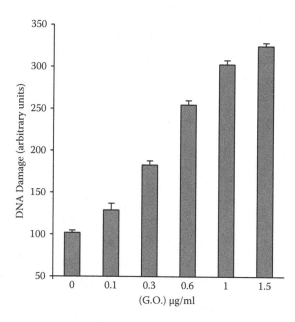

FIGURE 16.1   Dose dependent effect of H_2O_2 on the DNA of HeLa cells. HeLa cells (1×10^5 cells per ml) were exposed to increasing concentrations of H_2O_2 by the direct addition of increasing amounts of the enzyme glucose oxidase into the culture medium. Ten minutes later cells were collected by trypsinization and the cellular DNA damage was estimated by comet assay. Each point represents the mean value of three different samples. This experiment was repeated once more with essentially the same results.

(N,N,N′,N′-tetrakis(2-pyridyl-methyl)ethylenediamine) were also shown to offer protection (Figure 16.3).

EDTA (ethylenediaminetetraacetic acid) and DTPA (diethylenetriaminepenta-acetic acid) were also checked for their ability to prevent H_2O_2-induced formation of single strand breaks in the cellular DNA. Whereas the first was ineffective, the latter was able to protect (Figure 16.4). However, the protection offered by DTPA was pre-incubation time dependent with its capacity increasing with the time up to 6 h (Figure 16.4b). These results indicate that the iron binding by the respective chelator must inhibit its interactions with H_2O_2 in order to be effective. EDTA binds iron in such a way that allows its interaction with H_2O_2 so it does not offer any protection. The time-dependent effect of DTPA may be related to its slower uptake by the cells compared to the other iron chelators used above. DTPA (and EDTA), due to their negative charge, seem to be membrane impermeable and their intracellular action, it is possible to assume, is related to endocytotic mechanisms.

THE ACTION OF DESFERRIOXAMINE

Desferrioxamine (DFO), an effective iron chelator, is currently used extensively for the treatment of iron overload diseases such as β-thalassemia. We used this compound in order to check its ability to protect cellular DNA after exposure of HeLa cells to

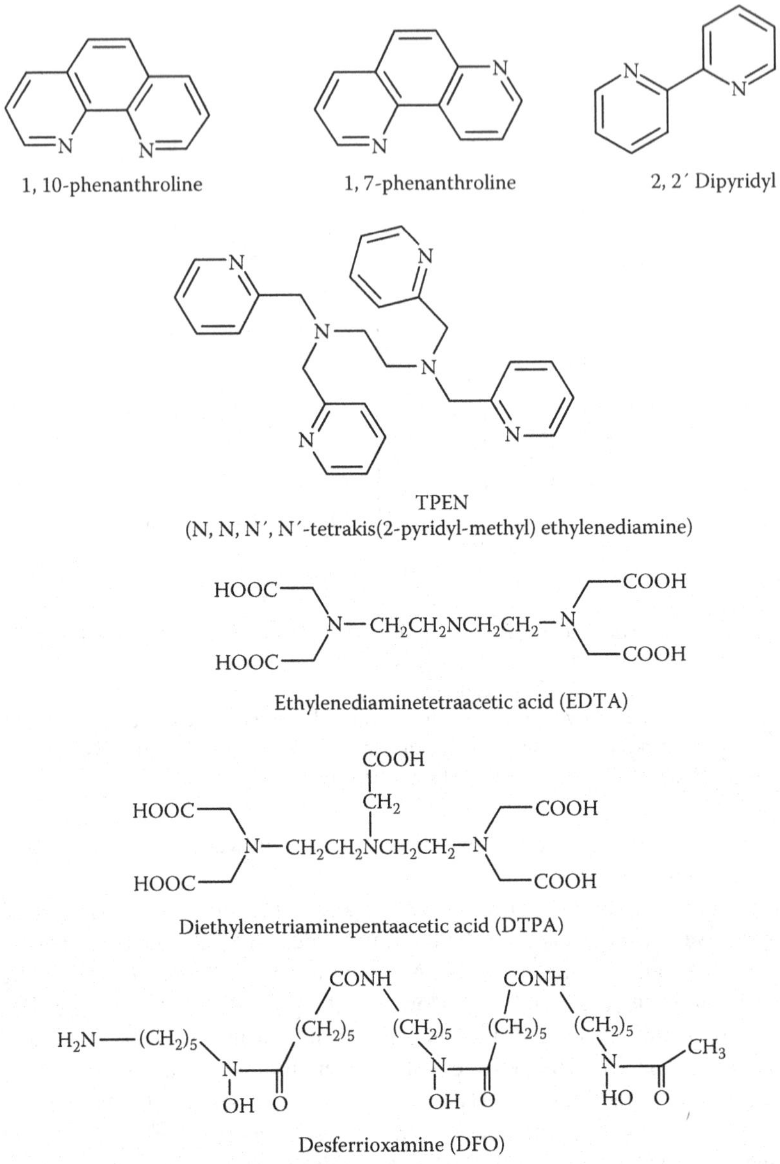

1, 10-phenanthroline

1, 7-phenanthroline

2, 2′ Dipyridyl

TPEN
(N, N, N′, N′-tetrakis(2-pyridyl-methyl) ethylenediamine)

Ethylenediaminetetraacetic acid (EDTA)

Diethylenetriaminepentaacetic acid (DTPA)

Desferrioxamine (DFO)

FIGURE 16.2   Chemical structures of iron chelators

H_2O_2. It was shown that DFO was able to protect cells in a dose- and time-dependent manner.[31,33] Interestingly, the degree of protection was significantly increased even when DFO was present in the culture medium for 1 h and then the cells were kept at normal conditions (without DFO) for another hour before being subjected to oxidative stress. In fact, whether DFO was present or absent during the second hour before oxidative stress did not make any difference for the degree of protection ($26.7 \pm 0.9\%$ and $28.9 \pm 2.4\%$ of the initial DNA damage, respectively, down from $42.7 \pm 2.4\%$

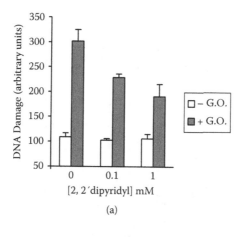

(a)

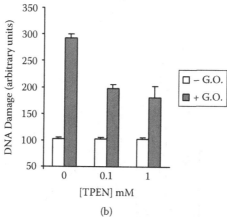

(b)

FIGURE 16.3 Protective effects of iron chelators on H_2O_2-induced DNA damage. Jurkat cells (1.5×10^5 cells per 100 μl) were incubated with the indicated concentrations of 2,2' dipyridyl (a) or TPEN (b) for 15 min before the addition of the H_2O_2 generating enzyme "glucose oxidase" (60 ngGO). Ten minutes later cellular DNA was analyzed for the formation of single strand breaks by using comet assay as in Figure 16.1.

following only 1 h of DFO pretreatment).[33] These experiments may be explained if we assume that DFO is taken up by the cells by endocytosis as already proposed.[34-36] This possibility is further supported by the results of the experiment presented in Figure 16.5. Both 1,10 phenanthroline and DFO were protective at 37°C, whereas only 1,10 phenanthroline but not DFO affected DNA damage at 4°C, indicating that an enzymatic process is needed in the case of DFO.

Endocytosis is a process by which extracellular material is internalized by different mechanisms initiating from the plasma membrane. The best studied of these mechanisms are clathrin coated vesicles, and caveolae-mediated and fluid phase endocytosis. To get insights into the precise mechanism by which DFO exerts its protective activity, molecular tools that specifically modulate these internalization

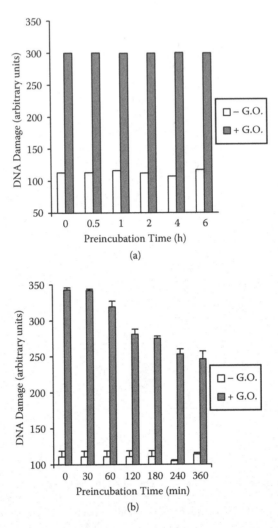

FIGURE 16.4 Effects of iron chelators on H_2O_2-induced DNA damage. Jurkat cells (1.5×10^5 cells per 100 μl) were incubated for the indicated periods of time with 2.5 mM EDTA (a) or 1 mM DTPA (b) before the addition of 60 ng of glucose oxidase. Ten minutes later cellular DNA damage was analyzed by comet assay as described in Figure 16.1.

pathways were used. HeLa cells were transiently transfected with plasmids carrying genes that encode the dominant negative or positive mutants of the GTPases, dynamin, and Rab5. Dynamin is known to be involved in the inward pinching of clathrin-coated vesicles[37] and caveolae[38] from the plasma membrane, but it is not involved in fluid phase endocytosis (Figure 16.5).[37,39] On the other hand, Rab5 plays a critical role in the mechanism of both fluid phase and clathrin-coated vesicle endocytosis.[40,41] Experiments with these cells show that transfection with vectors carrying a dominant positive or negative form of dynamin (K694A) and (K44A), respectively, which modulate clathrin- and caveolae-mediated endocytosis,[37,42,43] did not affect

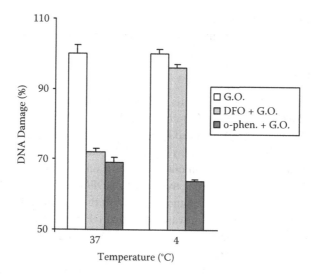

FIGURE 16.5 Effects of temperature on the protective action of desferrioxamine. HeLa cells (1 × 10^5 cells per ml) were preincubated at the indicated temperatures either with 1 mM desferrioxamine for 60 min or with 0.5 mM 1,10-phenanthroline for 15 min. Cells were then exposed to continuously generated H$_2$O$_2$ by the direct addition into the culture medium of 1 µg/ml of glucose oxidase (generating about 20 µM H$_2$O$_2$ per minute). Ten minutes later cells were collected by trypsinization and DNA damage was evaluated by comet assay.

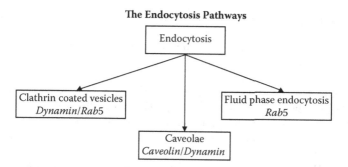

FIGURE 16.6 The different endocytotic pathways and the proteins that regulate these pathways.

the ability of DFO to protect cells from H$_2$O$_2$-induced DNA damage. On the contrary, transfection with inactive or overactive mutants of Rab5 (S34N and Q79L, respectively), which have been shown to modulate fluid phase and clathrin-coated vesicle endocytosis,[44,45] resulted in a much decreased or increased ability of DFO to prevent DNA damage (Figure 16.6).[33]

Collectively, the above results clearly indicate that DFO has to be endocytosed by the fluid phase pathway in order to protect against oxidative stress-induced DNA damage. Moreover, these results highlight the role of lysosomes in general and lyso-somal iron in particular in the mechanisms of H_2O_2-induced DNA damage.

CONCLUSIONS

In conclusion, preincubation of the cells with several chelating agents of varying spec-ificity, before the challenge with H_2O_2, revealed that the redox active pool of intracel-lular iron is pertinent to H_2O_2-induced DNA damage. Of particular interest were the protective effects that were exerted by DFO. DFO, due to its hydrophilic character and its molecular weigh (more than 600 Da), seems unlikely to be able to penetrate the cell membrane. Thus, the protective effect of DFO seems to be dependent on the pathway of its uptake. By using molecular tools that specifically modulated the various internalization pathways, it was shown that DFO is taken up by the cells by fluid phase endocytosis. As DFO most probably is not released into the cytosol, these results also indicate the importance of endosomal and lysosomal redox-active iron in the above process. This notion is substantiated by a number of recently published works,[46,47] but needs further investigations in order to be totally clarified.

REFERENCES

1. Chance, B., Sies, H., and Boveris, A., Hydroperoxide metabolism in mammalian organs, *Annu. Rev. Physiol.*, 59, 527, 1979.
2. Metodiewa, D. and Koska, C., Reactive oxygen species and reactive nitrogen species: relevance to cyto(neuro)toxic events and neurologic disorders: An overview, *Neurotoxi-col. Res.*, 3, 197, 2000.
3. Cauwels, A. et al., Caspase inhibition causes hyperacute tumor necrosis factor-induced shock via oxidative stress and phospholipase A2, *Nat. Immunol.*, 4, 387, 2003.
4. Kim, S.H. et al., Brain-derived neurotrophic factor can act as a pronecrotic factor through transcriptional and translational activation of NADPH oxidase, *J. Cell Biol.*, 159, 821, 2002.
5. Lenaz, G., The mitochondrial production of reactive oxygen species: mechanisms and implications in human pathology, *IUBMB Life*, 52, 159, 2001.
6. Sies, H., in *Oxidative Stress*, Academic Press, London, 1985.
7. McCord, J.M., The evolution of free radicals and oxidative stress, *Am. J. Med.*, 108, 652, 2000.
8. Maulik, N. and Das, D.K., Redox signaling in vascular angiogenesis, *Free Radic. Biol. Med.*, 33, 1047, 2002.
9. Rojkind, M. et al., Role of hydrogen peroxide and oxidative stress in healing responses, *Cell Mol. Life Sci.*, 59, 1872, 2002.
10. Forman, H.J. and Torres, M., Reactive oxygen species and cell signaling: respiratory burst in macrophage signalling, *Am. J. Respir. Crit. Care Med.*, 166, S4, 2002.
11. Halliwell, B. et al., Hydrogen peroxide. Ubiquitous in cell culture and *in vivo*? *IUBMB Life*, 50, 251, 2000.
12. Sundaresan, M. et al., Requirement for generation of H_2O_2 for platelet-derived growth factor signal transduction, *Science*, 270, 296, 1995.
13. Forman, H.J. and Torres, M., Redox signaling in macrophages, *Mol. Asp. Med.*, 22, 189, 2001.

14. Wolin, M.S. and Mohazzab, K.M., Mediation of signal transduction by oxidants, in *Oxidative Stress and the Molecular Biology of Antioxidant Defenses,* Scandalios, J.G. Ed., Cold Spring Harbor Laboratory Press, New York, 1997, pp. 21–48.

15. Hansen, L.L. et al., Insulin signaling is mediated by micromolar concentrations of H_2O_2: evidence for a role of H_2O_2 in tumor necrosis factor alpha-mediated insulin resistance, *J. Biol. Chem.*, 274, 25078, 1999.

16. Hampton, M.B. and Orrenius, S., Dual regulation of caspase activity by hydrogen peroxide: implications for apoptosis, *FEBS Lett.*, 414, 552, 1997.

17. Zhang, W.J. and Frei, B., Intracellular metal ion chelators inhibit TNFalpha-induced SP-1 activation and adhesion molecule expression in human aortic endothelial cells, *Free Radic. Biol. Med.*, 34, 674, 2003.

18. Sandstrom, B.E., Effects of quin2 acetoxymethyl ester on H_2O_2-induced DNA single-strand breakage in mammalian cells: H_2O_2-concentration-dependent inhibition of damage and additive protective effect with the hydroxyl-radical scavenger dimethyl sulphoxide, *Biochem. J.*, 305, 181, 1997.

19. Cantoni, O. et al., Calcium chelator Quin 2 prevents hydrogen-peroxide-induced DNA breakage and cytotoxicity, *Eur. J. Biochem.*, 182, 209, 1989.

20. Panayiotidis, M., Tsolas, O., and Galaris, D., Glucose oxidase-produced H_2O_2 induces Ca 2+-dependent DNA damage in human peripheral blood lymphocytes, *Free Radic. Biol. Med.*, 26, 548, 1999.

21. Imlay, J., Chin, S., and Linn, S., Toxic DNA damage by hydrogen peroxide through the Fenton reaction *in vivo* and *in vitro, Science*, 240, 640, 1998.

22. Meneghini, R., Iron homeostasis, oxidative stress, and DNA damage, *Free Radic. Biol. Med.*, 23, 783, 1997.

23. Halliwell, B. and Gutteridge, J.M.C., Role of free radicals and catalytic metal ions in human disease, *Methods Enzymol.*, 186, 1, 1992.

24. Chevion, M., A site specific mechanism for free radical induced biological damage: The essential role of redox-active transition metals, *Free Radic. Biol. Med.*, 5, 27, 1988.

25. Roveri, A. et al., Effects of hydrogen peroxide on calcium homeostasis in smooth muscle cells, *Arch. Biochem. Biophys.*, 297, 265, 1992.

26. Antunes, F. and Cadenas, E., Estimation of H_2O_2 gradients across biomembranes, *FEBS Lett.*, 475, 121, 2000.

27. Singh, N.P. et al., A simple technique for quantitation of low levels of DNA damage in individual cells, *Exp. Cell Res.*, 184, 461, 1988.

28. Doulias, P.T. et al., SIN-1-induced DNA damage in isolated human peripheral blood lymphocytes as assessed by single cell gel electrophoresis (comet assay), *Free Radic. Biol. Med.*, 30, 679, 2001.

29. Tselepis, A. et al., Trimetazidine protects low-density lipoproteins from oxidation and cultured cells exposed to H_2O_2 from DNA damage, *Free Radic. Biol. Med.*, 30, 1357, 2001.

30. Barbouti, A., et al., DNA damage and apoptosis in hydrogen peroxide-exposed Jurkat cells: bolus addition versus continuous generation of H_2O_2, *Free Radic. Biol. Med.*, 33, 691, 2002.

31. Barbouti, A., et al., Intracellular iron, but not copper, plays a critical role in hydrogen peroxide-induced DNA damage, *Free Radic. Biol. Med.*, 31, 490, 2001.

32. Boumans, H. et al., Differential inhibition of the yeast bc1 complex by phenanthrolines and ferroin. Implications for structure and catalytic mechanism, *J. Biol. Chem.*, 272, 16753, 1999.

33. Doulias, P.T. et al., Endosomal and lysosomal effects of desferrioxamine: protection of HeLa cells from hydrogen peroxide-induced DNA damage and induction of cell-cycle arrest, *Free Radic. Biol. Med.*, 35, 719, 2003.

34. Cable, H. and Lloyd, J.B., Cellular uptake and release of two contrasting iron chelators, *J. Pharm. Pharmacol.*, 51, 131, 1999.

35. Lloyd, J.B., Cable, H., and Rice-Evans, C., Evidence that desferrioxamine cannot enter cells by passive diffusion, *Biochem. Pharmacol.*, 41, 1361, 1991.

36. Öllinger, K. and Brunk, U.T., Cellular injury induced by oxidative stress is mediated through lysosomal damage, *Free Radic. Biol. Med.*, 19, 565, 1995.

37. Damke, H. et al., Induction of mutant dynamin specifically blocks endocytic coated vesicle formation, *J. Cell Biol.*, 127, 915, 1994.

38. Henley, J.R. et al., Dynamin-mediated internalization of caveolae, *J. Cell Biol.*, 141, 85, 1998.

39. Damke, H. et al., Clathrin-independent pinocytosis is induced in cells overexpressing a temperature-sensitive mutant of dynamin, *J. Cell Biol.*, 131, 69, 1995.

40. Bucci, C. et al., The small GTPase rab5 functions as a regulatory factor in the early endocytic pathway, *Cell*, 70, 715, 1992.

41. McLauchlan, H. et al., A novel role for Rab5-GDI in ligand sequestration into clathrin-coated pits, *Curr. Biol.*, 8, 34, 1998.

42. Oh, P., McIntosh, D.P., and Schnitzer, J.E., Dynamin at the neck of caveolae mediates their budding to form transport vesicles by GTP-driven fission from the plasma membrane of endothelium, *J. Cell Biol.*, 141, 101, 1998.

43. Sever, S., Muhlberg, A.B., and Schmid, S.L., Impairment of dynamin's GAP domain stimulates receptor-mediated endocytosis, *Nature*, 398, 481, 1999.

44. Lanzetti, L. et al., The Eps8 protein coordinates EGF receptor signalling through Rac and trafficking through Rab5, *Nature*, 408, 374, 2000.

45. Simonsen, A. et al., EEA1 links PI(3)K function to Rab5 regulation of endosome fusion, *Nature*, 394, 494, 1998.

46. Stoka, V. et al., Lysosomal protease pathways to apoptosis, *J. Biol. Chem.*, 276, 3149, 2001.

47. Madge, L.A. et al., Inhibition of phosphatidylinositol 3-kinase sensitizes vascular endothelial cells to cytokine-initiated cathepsin-dependent apoptosis, *J. Biol. Chem.*, 278, 21295, 2003.

17 LNA (Locked Nucleic Acid) and Functionalized LNA: *Toward Efficient Gene Targeting*

Jesper Wengel

CONTENTS

ABSTRACT

Locked nucleic acid (LNA) is a class of nucleic acid analogues possessing unprecedented binding affinity towards complementary RNA. LNA-type oligonucleotides are commercially available, are nontoxic, display increased RNA target accessibility, and can be designed to activate RNaseH or to function in steric block approaches. LNA-antisense, in particular, gapmer LNA, containing a central DNA or phosphorothioate-DNA segment flanked by LNA-gaps, rivals siRNA as the technology of choice for target validation and therapeutic applications.

INTRODUCTION

Two strategies in general are considered for sequence-specific down regulation of gene expression using synthetic oligonucleotides, i.e., antisense-DNA and siRNA (short interfering RNA).[1,2] The easy availability and high potency of siRNA *in vitro* are reasons for the current popularity of the siRNA approach relative to the antisense-DNA approach. However, it should be kept in mind that the effect of antisense-DNA

239

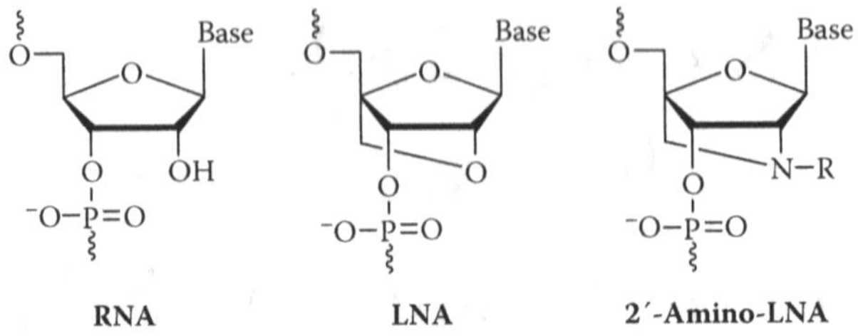

FIGURE 17.1   Structure of RNA, LNA, and 2′-amino-LNA monomers. Base = nucleobases.

in vitro and *in vivo* strongly depends on the nature of the nucleotide modifications used. Thus, phosphorothioate-DNA, the commonly used first-generation antisense-DNA chemistry, is rather inefficient and toxic, and should not be used to compare in a general way the antisense-DNA and siRNA technologies. Instead, optimized antisense-DNA should be considered, and one candidate molecular architecture in this relation is LNA-type oligonucleotides, e.g., LNA or 2′-amino-LNA (Figure 17.1). In this account, design and characteristics of various LNA constructs for antisense applications ("LNA-antisense") are described.

LNA (LOCKED NUCLEIC ACID)

We have defined LNA as an oligonucleotide containing at least one conformationally locked 2′-O,4′-C-methylene-β-D-ribofuranosyl nucleotide monomer (LNA monomer; Figure 17.2).[3,4] LNA oligonucleotides (fully modified LNA, LNA/DNA mixmers, LNA/RNA mixmers, LNA/phosphorothioate-DNA mixmers, etc.; Figure 17.2) possess unsurpassed affinity to complementary DNA and RNA obeying the Watson–Crick base-pairing rules (Table 17.1).[3-7] LNA oligonucleotides induce resistance towards nucleolytic degradation, display low toxicity in biological systems, and are efficiently transfected into mammalian cells using standard procedures.[7] LNA contains phosphordiester linkages and is therefore soluble in aqueous media, and can be synthesized by standard automated methods using commercial DNA synthesizers.

LNA-ANTISENSE FOR mRNA TARGETING

LNA-antisense studies were first reported by Wahlestedt et al. who demonstrated potent and nontoxic antisense effects of LNA oligonucleotides *in vivo*, using LNA-

LNA1:   5′- CAcGggCaGaAGgCA   [LNA/DNA mixmer design]

LNA2:   5′-ATCTTgttgacggTCTCA   [LNA-DNA-LNA gapmer design]

FIGURE 17.2   Examples of LNA/DNA mixmer (**LNA1**) and LNA-DNA-LNA gapmer (**LNA2**) designs; **A**, **C**, **G**, and **T** denote LNA monomers (i.e., adenin-9-yl, 5-methylcytosin-1-yl, guanin-9-yl, and thymin-1-yl derivatives) and a, c, g, and t denote DNA monomers. LNA oligonucleotides can activate RNase H if designed as gapmers.

TABLE 17.1

Examples of Thermal Denaturation Temperatures (T_m Values) of LNA Oligonucleotides Measured in Medium Salt Buffer

LNA	DNA Target		RNA Target	
	$T_m/°C$	$\Delta T_m/°C$	$T_m/°C$	$\Delta T_m/°C$
5'-d(gtgatatgc)	28	—	28	—
5'-(GTGATATGC)^L	64	+4.0	74	+5.1
5'-d(gT^{L}gaT^{L}aT^{L}gc)	44	+5.3	50	+7.3
5'-d(gtgaT^{L}atgc)	35	+7.0	37	+9.0

Note: 100 mM sodium chloride, 10 mM sodium phosphate, pH 7.0. ΔT_m values = increase in T_m values per LNA monomer calculated relative to the T_m values of the reference duplexes formed with the DNA strand (the oligonucleotide shown on top).

Source: Koshkin, A.A. et al., LNA (locked nucleic acid): An RNA mimic forming exceedingly stable LNA:LNA duplexes, *J. Am. Chem. Soc.*, 120, 13252, 1998. With permission.

DNA-LNA gapmers or LNA/DNA mixmers targeting the delta opioid receptor mRNA.[8] LNA-antisense was clearly superior to the isosequential DNA oligonucleotides with respect to dose-dependent and sequence-specific gene silencing. Braasch et al. studied the inhibition of luciferase activity by LNA-antisense.[9] Fully modified LNA, LNA/DNA mixmers, and LNA-DNA-LNA gapmers displayed potent and sequence-specific inhibition of luciferase activity in CV-1 cells, whereas other LNA-DNA-LNA gapmers and LNA/DNA mixmers were less efficient. These results indicate that the exact position within the mRNA target sequence can be of importance for LNA-antisense. Phosphorothioate LNA-DNA-LNA gapmers have also been demonstrated to induce potent inhibition of luciferase activity in HeLa cells with constitutive expression of luciferase.[10]

A fully modified LNA oligonucleotide caused down-regulation of the POLR2A protein in 15PC3 prostate cells, although not in a stringent sequence-dependent manner.[11] This LNA oligonucleotide was applied as an antitumor agent in a murine model system with 15PC3 xenografts (continuous administration for 14 d). At a dose of 1 mg/kg/d, potent and sequence-specific suppression of 15PC3 tumor xenograft growth was observed, whereas the isosequential phosphorothioate-DNA oligonucleotide displayed its optimum activity at a higher dose (5 mg/kg/d).[12] This study shows that LNA-antisense is superior to traditional phosphorothioate-DNA antisense for tumor suppression in this *in vivo* model system.

In a comparative study with other antisense chemistries and siRNA, the ability of LNA-DNA-LNA gapmers to down-regulate VR1 protein has been studied in a transient system employing Cos-7 cells.[13] The IC_{50} value for protein down-regulation by an LNA-DNA-LNA gapmer was approximately 175- and 550-fold lower than the value of the corresponding phosphorothioate-DNA and 2'-O-methyl-RNA gapmer, respectively, and the IC_{50} values for the optimized LNA-DNA-LNA gapmer and the optimized siRNA were 0.4 nM and 0.06 nM, respectively. Thus, LNA/DNA/LNA gapmers are more potent than other types of standard antisense oligonucleotides and are almost as efficient as optimized siRNA.

LNA-ANTISENSE FOR TARGETING OF NONCODING RNA

Targeting of noncoding RNA by so-called steric block antisense oligonucleotides that do not activate RNase H is an attractive approach for inhibition of gene-expression or prevention of interactions between RNA and other molecules. As noncoding RNAs are promising therapeutic targets, the use of LNA-antisense is very appealing in this context by virtue of its unprecedented RNA binding affinity. Notably, very efficient targeting of microRNAs was recently achieved employing LNA oligonucleotides; this study demonstrated antiviral effects of microRNAs in addition to their regulatory functions.[14]

By so-called oligonucleotide-directed misfolding of RNA (ODMiR), LNA was shown to inhibit *Candida albicans* group I intron splicing sequence-specifically.[15] An 8-mer fully modified LNA and a 12-mer LNA/DNA mixmer displayed IC_{50} values of 150 nM and 30 nM, respectively, whereas the corresponding 12-mer DNA and 2′-O-methyl-RNA displayed higher IC_{50} values.

Interaction between the HIV-1 Tat protein and its RNA recognition sequence TAR is an important step in viral replication of HIV-1 virus. In a cellular experiment, an LNA/2′-O-methyl-RNA mixmer displayed dose-dependent and sequence-specific inhibition of HIV-LTR/Tat dependent transcription of firefly luciferase whereas the corresponding 2′-O-methyl-RNA and PNA oligomers were inefficient.[16] This study underlines the importance and biological relevance of apparently minor changes in chemical constitution, here the exchange of five 2′-O-methyl-RNA nucleotides with five LNA nucleotides.

CONCLUSIONS

The key points characterizing LNA-antisense are shown in Figure 17.3. LNA nucleotides are compatible with other commercially available nucleotide monomers, e.g., DNA, phosphorothioate-DNA, RNA, and 2′-O-methyl-RNA monomers.[6] Both LNA oligonucleotides and LNA phosphoramidites are commercially available from several suppliers and, as discussed above, fully modified LNA, mixmer-LNA (steric block approach), and LNA-DNA-LNA gapmers (RNaseH approach) are useful LNA-antisense designs. Recent research conducted by the authors[17] and others[18] has convincingly shown that the strong RNA binding is translated into excellent RNA

- Strong RNA-binding: RNA-target accessibility
- LNA-type gapmers: RNase H activity
- Steric block LNAs: Targeting of non-coding RNA
- Active in cell cultures and *in vivo*
- Low acute toxicity
- Preclinical studies ongoing (Santaris Pharma A/S)

FIGURE 17.3 Characteristics of LNA-antisense.

accessibility, even with highly structured RNA targets. The first preclinical studies with LNA-antisense (LNA-DNA-LNA gapmers) have been successfully completed, and the first clinical trails were to be initiated in summer 2005, aiming at cancer treatment (www.santaris.com).

We are currently focusing our research towards functionalized LNA, in particular N2'-functionalized 2'-amino-LNA (Figure 17.1).[19,20] The high-affinity targeting of RNA that is characteristic also for 2'-amino-LNA oligonucleotides[20,21] enables exploratory oligonucleotide chemistry that likely will lead to novel applications within molecular diagnostics, biotechnology, nanotechnology, and biomedicine.[7,22,23] With these molecular building blocks at hand, the medicinal chemists have unique opportunities for designing the optimized nucleic acid therapeutics of the future.

REFERENCES

1. Achenbach, T.V., Brunner, B., and Heermeier, K., Oligonucleotide-based knockdown technologies: antisense versus RNA interference, *Chembiochem*, 4, 928, 2003.
2. Kurreck, J., Antisense technologies. Improvement through novel chemical modifications, *Eur. J. Biochem.*, 270, 1628, 2003.
3. Koshkin, A.A. et al., LNA (Locked Nucleic Acids): Synthesis of the adenine, cytosine, guanine, 5-methylcytosine, thymine and uracil bicyclonucleoside monomers, oligomerisation, and unprecedented nucleic acid recognition, *Tetrahedron*, 54, 3607, 1998.
4. Koshkin, A.A. et al., LNA (Locked Nucleic Acid): An RNA mimic forming exceedingly stable LNA:LNA duplexes, *J. Am. Chem. Soc.*, 120, 13252, 1998.
5. Obika, S. et al., Stability and structural features of the duplexes containing nucleoside analogues with a fixed N-type conformation, 2'-O,4'-C-methyleneribonucleosides, *Tetrahedron Lett.*, 39, 5401, 1998.
6. Wengel, J., Synthesis of 3'-C- and 4'-C-branched oligodeoxynucleotides and the development of locked nucleic acid (LNA), *Acc. Chem. Res.*, 32, 301, 1999.
7. Petersen, M. and Wengel, J., LNA: a versatile tool for therapeutics and genomics, *Trends Biotechnol.*, 21, 74, 2003.
8. Wahlestedt, C. et al., Potent and nontoxic antisense oligonucleotides containing locked nucleic acids, *Proc. Natl. Acad. Sci. U.S.A.*, 97, 5633, 2000.
9. Braasch, D.A., Liu, Y., and Corey, D.R., Antisense inhibition of gene expression in cells by oligonucleotides incorporating locked nucleic acids: effect of mRNA target sequence and chimera design, *Nucl. Acid. Res.*, 30, 5160, 2002.
10. Frieden, M. et al., Expanding the design horizon of antisense oligonucleotides with alpha-L-LNA, *Nucl. Acid. Res.*, 31, 6365, 2003.
11. Fluiter, K. et al., *In vivo* tumor growth inhibition and biodistribution studies of locked nucleic acid (LNA) antisense oligonucleotides, *Nucl. Acid. Res.*, 31, 953, 2003.
12. Fluiter, K. et al., Tumor genotype-specific growth inhibition *in vivo* by antisense oligonucleotides against a polymorphic site of the large subunit of human RNA polymerase II, *Cancer Res.*, 62, 2024, 2002.
13. Grünweller, A. et al., Comparison of different antisense strategies in mammalian cells using locked nucleic acids, 2'-O-methyl RNA, phosphorothioates and small interfering RNA, *Nucl. Acid. Res.*, 31, 3185, 2003.
14. Lecellier, C.H. et al., A cellular microRNA mediates antiviral defense in human cells, *Science*, 308, 557, 2005.
15. Childs, J.L., Disney, M.D., and Turner, D.H., Oligonucleotide directed misfolding of RNA inhibits *Candida albicans* group I intron splicing, *Proc. Natl. Acad. Sci. U.S.A.*, 99, 11091, 2002.

16. Arzumanov, A. et al., Inhibition of HIV-1 Tat-dependent trans activation by steric block chimeric 2'-O-methyl/LNA oligoribonucleotides, *J. Biochem.*, 40, 14645, 2001.

17. Vester, B. et al., LNAzymes: incorporation of LNA-type monomers into DNAzymes markedly increases RNA cleavage, *J. Am. Chem. Soc.* 124, 13682, 2002.

18. Schubert, S. et al., RNA cleaving '10-23' DNAzymes with enhanced stability and activity, *Nucl. Acid. Res.*, 31, 5982, 2003.

19. Wengel, J., Nucleic acid nanotechnology—towards Angstrom-scale engineering, *Org. Biomol. Chem.*, 2, 277, 2004.

20. Sørensen, M.D., Petersen, M., and Wengel, J., Functionalized LNA (locked nucleic acid): high-affinity hybridization of oligonucleotides containing N-acylated and N-alkylated 2'-amino-LNA monomers, *Chem. Commun.*, 2130, 2003.

21. Singh, S.K., Kumar, R., and Wengel, J., Synthesis of novel bicyclo[2.2.1] ribonucleosides: 2'-amino- and 2'-thio-LNA monomeric nucleosides, *J. Org. Chem.*, 63, 10035, 1998.

22. Jepsen, J.S. and Wengel, J., LNA-antisense rivals siRNA for gene silencing, *Curr. Opin. Drug Disc. Dev.*, 7, 188, 2004.

23. Jepsen, J.S., Sørensen, M.D., and Wengel, J., Downregulation of p21(WAF1/CIP1) and estrogen receptor alpha in MCF-7 cells by antisense oligonucleotides containing locked nucleic acid (LNA), *Oligonucleotides*, 14, 130, 2004.

Part IV

Drug – Xenobiotic Metabolism

18 The Effect of Diet on Drug Metabolism

K.J. Netter

CONTENTS

ABSTRACT

This chapter consists of two parts. First, the different levels of food interaction with drug pharmacokinetics are listed and broadly characterized, mainly by referring to existing review articles. Second, two examples of recent interesting new and somewhat unexpected interactions are described. These are the acute inhibition of hepatic oxidative drug conversion by a protein-rich meal and the enhancement of the CYP3A4-related intestinal hydroxylation of quinidine by a chronic high-salt diet.

INTRODUCTION

The question whether a certain diet may affect the clinical efficiency of drugs has attracted much interest in the last decades. This has increased in parallel with methodology improvements in analytical techniques to measure drugs and their metabolites in biological materials. Therefore, a number of papers and reviews have appeared that describe the effects of food intake, or the lack of it, on the pharmacokinetic characteristics of many drugs. It is understandable that primarily those drugs have been in the focus of this particular interest, which can easily be quantitatively analyzed. Studying the various reviews also reveals that groups of drugs have been in the focus of nutritional considerations simultaneously with their general popularity as new drugs or otherwise important therapeuticals. A similar parallelism can be observed with the progress in the elucidation of the enzymatic mechanisms of drug metabolism. Thus, one can say that studying the literature on nutrition and drug action yields a mirror image of contemporary heuristic achievements.

In principle, food can influence the pharmacokinetic behavior of drugs at several levels of their way through the organism. Food constituents can interact with drugs before reaching the systemic circulation simply by chemical interaction such as complex formation; an example of this is the reduced gastric absorption of tetracycline antibiotics when given with calcium-ion-containing food. Similarly, in intravenous nutrition, nutrients (amino acids, lipids) may interact with simultaneously applied drugs and lead to pharmacokinetic changes.

Food also influences the absorption of drugs from the gastrointestinal tract. There are many observations and some reviews on drugs, the absorption of which is enhanced, decreased, or unaffected by simultaneous eating. It is noteworthy that under these conditions impressive differences in the blood level curves of drugs are observed at the beginning of such experiments, but that long-term observation shows that the respective curves signify identical amounts to have been finally taken up into the systemic circulation; thus, food also delays or accelerates intestinal absorption, but does not change the ultimate amount.

The chemical conversion of drugs begins during the absorption phase by catalysis through drug-metabolizing enzymes present in the gastric and intestinal mucosa. These enzymes can be altered in their activity by food constituents, particularly salt, as will be discussed later.

It is also obvious that the absence of ingested food, i.e., starvation, will have its particular influences, although these are localized more prominently in the liver, where the activities of drug-metabolizing enzymes are changed by the nutritional status. Thus, protein-rich nutrition generally increases the respective activities, whereas starvation decreases the metabolic rate.

An increase in blood flow is an important physiological consequence of food intake. The increased translocation of drugs from the presystemic region into the liver will enhance the hepatic drug concentration and hence the metabolic conversion, particularly of drugs with a marked first-pass effect.

Foodstuffs may contain constituents that directly either inhibit or induce oxidative drug metabolism. Grapefruit juice has attracted much interest because it inhibits the metabolism of coumarins, for example. On the other hand, Brussels sprouts as well as other cruciferous vegetables and charcoal-grilled food contain inducers of cytochromes P450. It should prove interesting to investigate whether reductive enzymes also are influenced *in vivo* by plant materials such as quercetin.

Nutrition up to a certain degree furthermore influences the drug-metabolizing enzymes in the liver as well as in other organs. This concerns the oxidative phase I reactions that are catalyzed by the cytochrome P450-dependent monooxygenases that require molecular oxygen as well as NADPH for activity. It is obvious that there are various areas of interference affecting the enzyme activity as well as the availability of the cofactors. Conjugating enzymes (phase II) are also affected by nutritional factors. After drugs or their metabolites have reached the systemic circulation, they are, for obvious reasons, not subject to nutritional manipulation.

The final elimination of drugs, and hence the termination of their action, occurs mainly via renal excretion; this again can be influenced by nutrition. The amount of fluid intake, and hence urine flow, is a determining factor for the rate of excretion as well as the pH of the urine, which in turn can depend on the prevailing diet to a certain degree; it is easily conceivable that the latter argument applies only to drugs that are excreted as weak acids or bases.

Hormonal actions of plant constituents probably are rare, but there is the example of the aldosterone-like action of liquorice mediated by its constituent glycyrrhetinic acid.

The considerations mentioned previously are the basic principles of a thus far rather amorphous subdiscipline of pharmacology, namely *pharmacodietetics*. It has attracted much interest and produced an enormous literature, yet the generally applicable observations and dogmas have not led to a unified picture. This seems to be due to the biological fact that the affected systems have more than one direction in which to respond. It may well be that this heterogeneity in response may be of great Darwinian advantage, particularly for omnivorous mammals, especially humans.

In view of the multitude of possible food–drug interactions, it will hardly be possible to give a comprehensive account of this phenomenon. Therefore, it is intended to aid the reader by just mentioning a number of respective reviews and then dwell more extensively on the actions of amino acids on drug metabolism in isolated perfused livers and on the action of dietary salt on intestinal cytochrome P450 3A4.

SURVEY OF SURVEYS

It was about 25 years ago that enough material had accumulated to warrant a review of the role of nutrition in drug-metabolizing enzymes.[1] This was followed by a monograph[2] on nutrition and drug interrelations, and related symposium.[3] At the same time, the scope of these interactions was widened to toxicological aspects, in particular, also carcinogenicity[4] The term "pharmacodietetics" was introduced in 1979;[5] this was followed by applying the same terminology to toxic effects.[6] The influence of food on drug absorption has been authoritatively reviewed within the last 20 years.[7,8] A monograph, *Food, Nutrition, and Chemical Toxicity*, in 1993 directed attention to these considerations, which also include the general nutritional status with regard to micronutrients, etc.[9] The general points of interaction between biosynthesis and catalytic action of drug-metabolizing enzymes were pointed out in 1994.[10] More recently, the subject has found its way into advanced textbooks of nutrition.[11] A useful tabular presentation of the nonmetabolic interactions, particularly with respect to drug absorption, is to be found in *Advances in Pharmacology*,[12] which complements a review by Welling.[8] A further review can be found in *Clinical Pharmacokinetics*.[13] Contemporary explanations about the mechanisms of the influence of food on drug metabolism were given by Yang and Yoo in 1988.[14]

Based on these reviews and the knowledge represented in them, we now turn to two specific aspects.

SPECIAL PROBLEMS IN FOOD–DRUG INTERACTION

With the upcoming interest in antihypertensive blockers of adrenergic beta receptors, many pharmacokinetic studies were carried out with these drugs. Thus, in 1977, Melander et al. observed an increase in bioavailability of propanolol and metoprolol by food.[15] This general observation was complemented by the finding that, in particular, a protein-rich (84% protein plus lipid) diet would dramatically change the plasma propanolol time curve after 1 mg/kg of propranolol given orally. Whereas in control fasting human subjects the plasma curve was low and reached a shallow maximum 5 h after oral application, after a protein-rich meal, there was a dramatic initial rise after 1 h, which then reverted and reached the same values as the controls after about 5 h.[16] The authors suspect that the increased hepatic blood flow may be responsible, in conjunction with a possible inhibition of drug-metabolizing activity by the influx of amino acids. There is, however, one uncertainty in the latter suspicion because a 93% carbohydrate meal produces an almost identical rise in propranolol bioavailability. As the propranolol analysis in plasma was carried out with the aid of radioactive labeling, no data about possible changes in the metabolism could be furnished in the 1981 study. Subsequently,[17] an even more dramatic influence of eating a 250-g steak on the elimination of indocyanine green, injected intravenously, was noted; however, the measurement of propranolol steady-state plasma levels during an infusion in 12 human subjects showed a decrease from 46 to 30 ng/mL within 5 min after eating a steak of about 160 g protein (a steak of about 50 g).

Propranolol was measured by high-pressure liquid chromatography. This finding strongly indicates a metabolic interference with the steady-state elimination of propranolol, perhaps in conjunction with increased hepatic blood flow, which according to other authors is estimated to be between 10 and 40% of the original rate.

To further elucidate the situation, another approach was taken, namely, the perfusion of isolated rat livers with mixtures of amino acids.[18] In a single-path nonrecirculating perfusion system, propranolol was offered, after a meal of 40 mm, leading to a steady-state concentration of 12.5 mg/mL; when adding a balanced mixture of amino acids mimicking those concentrations reached in the portal vein after a high-protein meal of again 40 mm, the level of propranolol rises to 16 mg/mL in the effluent.

This strongly indicates that the amino acids in some way inhibit the oxidative and conjugative metabolism of propranolol. Therefore, the authors have measured the main metabolites in the same effluent and found a congruent fall in the main metabolites, namely, 4-hydroxypropranolol and N-desisopropylpropranolol, along with a small decrease in the minor metabolites. This applies also to propranolol conjugates. From these *in vitro* experiments it has become clear that the presence of amino acids in the liver interferes with the metabolic conversion of the test substance. The authors have come to the conclusion that there is *in vitro* as well as *in vivo* evidence that amino acids derived from protein digestion lead to an inhibition of phase I and phase II hepatic metabolism, at least of propranolol. Subsequently, the same research group[19] studied the inhibition of metoprolol metabolism in perfused rat livers by amino acids. Also, with this substance, amino acid infusion very markedly lowers the rate of metabolism to the demethylated and alpha hydroxylated product, whereby the difference becomes less pronounced at high-steady-state metoprolol concentrations in the effluent fluid. This effect is largely reversible. When characterizing amino acid inhibition in microsome suspensions, it was found that the V_M was reduced to about half, whereas the K_M, i.e., the affinity to cytochrome P450, did not change. From ancillary experiments, it became apparent that the cofactors NADPH and oxygen also do not seem to change. Thus, it seems as if amino acids in some way interfere with the function of the reaction cycle for cytochrome P450-catalyzed drug oxidations. It remains to be seen whether this effect is restricted to a single amino acid only or constitutes a general phenomenon. Also, the determination of a threshold concentration would be of interest.

The last aspect to be treated in this chapter concerns the presystemic oxidative metabolism of orally ingested drugs. An illustrative example of food-related modification of this biological process has been published,[20] which concerns not nutrients per se but rather dietary salt, investigating possible effects in the salt water balance on quinidine pharmacokinetics. The authors serendipitously found that dietary salt increases the first-pass elimination of oral quinidine. Comparison in healthy volunteers of the quinidine plasma concentrations at a low-salt (10 mEq/d) and a high-salt (400 mEq/d) diet shows that the high-salt diet leads to definitely lower blood levels of quinidine for the first 4 h after intake of 600 mg quinidine sulfate immediate-release tablets with 100 mL water. The diets were maintained for one week prior to quinidine application.

From other sources,[21] it was known that semisynthetic diets markedly reduced the cytochrome P450-dependent coumarin-7-ethoxy-O-deethylase activity in human volunteers. This was shown in serial intestinal biopsies taken from these volunteers on various forms of diet. It is also known that CYP3A4 is the predominant cytochrome in the intestinal mucosa.[22] When 300 mg quinidine is given intravenously, there is no difference in blood level pharmacokinetics between both diets. From this it seems obvious that a high-salt diet affects the oxidative metabolic transformation in the intestinal wall; this conclusion is corroborated by the finding of higher 3-hydroxyquinidine excretion in urine under high-salt conditions. Although the described phenomenon certainly needs further elucidation, the effect of diets on intestinal drug-metabolizing systems could be of heuristic interest.

CONCLUSIONS

The influence of diet on drug pharmacokinetics is a multifaceted phenomenon, which has attracted the interest of many investigators and has led to many, though often diverse, results. The attempt has been made to crudely classify the levels on which diets may interfere with drug action.

Two phenomena have been highlighted in detail: the effect of high-protein diets on the oxidative hepatic metabolism of beta blockers and the enhancing effect of a high-salt diet on the intestinal oxidative conversion of quinidine. It remains to be seen whether these findings represent general phenomena, which also apply to other drugs and chemical substances.

REFERENCES

1. Campbell, T.C. and Hayes, J.R., Role of nutrition in the drug metabolizing enzyme system, *Pharmacol. Rev.*, 26, 171, 1974.
2. Hathcock, J.N. and Coon, J., Eds., *Nutrition and Drug Interactions*, Academic Press, New York, 1978.
3. Hathcock, J.N., Ed., Symposium: metabolic interactions of nutrition and drugs, *Fed. Proc.*, 44, 123, 1985.

4. Parke, D.V. and Ioannides, C., The role of nutrition in toxicology, *Annu. Rev. Nutr.*, 1, 207, 1981.

5. Netter, K.J., Pharmacodietetics: perspectives in the nutritional manipulation of toxicity or drug effects, in *Drug-Action Modifications, Comparative Pharmacology*, Olive, G., Ed., Pergamon Press, Oxford, 1979, pp. 65–68.

6. Netter, K.J., Toxicodietetics: dietary alterations of toxic action, in *New Concepts and Developments in Toxicology*, Chambers, P.L., Gehring, P., and Sakai, F., Eds., Elsevier, Amsterdam, 1986, pp. 139–144.

7. Welling, P.G., Influence of food and diet on gastrointestinal drug absorption: a review, *J. Pharmacokinet. Biopharm.*, 5, 291, 1977.

8. Welling, P.G., Effects of food on drug absorption, *Annu. Rev. Nutr.*, 16, 383, 1996.

9. Netter, K.J., Nutrition and chemical toxicity, in *Food, Nutrition, and Chemical Toxicity*, Parke, D.V., Ioannides, C., and Walker, R., Eds., Smith-Gordon, London, 1993, pp. 17–26.

10. Netter, K.J., The role of nutrients in detoxification mechanisms, in *Nutritional Toxicology*, Kotsonis, F.N., Mackey, M., and Hjelle, J.J., Eds., Raven Press, New York, 1994, pp. 1–18.

11. Knapp, F.I.R., Nutrient–drug interactions, in *Present Knowledge in Nutrition*, 7th ed., Ziegler, E.E. and Filer, U., Jr., Eds., ILSI Press, Washington, 1996, pp. 540–546.

12. Tschanz, C., Stargel, W.W., and Thomas, J.A., Interactions between drugs and nutrients, in *Advances in Pharmacology*, Vol. 35, August, J.T., Anders, I.V.I.W., Murad, F., and Coyle, J.T., Eds., Academic Press, San Diego, CA, 1996, pp. 1–26.

13. Walter-Sack, I. and Klotz, U., Influence of diet and nutritional status on drug metabolism, *Clin. Pharmacokinet.*, 31, 47, 1996.

14. Yang, C.S. and Yoo, J.S.H., Dietary effects on drug metabolism by the mixed function oxidase system, *Pharmacol. Ther.*, 38, 53, 1988.

15. Melander, A. et al., Enhancement of the bioavailability of propranolol and metoprolol by food, *Clin. Pharmacol. Ther.*, 22, 108, 1977.

16. McLean, A.J. et al., Reduction of first-pass hepatic clearance of propranolol by food, *Clin. Pharmacol. Ther.*, 30, 31, 1981.

17. Feely, J., Nadeau, J., and Wood, A.J.J., Effects of feeding on the systemic clearance of indocyanine green and propranolol blood concentrations and plasma binding, *Br. J. Clin. Pharmacol.*, 15, 383, 1983.

18. Semple, H.A. and Xia, F., Interaction between propranolol and amino acids in the single-pass isolated perfused rat liver, *Drug Metab. Dispos.*, 23, 794, 1995.

19. Wang, B. and Semple, H.A., Inhibition of metoprolol metabolism by amino acids in perfused rat livers. Insights into the food effect? *Drug Metab. Dispos.*, 25, 287, 1997.

20. Darbar, D. et al., Dietary salt increases first-pass elimination of oral quinidine, *Clin. Pharmacol. Ther.*, 61, 292, 1997.

21. Hoensch, H.P. et al., Effects of semisynthetic diets on xenobiotic metabolizing enzyme activity and morphology of small intestinal mucosa in humans, *Gastroenterology*, 86, 1519, 1984.

22. de Waziers, I. et al., Cytochrome P450 isoenzymes, epoxide hydolase and glutathione transferases in rat and human hepatic and extrahepatic tissues, *J. Pharmacol. Exp. Ther.*, 253, 387, 1990.

19 Cytochromes P450 in the Metabolism and Bioactivation of Chemicals

Costas Ioannides

CONTENTS

ABSTRACT

To survive and thrive in the chemical environment in which it exists, the living organism has developed a number of versatile enzyme systems that facilitate the complete elimination of xenobiotics to which it is inescapably exposed by converting them metabolically to hydrophilic, readily excretable, metabolites. Cytochromes P450, a superfamily of enzymes, are undoubtedly the most prolific enzyme system in xenobiotic metabolism, capable of catalyzing efficiently the metabolism of structurally very diverse compounds. Paradoxically, however, such metabolism may also lead to the generation of reactive intermediates that interact covalently with cellular macromolecules, giving rise to various forms of toxicity, including carcinogenicity, a process referred to as *bioactivation*. As a defense mechanism, the body can deactivate these enzymically, so that production of reactive intermediates, and hence toxicity, is dependent on the balance of bioactivation/deactivation. Any factor that disturbs this balance will profoundly influence the appearance and severity of toxicity.

INTRODUCTION

The human body is continuously and inescapably exposed to a wide diversity of chemicals, both manmade and naturally occurring, frequently referred to as *xenobiotics* (Gr., foreign to life). Although most of the emphasis, at least regarding safety, is focused primarily on anthropogenic chemicals, more that 99% of human exposure is to naturally occurring chemicals,[1] which can be as biologically active as their synthetic counterparts. Indeed, what has recently sparked interest in naturally occurring chemicals, largely phytochemicals, was the realization that these have a major role to play in affording protection against major high-morbidity diseases in humans such as cancer and cardiovascular disease. Chemicals have been identified in food that, at least in animal models, can effectively antagonize chemical carcinogens. Such anticarcinogenic chemicals are polyphenols,[2] isothiocyanates,[3] organosulfates,[4] and many others.

Although some chemicals are taken voluntarily, such as drugs and food additives, the vast majority are taken involuntarily—for example, through the diet, undoubtedly the principal source of human exposure to chemicals. Such chemicals may be inherent in the diet, may be contaminants such as pesticides or microbial toxins, or may be generated during domestic cooking conditions. Exposure to chemicals is, therefore, continual and inevitable.

The body cannot exploit these xenobiotics beneficially either to generate energy, build new tissue, or serve as chemical messengers, and, because they cannot in any way assist in the normal physiological function of the body, they are recognized as foreign and potentially deleterious; the body's immediate reaction is to eliminate them. Chemicals that find their way into the body and are distributed into the various tissues, by necessity, have a lipophilic character so as to traverse lipid membranes. However, the excretory mechanisms of the body are unable to eliminate efficiently lipophilic compounds so that these must be rendered hydrophilic to ensure their elimination. To attain this objective, the body has developed an array of nonspecific enzyme systems that can metabolically convert lipophilic compounds to hydrophilic, readily excretable metabolites.[5] Moreover, in most cases, such metabolism terminates the biological activity of chemicals because the metabolites, in contrast to the parent compounds, are unable to reach, and interact with, the appropriate receptors to elicit a biological effect; thus, metabolism has a very profound effect on the biological activity of a chemical. It is unlikely that the human body could withstand the constant onslaught of the chemicals to which it is unavoidably exposed if these enzymes did not evolve to ensure their rapid and complete elimination.

METABOLISM OF XENOBIOTICS

The process of transforming lipophilic chemicals to polar entities occurs in two distinct metabolic phases.[6] During phase I, functionalization, an atom of oxygen is

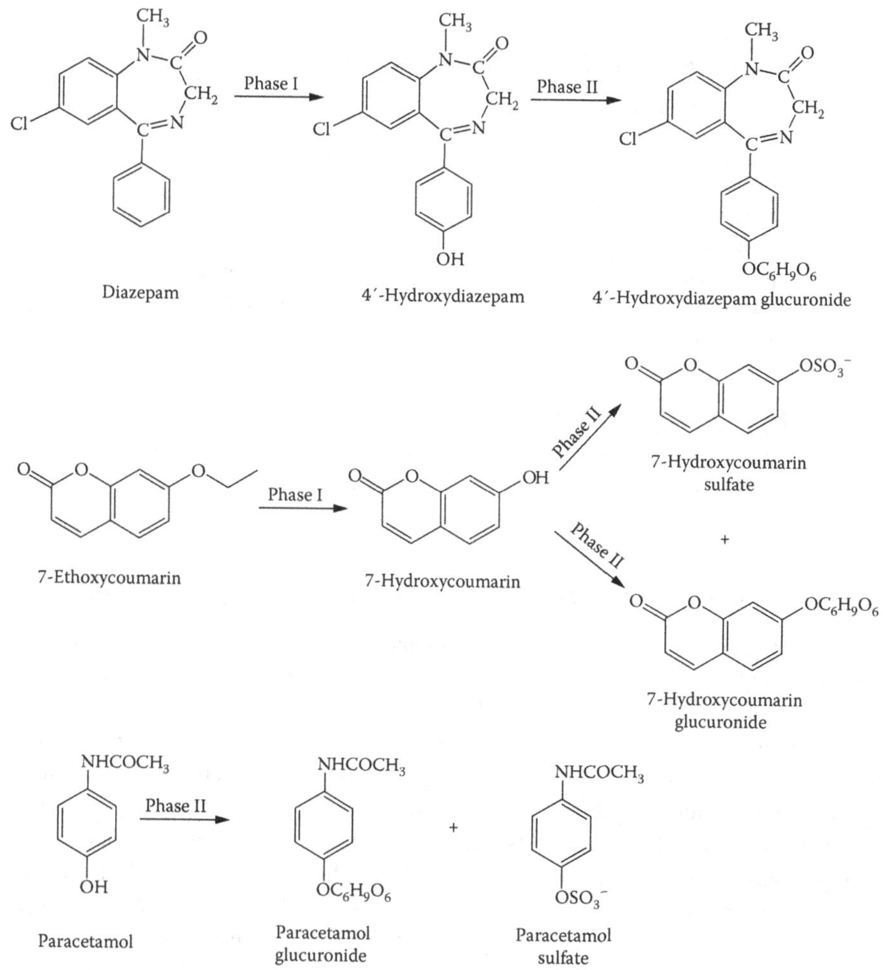

FIGURE 19.1   Phase I and phase II metabolism of xenobiotics.

incorporated into the chemical, and functional groups such as -OH, -COOH, etc. are generated, as in the hydroxylation of diazepam, in which a phenolic group is generated at the 4′-position (Figure 19.1). Alternatively, such a functional group may be unmasked, as in the deethylation of 7-ethoxycoumarin to 7-hydroxycoumarin. Such metabolites are not only more polar than the parent compound but, furthermore, are capable of undergoing phase II metabolism, namely, conjugation, in which endogenous substrates, e.g., glucuronic acid and sulfate, are introduced to form highly hydrophilic molecules, thus ensuring their elimination (Figure 19.1). Xenobiotics that already possess a functional group can bypass phase I metabolism and directly participate in conjugation reactions; the mild analgesic paracetamol (acetaminophen), because of the presence of a phenolic group, is readily metabolized by sulfate and glucuronide conjugation, and these constitute its principal pathways of metabolism (Figure 19.1).

TABLE 19.1

Enzyme Systems that Metabolize Xenobiotics

Enzyme System	Principal Localization Site
Cytochrome P450	Endoplasmic reticulum
FAD-monooxygenase	Endoplasmic reticulum
Monoamine oxidase	Mitochondria
Alcohol/aldehyde dehydrogenase	Cytosol
Epoxide hydrolase	Endoplasmic reticulum
Glucuronosyl transferase	Endoplasmic reticulum
Glutathione S-transferase	Cytosol
Sulfotransferase	Cytosol
Acetyltransferase	Cytosol
Methyltransferase	Cytosol
Oxidoreductase	Cytosol
Xanthine oxidase	Cytosol

CYTOCHROMES P450 IN THE METABOLISM OF XENOBIOTICS

A number of enzyme systems capable of metabolizing xenobiotics have been identified, primarily localized in the endoplasmic reticulum and cytosolic fraction of the
cell, and extensively investigated[5] (Table 19.1). Undoubtedly, by far the most important enzyme system in the phase I metabolism of xenobiotics is the cytochrome
P450-dependent mixed-function oxidases, a ubiquitous system of heme-thiolate
enzymes encountered in almost every human organ, but with the highest concentration in the liver, which consequently functions as the center of xenobiotic metabolism,
being capable of catalyzing both oxidation and reduction pathways.[7] Almost every
lipophilic chemical that finds its way into the body is, at least to some extent, subject
to metabolism catalyzed by this enzyme system. The cytochrome P450-dependent
mixed-function oxidase system comprises an electron transport chain consisting of
the flavoprotein cytochrome P450 reductase and the cytochrome P450 that functions
as a terminal oxidase, and it catalyzes the incorporation of one atom of molecular
oxygen to the substrate (RH) while the second atom forms water.

$$RH + NADPH + H^+ + O_2 \rightarrow ROH + NADP^+ + H_2O$$

The cytochrome P450 system can catalyze a wide variety of reactions, e.g., aliphatic
and aromatic hydroxylations, dealkylations, and even reductions when oxygen ten-
sion is low, which explains its predominant role in the metabolism of xenobiotics.

A principal attribute of the cytochrome P450 system is the unprecedented broad
substrate specificity it displays; it catalyzes efficiently the metabolism of chemicals
of immense structural diversity.[7] Substrates include small-molecular-weight compounds such as the solvent methanol (MW = 42) as well as large-molecular-weight
compounds such as the immunosuppressant cyclosporin (MW = 1203); it affects the
metabolism of planar molecules such as benzo[a]pyrene, a carcinogenic polycyclic

aromatic hydrocarbon, as well as of globular molecules such as the anticonvulsant phenobarbitone. Naturally, no single enzyme can accommodate in its substrate-binding site such a diversity of substrates, and consequently it is not surprising that cytochrome P450 is not a single enzyme, but a superfamily comprising a number of enzymes, each with different, albeit overlapping, substrate specificity.[8,9] The cytochrome P450 superfamily is divided into a number of families, which in turn are subdivided into subfamilies, each of which may consist of one or more enzymes (isoforms). Classification of cytochrome P450 enzymes within families and subfamilies is carried out strictly on the basis of primary sequence homology, with no consideration of their substrate specificity. Cytochrome P450 enzymes belonging to the same family share at least a 40% structural homology whereas homology between enzymes within a subfamily is at least 55%. For example, CYP3A4 enzyme denotes a cytochrome P450 protein belonging to family 3, subfamily A and is protein 4. The families responsible for the metabolism of xenobiotics are CYP1 to CYP3 and, to a much lesser extent, CYP4, whereas cytochrome P450 enzymes belonging to other families are concerned with the metabolism of endogenous substrates. Indeed, the function of cytochrome P450 enzymes is not confined to the metabolism of xenobiotics, but they play a crucial role in the metabolism, both biosynthesis and degradation, of vital endogenous chemicals, including hormones, such as steroid hormones and melatonin. Cytochromes P450 are believed to have evolved from a common ancestor many millions of years ago. The function of the earliest forms is considered to have been in the metabolism of essential endogenous chemicals, such as steroid hormones, then evolved to enzymes capable of facilitating the elimination of xenobiotics. As a result, the cytochrome P450 proteins lost their narrow substrate specificity toward steroids and, in order to cope with the new, increasingly chemical environment, they developed into broad-specificity enzymes that could metabolize xenobiotics, thus facilitating their elimination. It has been proposed that what forced the evolution of these proteins is the necessity to develop defense mechanisms to protect against plant toxins present in the food chain that were produced in order to discourage predators.[10] Generally there is little overlap between the cytochrome P450 enzymes that metabolize endogenous and exogenous compounds.

BIOCHEMICAL ACTIVATION OF CHEMICALS

In the 1970s, it became apparent that cytochromes P450 could metabolize chemicals to form highly reactive, electrophilic metabolites that could readily interact irreversibly with vital cellular macromolecules with deleterious consequences, and thus had a role in the etiology of chemically induced disease.[10] Therefore, the initial view that the function of cytochromes P450 was exclusively in the deactivation and elimination of chemicals was clearly incorrect. Relatively few chemicals have inherent reactivity to interact covalently with cellular macromolecules but may acquire it through metabolism. The process through which inert chemicals are biotransformed to reactive intermediates capable of causing cellular damage is referred to as *metabolic activation* or *bioactivation*.[11,12] For these chemicals, toxicity/carcinogenicity is inextricably linked to their metabolism. As these reactive species are generated intracellularly, they can readily and irreversibly interact with vital cellular macromolecules,

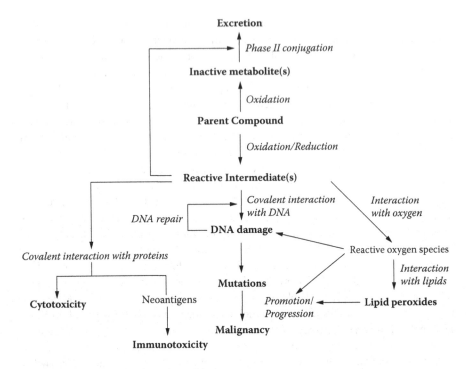

FIGURE 19.2 Bioactivation of xenobiotics.

such as DNA, RNA, and proteins, to provoke various types of toxicity; thus, in this case, metabolism confers on the chemical adverse biological activity (Figure 19.2). The generated reactive intermediates may interact with DNA to form adducts that, if they escape the repair mechanisms of the cell, may be fixed and passed to the progeny, thus giving rise to a mutation. Reactive intermediates of chemicals may also induce DNA damage through an alternative mechanism that involves interaction with molecular oxygen to generate reactive oxygen species that can cause cellular damage similar to that resulting from the covalent interaction of the reactive species of chemicals with cellular components; they oxidize (1) DNA to induce mutations, (2) lipids to form lipid peroxides, which appear to play an important role in the promotion and progression stages of chemical carcinogenesis, and also (3) proteins.[13] The reactive intermediates of chemicals may also interact covalently with proteins, disturbing physiological homeostasis, leading to cell death. Alternatively, they may function as haptens, conferring on proteins antigenic potential and eliciting immunotoxicity.[13] Drugs such as tienilic acid, dihydralazine, and halothane are metabolically converted to metabolites that bind covalently to proteins to generate neoantigens resulting in the production of autoantibodies. Subsequent exposure to these drugs may provoke an autoimmune response leading to hepatitis. An example of a cytotoxic agent is the analgesic drug paracetamol, which, in addition to the conjugation reactions already described (see previous text), undergoes oxidation to form a benzoquinoneimine capable of interacting with the −SH groups of proteins, thus constituting an essential step of the mechanism leading to its hepatotoxicity

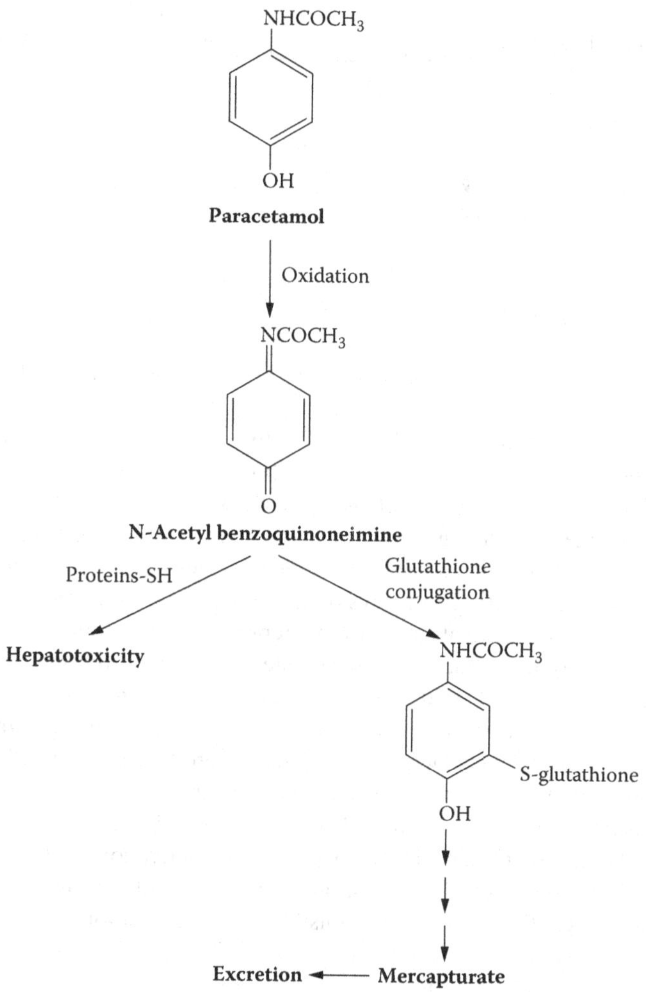

NHCOCH$_3$

OH

Paracetamol

Oxidation

NCOCH$_3$

O

N-Acetyl benzoquinoneimine

Proteins-SH

Glutathione
conjugation

Hepatotoxicity

NHCOCH$_3$

S-glutathione

OH

Excretion ◄——— Mercapturate

FIGURE 19.3   Bioactivation of paracetamol.

(Figure 19.3). The antimalarial drug amodiaquine, a 4-aminoquinoline, is also oxidized to a quinoneimine, which in this case functions as a hapten; it interacts with
proteins to produce neoantigens that can provoke an immune response.

BALANCE OF ACTIVATION AND DEACTIVATION

The body has a number of defensive mechanisms to protect itself against the production of reactive intermediates (Figure 19.4). First, it can direct the metabolism
through pathways that generate inactive, readily excretable metabolites. Alternatively, it can detoxicate the reactive intermediates, for example, neutralizing them
as a result of conjugation with the nucleophilic tripeptide glutathione, which is characterized by high intracellular concentration, catalyzed by the cytosolic glutathione

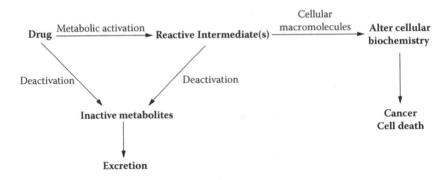

FIGURE 19.4   Balance of activation/deactivation in the metabolism of xenobiotics.

S-transferases.[14] Thus, a chemical is subject to metabolism through a number of pathways, the majority of which will bring about its deactivation and facilitate its excretion. However, some routes of metabolism will transform the chemical to a metabolite capable of inducing toxicity. Clearly, the amount of reactive intermediate produced, and hence the incidence and degree of toxicity, will be largely dependent on the rates of the competing pathways of activation and deactivation.

In most cases, the activation pathways represent minor routes of metabolism, so that the generation of reactive intermediates is minimal, the low levels formed are effectively deactivated by the defensive mechanisms, and no toxicity is apparent. The rates of activation and deactivation, however, vary among animal species, so that if an animal species lacks the enzyme system that catalyzes the activation pathway, then it will be resistant to its toxicity and carcinogenicity. For example, the guinea pig is unable to carry out the N-hydroxylation of the carcinogen 2-acetylaminofluorene, the initial and rate-limiting step in its bioactivation, and is consequently resistant to its carcinogenicity.[15] Similarly, the cynomolgus monkey lacks the enzyme, namely CYP1A2, that is responsible for the activation of the heterocyclic amine MeIQx (2-amino-3,8-dimethylimidazo[4,5-*f*]quinoxaline), a food carcinogen, and is consequently refractive to its carcinogenicity,[16] whereas rodents such as mice and rats, which catalyze the activation of this carcinogen, are susceptible. However, in certain situations, the activation pathways may assume a greater role, leading to enhanced production of reactive intermediates, overwhelming the deactivation pathways, resulting in their accumulation in the body, thus increasing the likelihood of an interaction with cellular components with ensuing toxicity.

Whatever alters this balance of activation/deactivation will have a very profound effect on the toxicity of a compound. For example, if the activity of a cytochrome P450 enzyme that functions as the major catalyst of the bioactivation of a chemical is elevated, then increased production of reactive intermediates will lead to enhanced toxicity. An established clinical fact is that chronic alcoholics are susceptible to the hepatotoxicity of paracetamol; these people display symptoms of liver disease even when they consume therapeutic doses of the drug.[17,18] As already discussed (see previous text), paracetamol is largely metabolized by conjugation with sulfate and glucuronic acid, and these are the principal pathways of its elimination (Figure 19.1). To a minor extent, it undergoes cytochrome P450-catalyzed oxidation to generate the quinoneimine, an electrophile that can interact with hepatic proteins to precipitate hepatotoxicity (Figure 19.3). However, the body protects itself by neutralizing this metabolite through conjugation with glutathione; the glutathione conjugate is further processed and is eventually excreted in the urine and bile as mercapturate. One of the cytochrome P450 proteins that catalyzes the oxidation of paracetamol to quinoneimine is CYP2E1, an enzyme inducible by exposure to alcohol. For this reason, alcoholics have higher CYP2E1 activity, and as a result convert more of paracetamol to quinoneimine, leading to hepatotoxicity, which explains their sensitivity to paracetamol. A similar situation of elevated levels of reactive intermediates can arise when the deactivation process is impaired. Bromobenzene is a toxic and carcinogenic compound; it is metabolized by cytochromes P450 to form a number of epoxides that are responsible for its toxicity. The epoxides may rearrange to bromophenols or be deactivated by hydration, catalyzed by epoxide hydrolase, by conjugation with glutathione, catalyzed by the glutathione S-transferases. However, in situations in which the levels of glutathione are limiting, the toxicity of bromobenzene rises dramatically. When the toxicity of bromobenzene was evaluated in fed rats and in rats that had been starved for two days, so that glutathione levels were very low, compromising the deactivation pathway, toxicity was very low in the well-nourished animal but dramatically higher in the starved animals.[19]

BIOACTIVATION OF CHEMICAL CARCINOGENS

It is in the etiology of chemically induced cancer, however, that cytochrome P450 proteins appear to play a pivotal role. The vast majority of chemical carcinogens require bioactivation to manifest their carcinogenicity, and this is largely catalyzed by the cytochromes P450, although other enzyme systems may also contribute to the activation process. Bioactivation of chemical carcinogens can take place in almost every tissue, but by far the most active is the liver, because it contains high levels of most xenobiotic-metabolising systems, including cytochromes P450. In contrast, extrahepatic tissues are characterized by a more restricted number of enzymes, generally present at low levels. Consequently, it would be logical to expect that the liver, being the primary site of chemical activation, would be the major target of toxic and carcinogenic manifestations. In reality, however, tissues such as breast, lung, and colon, which have minimal ability to activate chemicals by oxidation,[20] are far more frequent sites of tumorigenesis, despite their limited metabolic competence. It is conceivable that the liver functions as the principal manufacture center of reactive intermediates, which are then exported to other tissues, where they may provoke adverse effects. Such a transport mechanism has been invoked to explain the bladder carcinogenicity of aromatic amines (Figure 19.5). The amine is N-hydroxylated in the liver either by cytochromes P450 or the flavin monooxygenase system; the hydroxylamine is trapped as *N*-glucuronide and is transported in this form to the bladder. Under the acidic pH conditions that prevail in the bladder, the glucuronide breaks down to release the free hydroxylamine, which is then transformed to the nitrenium ion, the entity that interacts with DNA, the ultimate carcinogen.[21] So in this case, the reactive intermediate is generated in the liver and exported to the bladder. The nitrenium ion can also be formed *in situ* because the bladder has low levels

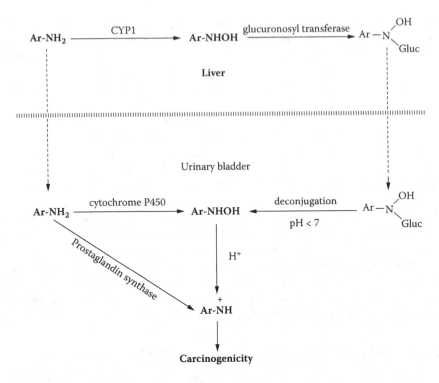

FIGURE 19.5 Transfer of reactive intermediates between tissues.

of cytochrome P450 activity and substantial levels of prostaglandin synthetase activity that may contribute to the production of the nitrenium ion. The prostaglandin synthetase enzyme system has the potential to make a substantial contribution to the bioactivation of aromatic amines in extrahepatic tissues.[22,23] Similarly, the breast is a frequent site of tumorigenesis despite the fact that its capacity to bioactivate chemicals is poor.[20] It has been proposed that heterocyclic carcinogenic amines occurring in food, such as PhIP (2-amino-1-methyl-6-phenylimidazo[4,5-b]pyridine), are N-oxidized in the liver by cytochromes P450 and, in the form of either the hydroxylamine or following esterification to the acetoxyester, are then transported to extrahepatic tissues such as the colon to yield DNA adducts.[24] Extrahepatic tissues appear capable of catalyzing the esterification of hydroxylamines, another important step in their activation process but poor in catalyzing the initial oxidation of the heterocyclic amine as a result of the very low levels of cytochrome P450.[25]

ROLE OF CYTOCHROME P450 PROTEINS IN THE BIOACTIVATION OF CHEMICAL CARCINOGENS

Most xenobiotic-metabolizing cytochrome P450 enzymes can contribute to the bioactivation process of chemicals, but the most closely linked to chemical carcinogenesis are the CYP1 family and the CYP2E subfamily.

TABLE 19.2
Chemicals Activated by the CYP1 Family

Class	Example
Polycyclic aromatic hydrocarbons	Benzo[a]pyrene
Aromatic amines	4-Aminobiphenyl
Heterocyclic amines	2-Amino-3-methylimidazo-4,5-[4,5-f]quinoline
Aromatic amides	2-Acetylaminofluorene
Azocompounds	Dimethylaminoazobenzene
Polyhalogenated biphenyls	3,4,3',4'-Tetrachlorobiphenyl
Nitrosamines	4-(Methylnitrosamino)-1-(3-pyridyl)-1-butanone (NNK)
Mycotoxins	Aflatoxin B_1
Certain drugs	Paracetamol
Steroids	Estradiol
Synthetic steroids	Diethylstilboestrol
Furans	Ipomeanol

The CYP1 family consists of two subfamilies, 1A and 1B; the former comprises two proteins, 1A1 and 1A2, whereas only a single protein belonging to the latter, 1B1, has thus far been identified. CYP1A2 is almost exclusively a hepatic protein and appears not to be expressed at the protein level in most extrahepatic tissues, whereas CYP1A1 is largely expressed in extrahepatic tissues and is readily inducible in the liver.[26] The CYP1B1 isoform is distributed in many tissues, and high constitutive expression was noted in the adrenals. CYP1 is probably the most conserved family within the phylogenetic tree so that the human proteins share extensive structural similarity and display similar substrate specificity to the orthologous rodent proteins, alluding to an important role for this family in a vital biological function.[27]

The CYP1 family, originally known as cytochrome P448, was one of the first to be identified, and consequently it has been extensively studied, so that its preferred substrates have been well defined[28] (Table 19.2). The substrates of the CYP1 family are essentially lipophilic planar compounds characterized by a small depth and a large area/depth^2 ratio, an index of planarity.[29,30] It plays a limited role in the metabolism of most drugs,[31] but it is the principal catalyst of drugs like theophylline, caffeine, and phenacetin, and participates significantly in the metabolism of drugs such as propranolol, imipramine, and dantrolene. In contrast, it is the most important family in the bioactivation of chemical carcinogens, including many major and ubiquitous classes of relevance to humans, such as polycyclic aromatic hydrocarbons, heterocyclic and aromatic amines, and mycotoxins.[28,31,32] The CYP1 family is believed to be responsible for the activation of more than 90% of the known carcinogenic chemicals.[33] Most ubiquitous environmental and dietary carcinogens to which humans are frequently exposed are planar in nature, and therefore favored substrates of the CYP1 family. The CYP1 family through arene oxidation and N-oxidation activates polycyclic aromatic hydrocarbons, aromatic amines, aflatoxins, and heterocyclic amines. Generally, CYP1A1 is more effective in catalyzing arene oxidation whereas CYP1A2 is more active in N-oxidation.[28,34] The latter isoform has also been shown to play a prominent role in the activation of the mycotoxin aflatoxin B_1 through arene oxidation.[35] Recent studies have shown that CYP1B1 is also effective in catalyzing the metabolism and activation of polycyclic aromatic hydrocarbons such as benzo(a)pyrene, but less effectively compared with CYP1A1; it is also capable of activating aromatic amines.[32,36]

CYP2E is also an important contributor to the bioactivation of chemical carcinogens, and in most animal species comprises a single enzyme, 2E1. In common with CYP1, CYP2E1 is one of the most conserved proteins in animal species. It is very active in the metabolism of low-molecular-weight compounds, and its substrates are characterized by a small molecular diameter.[37] Its contribution to the metabolism of drugs is modest and is restricted to anesthetics such as enflurane and other drugs such as isoniazid and chlorzoxazone. CYP2E plays a dominant role in the metabolism of small carcinogenic compounds such as azoxymethane, benzene, 1,3-butadiene, nitrosamines such as dimethylnitrosamine and nitrosopyrrolidine, and halogenated hydrocarbons such as carbon tetrachloride and vinyl chloride.[38] A major and toxicologically important characteristic of the CYP2E subfamily is its high propensity to generate reactive oxygen species.[39] In the rat, CYP2E1 was the most active in comparison with other isoforms in producing hydrogen peroxide.[40] CYP2E may therefore facilitate carcinogenesis by two distinct mechanisms, namely, the oxidative activation of chemicals and the generation of reactive oxygen species.

REGIOSELECTIVITY IN THE METABOLISM OF CHEMICAL CARCINOGENS BY CYTOCHROME P450

Although a number of cytochrome P450 proteins can metabolize the same compound, they may display regioselectivity in that they metabolize it at different sites, so that some may direct its metabolism toward the formation of reactive intermediates and thus facilitate its bioactivation, whereas others produce inactive, readily excretable metabolites and promote its deactivation and elimination. For example, the human bladder carcinogen β-naphthylamine is metabolized selectively by CYP1A2 through N-hydroxylation, an activation step, whereas many isoforms catalyze its ring-oxidation at 1- and/or 6-positions, both being deactivation reactions, as these phenols are conjugated and excreted[41] (Figure 19.6). So, clearly, for a chemical to display genotoxicity in a living organism, at least two prerequisites must be fulfilled:

The chemical must be or must have the potential to be metabolically converted to a reactive intermediate capable of inducing DNA damage.
At the time of exposure, the living organism must possess the necessary enzymic apparatus to catalyze the activation pathways.

If the enzymic apparatus is not available to activate a compound, then a toxic manifestation will not be observed. This is one of the reasons that isomers, or structurally closely related compounds, display so markedly different carcinogenic activity. 4-Aminobiphenyl is an established human carcinogen, inducing tumors in a number of tissues, particularly the urinary bladder. In contrast, its 2-isomer is noncarcinogenic. The first and rate-limiting step in the hydroxylation of 4-aminobiphenyl is an

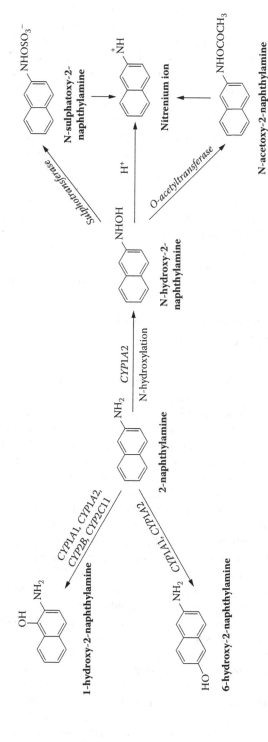

FIGURE 19.6 Metabolism of 2-naphthylamine.

N-hydroxylation that is catalyzed by the cytochrome P450 system. The hydroxyl-amine conjugates with sulfate, and the generated sulfatoxy ester breaks down spontaneously to form the nitrenium ion, the ultimate carcinogen. Of the two isomers only 4-aminobiphenyl elicits a mutagenic response in the presence of an activation system, implying that the hydroxylamine of 2-aminobiphenyl is either not formed or is not genotoxic.[42] The hydroxylamines of both isomers displayed direct genotoxicity in the Ames test, indicating that both hydroxylamines are capable of inducing DNA damage. This would suggest that the lack of carcinogenicity of 2-aminobiphenyl is due to the fact that it is not metabolically converted to the hydroxylamine.

Thus far, only a single isoform has been shown to catalyze the N-hydroxylation of 4-aminobiphenyl: CYP1A2. Although CYP1A2 was capable of catalyzing the N-hydroxylation of 4-aminobiphenyl, it is apparently not capable of catalyzing the N-hydroxylation of the 2-isomer. This is most likely due to the presence of the amino group at the 2-position, *ortho* to the ring, causing distortion and forming a dihedral angle of some 40°. The planarity index of 4-aminobiphenyl is 7.3, but this drops to 4.1 in the 2-isomer.[42] This loss of planarity provides a rational explanation as to why 2-aminobiphenyl is a poor substrate of the CYP1 family, in contrast to the 4-isomer. As it cannot be N-hydroxylated, the 2-isomer lacks carcinogenic activity.

CYTOCHROME P450 INDUCTION AND CHEMICAL CARCINOGENESIS

CYP1 is not a major family in the liver of animals and humans, comprising less than 5% of the cytochrome P450 content in the liver of rats and about 10% in the human liver. Under such conditions, activation would be expected to be a minor route of metabolism, and the low levels of reactive intermediates formed effectively detoxicated by other pathways such as conjugation with glutathione. When the genotoxic potential of carcinogenic chemicals is evaluated, as for example in the Ames test, when the activation is entrusted to hepatic preparations from untreated rats, many well-established and potent carcinogens either fail to register a mutagenic response or show only a weak one. If, however, the activation system has been derived from rats that have been treated with Aroclor 1254, a procedure that markedly increases the level of the CYP1 family in the liver from less than 5% to over 50%, then a clear mutagenic response is observed with all these carcinogens. Clearly, the control animal lacks the enzymic apparatus to significantly activate the carcinogen. However, it is probably the most inducible cytochrome P450 family.[34] In fact, a characteristic of many carcinogens that are activated by the CYP1 family is that, on repeated administration, they can markedly and selectively induce this enzyme, so that activation becomes a more predominant pathway. The consequent increased generation of reactive intermediates overwhelms the detoxication pathways, and the higher availability of these intermediates increases the likelihood of a DNA interaction.

It is relevant to point out that in animal studies to evaluate the carcinogenic potential of chemicals, not only are these administered at high doses but, moreover, are given daily. Such a treatment, after a single or just a few doses of the chemical, may lead to selective induction of the bioactivating enzymes, thereby exaggerating the production of reactive intermediates and markedly increasing the likelihood of a positive carcinogenic effect. Consequently, the ability of a chemical to induce its own activation may be an important factor in determining its carcinogenic activity.[43]

To illustrate this concept, 2-acetylaminofluorene will be used as an example. 2-Acetylaminofluorene is an established hepatocarcinogen whereas its 4-isomer is devoid of such activity. The activation of 2-acetylaminofluorene proceeds, as in the case of 4-aminobiphenyl (see previous text), through N-hydroxylation, the rate-limiting step in the bioactivation pathway, leading eventually to the formation of the nitrenium ion. The hydroxylation of the 2-isomer is catalyzed by the CYP1A family, and one would expect the same family to catalyze that of the 4-isomer, because both compounds are essentially planar. Two mechanisms may account for their difference in carcinogenic potential:

- Higher rate of generation and/or genotoxic potential of the nitrenium ion of 2-acetylaminofluorene compared with the 4-isomer
- Different potential in inducing the CYP1 family that activates them and, thus, their own genotoxic potential

Both compounds induced a mutagenic response in the Ames test in the presence of hepatic microsomes, but it was far higher in the case of the 2-isomer.[44] What these studies imply is that the nitrenium ion of the 2-isomer is either generated more readily or has higher genotoxic potential compared with that of the 4-isomer. The stability of the nitrenium ions of aromatic amines has been directly related to their mutagenicity;[45] 2-acetylaminofluorene nitrenium ion was found to be more stable than that of 4-acetylaminofluorene, which may explain the higher mutagenic potency of the 2-isomer.[44] These observations accord with previous studies in which N-acetoxy-2-acetylaminofluorene was shown to be a more potent carcinogen than N-acetoxy-4-acetylaminofluorene.[46]

A second possible explanation for the difference in the carcinogenic potential of these two isomers may lie in their ability to induce CYP1 activity and thus their own activation. Indeed, when rats were treated with the two isomers, 2-acetylaminofluorene was a more potent inducer of the hepatic CYP1A subfamily.[44] The consequence of the increased CYP1A activity was that 2-acetylaminofluorene stimulated markedly its own activation to genotoxic metabolites, whereas the 4-isomer had only a modest effect. It is, therefore, conceivable that the higher carcinogenic potential of the 2-isomer may be, at least partly, due to the fact that it can autoinduce its own activation. Furthermore, the 2-isomer interacted much more avidly than the 4-isomer with the cytosolic Ah receptor; this receptor, in addition to regulating a number of xenobiotic-metabolizing enzymes including the CYP1 family, also plays an important role in cellular proliferation and differentiation, which are critical steps in the promotion and progression stages of chemical carcinogenesis.[47] Consequently, this marked difference in affinity for the Ah receptor may also be a crucial contributory factor to the difference in carcinogenic potential. Similar studies have been conducted with other isomers and structurally related compounds, largely polycyclic aromatic hydrocarbons and aromatic amines, and the carcinogenic isomers were more potent inducers of CYP1 activity (Table 19.3).

To summarize, toxicity is not simply a consequence of the intrinsic molecular structure of the chemical but is also determined by the nature of the enzymes present at the time of exposure; these enzyme systems are in turn not only regulated

TABLE 19.3

CYP1 Induction in the Liver of Rats by Structural Isomers and Structural Analogs

Compound	CYP1A1 Induction (% of Control Activity[a])	CYP1A2 Induction (% of Control Activity[b])	Carcinogenic Activity
Benzo[a]pyrene	7400	100	Carcinogen
Benzo[e]pyrene	600	80	Noncarcinogen
15,16-Dihydro-2-methyl cyclopenta[a]phenahtnrene-17-one	500	70	Noncarcinogen
15,16-Dihydro-11-methyl cyclopenta[a]phenahtnrene-17-one	1300	50	Carcinogen
2-Aminobiphenyl	85	65	Noncarcinogen
4-Aminobiphenyl	205	245	Carcinogen
1-Naphthylamine	100	240	Noncarcinogen
2-Naphthylamine	910	210	Carcinogen
2-Acetylaminofluorene	465	335	Carcinogen
4-Acetylaminofluorene	300	145	Noncarcinogen
Diethylaminoazobenzene	110	115	Noncarcinogen
3-Methoxyaminoazobenzene	795	1015	Carcinogen
o-Aminoazotoluene	712	395	Carcinogen

[a] As exemplified by the O-deethylation of ethoxyresorufin and [b] by either the O-demethylation of methoxyresorufin or the metabolic activation of Glu-P-1 to mutagens in the Ames test. Both activities are expressed in terms of total cytochrome P450 activity.

genetically[48] but are also modulated by environmental factors such as previous exposure to chemicals[49], nature of the diet,[50] intake of herbal remedies,[51] and the presence of disease.[52] Toxicologists have the habit of referring to chemicals as being toxic or nontoxic, whereas in the strictest sense the vast majority of chemicals are innocuous, and it is the living organism that renders them toxic through metabolism. As we are becoming more competent in phenotyping humans for individual cytochromes P450, it may well become more appropriate to talk of toxicophilic and toxicophobic individuals rather than of toxic and nontoxic chemicals.

REFERENCES

1. Ames, B.N. and Gold, L.S., The prevention of cancer, *Drug Metab. Rev.*, 30, 201, 1998.
2. Moyers, S.B. and Kumar, N.B., Green tea polyphenols and cancer chemoprevention: multiple mechanisms and endpoints for phase III trials, *Nutr. Rev.*, 62, 204, 2004.

3. Hecht, S.S., Inhibition of carcinogenesis by isothiocyanates, *Drug Met. Rev.*, 32, 395, 2000.

4. Mori, H. and Nishikawa, A., Naturally occurring organosulphur compounds as potential anticarcinogens, in *Nutrition and Chemical Toxicity*, Ioannides, C., Ed., John Wiley & Sons, Chichester, 1998, pp. 285–299.

5. Ioannides, C., *Enzyme Systems that Metabolise Drugs and Other Xenobiotics*, John Wiley & Sons, Chichester, 2002.

6. Ioannides, C., Xenobiotic metabolism: an overview, in *Enzyme Systems that Metabolise Drugs and Other Xenobiotics*, Ioannides, C., Ed., John Wiley & Sons, Chichester, 2002, pp. 1–32.

7. Danielson, P.B., The cytochrome P-450 superfamily: biochemistry, evolution and drug metabolism in humans, *Curr. Drug Met.*, 3, 561, 2002.

8. Ioannides, C., *Cytochromes P-450: Metabolic and Toxicological Aspects*, CRC Press, Boca Raton, FL, 1996.

9. Ioannides, C. and Lewis, D.F.V., Cytochromes P-450 in the bioactivation of chemicals, *Curr. Top. Med. Chem.*, 4, 1767, 2004.

10. Gonzalez, F.J. and Nebert, D.W., Evolution of the P-450 gene superfamily animal plant "warfare," molecular drive and human differences in drug oxidation, *Trends Genet.*, 6, 182, 1990.

11. Hinson, J.A., Pumford, N.R., and Nelson, S.D., The role of metabolic activation in drug toxicity, *Drug Met. Rev.*, 26, 395, 1994.

12. Park, B.K. et al., The role of metabolic activation in drug-induced hepatotoxicity, *Annu. Rev. Toxicol.*, 45, 177, 2005.

13. Wiseman, H. and Halliwell, B., Damage to DNA by reactive oxygen and nitrogen species: role in inflammatory disease and progression to cancer, *Biochem. J.*, 313, 17, 1996.

14. Hayes, J.D., Flanagan, J.U., and Jowsey, I.R., Glutathione transferases, *Annu. Rev. Toxicol.*, 45, 51, 2005.

15. Kawajiri, K. et al., Biochemical basis for the resistance of guinea pigs to carcinogenesis by 2-acetylaminofluorene, *Biochem. Biophys. Res. Commun.*, 85, 959, 1978.

16. Ogawa, K. et al., Lack of carcinogenicity of 2-amino-3,8-dimethylimidazo[4,5-f]quinoxaline (MeIQx) in cynomolgus monkeys, *Jap. J. Cancer Res.*, 90, 622, 1999.

17. Seeff, L.B. et al., Acetaminophen toxicity in the alcoholic. A therapeutic misadventure, *Annals Int. Med.*, 104, 399, 1986.

18. Zimmerman, H.J. and Maddrey, W.C., Acetaminophen (paracetamol) hepatotoxicity with regular intake of alcohol: analysis of instances of therapeutic misadventure, *Hepatology*, 22, 767, 1995.

19. Pessayre, D. et al., Effect of fasting on metabolite-mediated hepatotoxicity in the rat, *Gastroenterology*, 77, 264, 1979.

20. Davis, C.D. et al., Metabolic activation of heterocyclic amine food mutagens in the mammary gland of lactating Fischer 344 rats, *Cancer Lett.*, 84, 67, 1994.

21. Kadlubar, F.F. et al., Alteration of urinary levels of the carcinogen N-hydroxy-2-naphthylamine, and its N-glucuronide in the rat by control of urinary pH, inhibition of metabolic sulfation and changes in biliary excretion, *Chem. Biol. Inter.*, 33, 129, 1981.

22. Eling, T.E. and Curtis, F.J., Xenobiotic metabolism by prostaglandin H synthase, *Pharmacol. Ther.*, 53, 261, 1992.

23. Degen, G.H., Vogel, C., and Abel, J., Prostaglandin synthases, in *Enzyme Systems that Metabolise Drugs and Other Xenobiotics*, Ioannides, C., Ed., John Wiley & Sons, Chichester, 2002, pp. 189–229.

24. Kaderlik, K.R. et al., Metabolic activation pathway for the formation of DNA adducts of the carcinogen 2-amino-1-methyl-6-phenylimidazo[4,5-b]pyridine (PhIP) in rat extrahepatic tissues, *Carcinogenesis*, 15, 1703, 1994.

25. Stone, E.M. et al., Interindividual variation in the metabolic activation of heterocyclic amines and their N-hydroxyderivatives in primary cultures of human mammary epithelial cells, *Carcinogenesis*, 19, 873, 1998.

26. Eaton, D.L. et al., Role of cytochrome P4501A2 in chemical carcinogenesis: Implications for human variability in expression and enzyme activity, *Pharmacogenetics*, 5, 259, 1995.

27. Kawajiri, K. and Hayashi S.-I., The CYP1 family, in *Cytochromes P-450: Metabolic and Toxicological Aspects*, Ioannides, C., Ed., CRC Press, Boca Raton, FL, 1996, pp. 77–97.

28. Ioannides, C. and Parke, D.V., The cytochrome P450I gene family of microsomal haemoproteins and their role in the metabolic activation of chemicals, *Drug Met. Rev.*, 22, 1, 1990.

29. Lewis, D.F.V., Ioannides, C., and Parke, D.V., Molecular dimensions of the substrate binding site of cytochrome P-448, *Biochem. Pharmacol.*, 35, 2179, 1986.

30. Lewis, D.F.V., Ioannides, C., and Parke, D.V., Structural requirements for substrates of cytochromes P-450 and P-448, *Chem. Biol. Inter.*, 64, 39, 1987.

31. Eaton, D.L. et al., Role of cytochrome P450 1A2 in chemical carcinogenesis: implications for human variability in expression and enzyme activity, *Pharmacogenetics*, 5, 259, 1995.

32. Shimada, T. et al., Activation of chemically diverse procarcinogens by human cytochrome P-450 1B1. *Cancer Res.*, 56, 2979, 1996.

33. Rendic, S. and Di Carlo, F.J., Human cytochrome P450 enzymes: a status report summarizing their reactions, substrates, inducers, and inhibitors, *Drug Met. Rev.*, 29, 413, 1997.

34. Ioannides, C. and Parke, D.V., Induction of cytochrome P4501 as an indicator of potential chemical carcinogenesis, *Drug Met. Rev.*, 25, 485, 1993.

35. Aoyama, T. et al., Five of 12 forms of vaccinia virus-expressed human hepatic cytochrome P450 metabolically activate aflatoxin B_1, *Proc. Natl. Acad. Sci. U.S.A.*, 87, 4790, 1990.

36. Kim, J.H. et al., Metabolism of benzo[a]pyrene and benzo[a]pyrene-7,8-diol by human cytochrome P450 1B1, *Carcinogenesis*, 19, 1847, 1998.

37. Lewis, D.F.V., Ioannides, C., and Parke, D.V., Validation of a novel molecular orbital approach (COMPACT) for the prospective safety evaluation of chemicals by comparison with rodent carcinogenicity and Salmonella mutagenicity data evaluated by the US NCI/NTP, *Mut. Res.*, 291, 61, 1993.

38. Guengerich, F.P. and Shimada, T., Oxidation of toxic and carcinogenic chemical by human cytochrome P-450 enzymes, *Chem. Res. Toxicol.*, 4, 391, 1991.

39. Ronis, M.J.J., Lindros, K.O., and Ingelman-Sundberg, M., The CYP2E subfamily, in *Cytochromes P450: Metabolic and Toxicological Aspects*, Ioannides, C., Ed., CRC Press, Boca Raton, FL, 1996, pp. 211–239.

40. Albano, E. et al., Role of ethanol-inducible cytochrome P450 (P450IIE1) in catalysis of the free radical activation of aliphatic alcohols, *Biochem. Pharmacol.*, 41, 1895, 1991.

41. Hammons, G.J. et al., Metabolic oxidation of carcinogenic arylamines by rat, dog, and human hepatic microsomes and by purified flavin-containing and cytochrome P-450 monooxygenases, *Cancer Res.*, 45, 3578, 1985.

42. Ioannides. C. et al., A rationale for the non-mutagenicity of 2- and 3-aminobiphenyls, *Carcinogenesis*, 10, 1403, 1989.

43. Ioannides, C., Induction of cytochrome P450 I and its influences in chemical carcinogenesis, *Biochem. Soc. Trans.*, 18, 32, 1990.

44. Ioannides. C. et al., The mutagenicity and interactions of 2- and 4-(acetylamino)fluorene with cytochrome P450 and the aromatic hydrocarbon receptor may explain the difference in their carcinogenic potency, *Chem. Res. Toxicol.*, 6, 535, 1993.

45. Ford, G.P. and Herman, P.S., Relative stabilities of nitrenium ions derived from polycyclic aromatic amines, Relationship to mutagenicity, *Chem. Biol. Inter.*, 81, 1, 1992.
46. Yost, Y., Gutmann, H.R., and Rydell, R.E., The carcinogenicity of fluorenylhydroxamic acids and N-acetoxy-N-fluorenylacetamides for the rat as related to the reactivity of the esters towards nucleophiles, *Cancer Res.*, 35, 447, 1975.
47. Hankinson, O., The aryl hydrocarbon receptor complex, *Annu. Rev. Pharmacol. Toxicol.*, 35, 307, 1995.
48. Wormhoudt, L.W., Commandeur, J.N.M., and Vermeulen, N.P.E., Genetic polymorphisms on human N-acetyltransferase, cytochrome P450, glutathione-S-transferase, and epoxide hydrolase enzymes: Relevance to xenobiotic metabolism and toxicity, *Crit. Rev. Toxicol.*, 29, 59, 1999.
49. Hollenberg, P.F., Characteristics and common properties of inhibitors, inducers and activators of CYP enzymes, *Drug Met. Rev.*, 34, 17, 2002.
50. Ioannides, C., Effect of diet and nutrition on the expression of cytochromes P450, *Xenobiotica*, 29, 109, 1999.
51. Ioannides, C., Pharmacokinetic interactions between herbal remedies and medicinal drugs, *Xenobiotica*, 32, 451, 2002.
52. Ioannides, C. et al., Expression of cytochrome P450 proteins in disease, in *Cytochromes P450: Metabolic and Toxicological Aspects*, Ioannides, C., Ed., CRC Press, Boca Raton, FL, 1996, pp. 301–327.

20 In Vitro Methods to Measure Drug Metabolism and Drug Interactions in Drug Discovery and Development

O. Pelkonen, M. Turpeinen, J. Uusitalo,
P. Taavitsainen, and H. Raunio

CONTENTS

KEYWORDS

Drug metabolism, cytochrome P450, CYP

ABBREVIATIONS

ADME: absorption, distribution, metabolism, excretion
CYP: cytochrome P450
DME: drug-metabolizing enzyme
HTS: high-throughput screening
NCE: new chemical entity

ABSTRACT

Drug metabolism profoundly affects drug action because almost all drugs are metabolized in the body, and thus their concentrations and retention times are dependent on metabolic activity. Drug metabolism contributes substantially to interindividual differences in drug response and is also often involved in drug interactions, resulting in either therapeutic failure or adverse effects. Knowledge about the metabolism of a new chemical entity (NCE) and its affinity to drug-metabolizing enzymes helps in the drug development process by providing important information for the selection of a lead compound from among a number of substances equally effective in their therapeutic response, pharmacologically. In drug development protocols, metabolism characteristics should be assessed very early during the development process. This has been made possible by the advances made especially in *in vitro* technologies used to predict *in vivo* drug metabolism, kinetic parameters, and interaction potential.

BACKGROUND

Elucidation of metabolic stability, metabolic routes, metabolizing enzymes and their kinetics, and metabolic interactions is obviously important for any chemicals to which humans are exposed. It is even more important for pharmaceuticals because this information is needed for selecting leads and candidate drugs during drug discovery and development. Furthermore, lack of detailed knowledge of metabolic fate and interactions of marketed drugs can result in morbidity or mortality, therapeutic failure, and toxicity from unanticipated overdose or metabolic reactions. In short, metabolism determines to a great extent the pharmacokinetic properties of most drugs and is often behind bioavailability problems, interindividual variations, metabolic interactions, idiosyncrasies, and so on.[1,2] This fact is the overpowering rationale for *in vitro* methods to measure drug metabolism and metabolic interactions in conjunction with drug discovery and development.

Before an NCE becomes a drug candidate and is subjected to clinical trials, it has to be tested for safety and the possibility of drug–drug interactions. These preclinical experiments must be reliable and give relevant information about the studied properties of an NCE. Safety and interactions are, even today, examined to a considerable extent by animal tests. Extrapolation of the results from animal studies to the human situation is usually difficult and contains many sources of errors. The most important reasons for this are the species-specific differences in drug-metabolizing enzymes, both qualitative (different metabolic pathways) and quantitative (different intrinsic clearances), between the human being and the test species. Therefore, "humanized" experimental systems are needed early in the drug development process.

There are currently several *in vitro* systems shown or claimed to be useful in drug development: human liver microsomes, liver slices, recombinant expressed enzymes, human hepatocytes, and permanent cell lines either as such or containing single or multiple transfected human enzymes.[3] Although most of them can give useful qualitative information to predict *in vivo* metabolic pathways and metabolites of a drug, there has been very little validation as to quantitative prediction.[1,4] However, these approaches are thought to provide crucial information for more in-depth studies during the preclinical and clinical phases of drug development. Examples of the integrative and comparative approach have been published.[5–7]

Important goals for using various human-liver-derived *in vitro* systems are the following: (1) to elucidate and determine principal metabolic routes of an NCE and tentatively identify principal metabolites; (2) to identify, with human liver homogenate or microsomal preparations, the CYP enzymes catalyzing the principal oxidation routes (primary metabolites) and to gain some quantitative data on their significance for the overall metabolic fate of an NCE; and (3) to provide useful background information for characterizing potential interactions and physiological, genetic, and pathological factors affecting the kinetics and variability of an NCE in an *in vivo* situation.

OVERVIEW OF DRUG METABOLIZING ENZYMES

The principal route of elimination of drugs from the body is biotransformation. The oxidative reactions are mainly catalyzed by cytochrome P450 (CYP) enzymes (phase I metabolism) and, after that, by conjugating enzymes (phase II metabolism). Glucuronidation, especially, catalyzed by the several UDP-glucuronosyltransferase isoforms is an important route of phase II drug metabolism in humans. Some prodrugs need to be metabolically activated before they are pharmacologically active. This activation usually occurs via CYP or hydrolytic enzymes.

CYP enzymes involved in drug metabolism belong to the superfamily of mainly microsomal hemoproteins that catalyze the oxidative, peroxidative, and reductive metabolism of a wide variety of endogenous and exogenous compounds.[8] The most important human hepatic CYPs and their relative amounts and model substrates and inhibitors are listed in Table 20.1. About 70% of the CYP enzymes in the human liver belong to the families that participate in drug metabolism. Of these, CYP3A4 represents about 30% and CYP2C about 20% of the total CYP enzymes.

Expression of CYP enzymes varies between individuals due to genetic and environmental factors and some diseases.[9] These factors produce inter-individual variation in the rate and metabolic pathways of xenobiotics. One example of genetic

TABLE 20.1

Some Characteristics of Xenobiotic-Metabolizing Human Hepatic CYPs

CYP	Relative Amount in Liver (%)	Substrates (Reaction Catalyzed)	Selective Inhibitors	Other Characteristics
1A2	~10	Ethoxyresorufin (O-deethylation) Caffeine (N-demethylation)	Furafylline	Inducible
2A6	~10	Coumarin (7-hydroxylation)	Tranylcypromine	Polymorphic
2B6	~1	Bupropion (hydroxylation)	Thio-tepa	Inducible
2C8	<1	Paclitaxel (6α-hydroxylation)	Quercetin	
2C9	~20	S-warfarin (7-hydroxylation)	Sulfaphenazole	Polymorphic Inducible
2C19	~5	Omeprazole (oxidation)		Polymorphic Inducible
2D6	~5	Dextromethorphan (O-demethylation)	Quinidine	Polymorphic
2E1	~10	Chlorzoxazone (6-hydroxylation)	Pyridine	Inducible
3A4	~30	Midazolam (1′- and 4-hydroxylation) Testosterone (6β-hydroxylation)	Azole antimycotics	Inducible

Note: For reviews, see Pelkonen, O. et al., Inhibition and induction of human cytochrome P450 (CYP) enzymes, *Xenobiotica*, 28, 1203, 1998; Rendic, S. and Di Carlo, F.J., Human cytochrome P450 enzymes: a status report summarizing their reactions, substrates, inducers, and inhibitors, *Drug Metab. Rev.* 29, 413, 1997; Nebert, D.W. and Russell, D.W., Clinical importance of the cytochromes P450, *Lancet*, 360, 1155, 2002.

factors influencing interindividual variation is the polymorphic expression of at least CYP2A6, CYP2C9, CYP2C19, and CYP2D6 enzymes in the population.[10] The frequency of poor metabolizers (PMs) varies between ethnic groups. Some dietary compounds, cigarette smoking, alcohol, and drugs may cause induction or diminution of the expression of certain CYPs.

METHODS FOR STUDYING *IN VITRO* METABOLISM: ENZYME SOURCES

There are several approaches to preclinical metabolism studies. The enzyme sources in these studies are usually human-liver-derived systems. These systems consist of liver microsomes, hepatocytes, and cell lines heterologously expressing drug-metabolizing enzymes, liver slices, and individually cDNA-expressed enzymes in

TABLE 20.2
Comparison of *In Vitro* Enzyme Sources Used in Preclinical Research

Enzyme Sources	Availability	Advantages	Disadvantages
Microsomes (Kremers 1999[29])	Relatively good, from transplantations or commercial sources.	Contains important rate-limiting enzymes. Relatively inexpensive technique.	Contains only phase I enzymes and UGTs. Requires strictly specific substrates and inhibitors or antibodies for individual DMEs.
Liver homogenates (Kremers 1999[29])	Relatively good. Commercially available.	Contains basically all hepatic enzymes.	Liver architecture lost.
cDNA-expressed individual CYPs (Rodrigues 1999[13])	Commercially available.	Can be utilized with HTS substrates. The role of individual CYPs in the metabolism can be easily studied.	The effect of only one enzyme at a time can be evaluated. Problems in extrapolation.
Primary hepatocytes (Guillouzo et al. 1995[30]; Gomez-Lechon et al. 2003[31])	Difficult to obtain, relatively healthy tissue needed. Commercially available. Cryopreservation possible.	Contains the whole complement of DMEs cellularly integrated. The induction effect of an NCE can be studied.	Requires specific techniques and well-established procedures. The levels of many DMEs decrease rapidly during cultivation.
Liver slices (Beamand et al. 1993[32])	Difficult to obtain, fresh tissue needed. Cryopreservation possible.	Contains the whole complement of DMEs and cell–cell connections. The induction effect of an NCE can be studied.	Requires specific techniques and well-established procedures.
Immortalized cell lines	Available at request, not many adequately characterized cell lines exist.	Nonlimited source of enzymes.	The expression of most DMEs is poor or absent, if characterized at all.

host cell microsomes. A summary of the major biological preparations for *in vitro* studies is provided in Table 20.2. Human liver samples obtained under proper ethical permission have been used as a golden standard. These preparations have usually been extensively characterized to be used for the primary screening (sufficient model activities, no known polymorphisms, expected effects of model inhibitors, quantitation of CYPs by Western blotting). Currently, *in silico* models are being increasingly used for predicting metabolism and interactions. A thorough treatment of these experimental systems can be found in recent reviewes.[3,11,12]

TABLE 20.3
In Vitro Studies for the Characterization of Metabolism and Metabolic Interactions of Potential Drugs

In Vitro Test	Preparations	Parameters	Extrapolations
Metabolic stability	Microsomes Homogenates Cells Slices	Disappearance of the parent molecule or appearance of (main) metabolites	Intrinsic clearance Interindividual variability
Metabolite identification	Same as above	Tentative identification of metabolites by, e.g., LC-TOF-MS	Metabolic routes Semiquantitative metabolic chart
Identification of metabolizing enzymes	Microsomes with inhibitors or inhibitory antibodies Recombinant individual enzymes	Assignment and relative ability of enzymes to metabolize a compound	Prediction of effects of various genetic, environmental, and pathological factors Interindividual variability
Enzyme inhibition	Microsomes Recombinant enzymes	Inhibition of model activities by a substance	Potential drug–drug interactions
Enzyme induction	Cells Slices Permanent cell lines (if available)	Induction of model activities (or mRNA)	Induction potential of a substance

Note: For reviews, see articles by Boobis, A.R., Kremers, P., Pelkonen, O., and Pithan, K., Eds., *EUR 18569—European Symposium on the Prediction of Drug Metabolism in Man: Progress and Problems,* Luxembourg, Office for Official Publications of the European Communities, 332 pp., 1999; Boobis, A.R., Prediction of inhibitory drug-drug interactions by studies *in vitro,* in *Advances in Drug Metabolism in Man,* Pacifici, G.M. and Fracchia, G.N., Eds., Office for the Official Publications of the European Communities, Luxembourg, 1995, pp. 513–539; and von Moltke, L.L. et al., *In vitro* approaches to predicting drug interactions *in vivo, Biochem. Pharmacol.,* 55, 113, 1998. See also Andersson, T.B. et al., An assessment of human liver-derived *in vitro* systems to predict the *in vivo* metabolism and clearance of almokalant, *Drug Metab. Dispos.,* 29, 712, 2001; Pelkonen, O. et al., Carbamazepine: a "blind" assessment of CYP-associated metabolism and interactions in human liver-derived *in vitro* systems, *Xenobiotica,* 31, 321, 2001; Salonen, J.S. et al., Comparative studies on the CYP-associated metabolism and interaction potential of selegiline between human liver-derived *in vitro* systems, *Drug Metab. Dispos.,* 31, 1093, 2003.

Liver homogenate with appropriate cofactors can be used for metabolic stability studies when the biotransformation pathways of a new compound are unknown. Liver homogenate fortified with appropriate cofactors exhibits enzyme activities that are present in intact tissue. In incubations with liver homogenate, the whole spectrum of *in vitro* metabolites is formed and identification of metabolites of an unknown compound is possible. When the transport of an NCE through cell membranes is under study, whole cell systems or liver slices are a good choice, because in these preparations the transport systems into and out of the cell are present, and even the cell–cell connections are preserved in liver slices. These preparations are also suitable for induction studies. Recombinant enzymes are used for the identification of enzymes metabolizing the compound under study, but they are increasingly studied as an alternative method to predict hepatic clearance.[13]

DETERMINATION OF METABOLISM IN *IN VITRO* SYSTEMS

During the development of an NCE, at least the following investigations are usually required—the earlier the better (Table 20.3):

- Measurement of "metabolic stability," i.e., the rate of disappearance of a chemical in human liver preparations
- Identification of principal metabolites—"metabolite profile" and proposed metabolic tree of a chemical
- Development of an analytical "routine" method for metabolites
- Identification of enzymes catalyzing metabolic routes with the aid of diagnostic inhibitors and antibodies, recombinant enzymes, and correlation studies
- Enzyme kinetic characterization of principal metabolic reactions, for scaling up and predicting *in vivo* kinetics of an NCE
- Affinity of a chemical (and/or its principal metabolites) for CYP (or other) enzymes, for predicting potential metabolic interactions

Metabolic stability of an NCE:[14] The metabolic stability of an NCE determines, to a great extent, its future as a drug candidate. If an NCE is rapidly metabolized in human liver preparations, its bioavailability *in vivo* is most probably too low for it to be a drug candidate. This naturally depends on the administration route of the drug. By determining the time and concentration dependence of metabolite formation from an NCE on the disappearance of the NCE *in vitro* in an appropriate system, its metabolic fate and half-life *in vivo* can be predicted. Similar studies performed in human and test species give valuable information for the selection of test species for pharmacokinetic and toxicological *in vivo* studies.

Identification of metabolites and metabolic routes: Metabolite identification can be developed from incubations with human liver preparations, homogenates or microsomes. For example, mass-spectrometric (MS) methods employing liquid chromatography (LC) as a separative tool have evolved into extremely sensitive and accurate techniques.[15,16] By these methods, it is possible to determine with high accuracy the exact molecular masses and metabolite structures. Sample preparation for MS studies is a critical step because the available chemical information of an NCE is often limited. The incubation conditions and the reaction-terminating reagent have to be chosen so as not to alter the parent compound or the metabolites chemically and to keep the recovery of the substrate and the metabolites close to 100%. Otherwise, it is impossible to predict the pathways for biotransformation.

Identification of CYPs metabolizing an NCE:[3] After characterizing the metabolic stability and metabolic routes of an NCE, the *in vivo* prediction requires clarification of the drug-metabolizing enzymes that participate in the *in vitro* biotransformation of the NCE. After determining the initial velocity conditions and enzyme kinetic parameters, CYPs involved in the metabolism of an NCE can be characterized by chemical and antibody inhibitors selective or specific for respective CYPs. In case the *in vivo* concentrations are known, it is recommendable to use substrate concentrations close to the therapeutic level, if at all feasible.

Some typical enzyme identification experiments and their outcomes are the following:

- Effects of various diagnostic CYP-selective inhibitors on NCE metabolism: These allow an estimation of the role of each CYP enzyme to contribute to the metabolism of a chemical. On the basis of this information, it is possible to perform predictive calculations to the *in vivo* behavior of a chemical.
- Ability of recombinant enzymes to catalyze metabolite formation of an NCE: This gives a direct indication of the potential of each recombinant CYP to catalyze the reaction. However, it is still somewhat uncertain whether the results from recombinant enzymes could be extrapolated to the whole microsomes and to the *in vivo* situation.
- Correlation of CYP-selective model activities with metabolite formation of an NCE: It gives an indirect indication of the participation of each CYP enzyme to catalyze the reaction. The results are, however, confirmatory and should be evaluated in the context of other approaches.

Measures of affinities of an NCE for CYPs:[3,11] From the therapeutic point of view, it is important to know to which drug-metabolizing enzymes the substance under development has affinity. The effect of an NCE on CYP-specific activities is studied by co-incubating series of dilutions of the study substance with a reaction mixture and a specific substrate. The effect of an NCE is described as the concentration of the studied compound causing 50% inhibition of the CYP-specific activities. If there is affinity toward some CYP-enzyme, the apparent Ki can be determined. By comparing the effects of an NCE on the CYP specific-activities to the respective effects of diagnostic inhibitors, a tentative prediction of the *in vivo* situation can be made.

The effects of an NCE on drug-metabolizing CYPs can be tested by using CYP isoform-specific model substrates and reactions.[11] The effects of the studied compound on metabolite formation in the selected model system, usually human liver microsomes, are evaluated by incubating the substrate and the studied compound with the enzymes and observing the metabolite formation in incubations with the studied compound. Testing the effects of an NCE on CYP-specific model activities and the effects of CYP-specific reference inhibitors on the metabolism of an NCE in human liver microsomes *in vitro* gives information about the affinity of an NCE for CYP enzymes. This approach also permits *in vivo* predictions about the behavior of the NCE in humans (metabolic pathways, intrinsic clearance, etc.), which helps to design *in vivo* studies for revealing possible interactions.

After the preceding studies, it is possible to plan further preclinical, molecular, toxicological, and clinicopharmacological studies, with focused consideration of those CYP enzymes that are important for the metabolism and kinetics of an NCE.

HIGH-THROUGHPUT SCREENING IN DRUG METABOLISM

There is a tremendous pressure in the pharmaceutical industry to screen for many parameters as early as possible in the drug development process.[17] The technology of higher-throughput ADME (absorption, distribution, metabolism, excretion) screening is developing fast, yet the fastest ADME methods do not come within two orders of magnitude of the rates routinely achieved by truly high-throughput discovery screening. The greatest need in ADME screening for discovery support is a better understanding of the integrated operation of *in vivo* absorption, first-pass metabolism, organ uptake and efflux mediated by transporters, plasma protein binding, competition among metabolic enzymes, and cellular determinants of intrinsic clearance.

As an affinity-screening tool, cDNA-expressed enzymes are highly valuable. It is also the most useful system allowing a high-throughput screening (HTS) technology for CYP studies today because of the difficulties and high costs in the detection of multiple substrates and metabolites produced in the HTS applications of other techniques, such as human liver microsomes. The detection of these multiple metabolites requires novel, highly sensitive mass spectrometry tools, whereas cDNA-expressed systems can utilize the conventional measurement of fluorescence metabolite production for multiple enzymes (see, for example, www.gentest.com).

IN VITRO–IN VIVO SCALING OF AN NCE

In general, *in vitro–in vivo* scaling consists of two approaches: prediction of the intrinsic clearance (Clint) of an NCE and prediction of drug–drug interactions. In the literature, methods for both purposes are extensively presented and discussed in the light of actual experimental or clinical studies.[4,14,18] It has to be stressed that these extrapolations contain a number of uncertainties, which have to be acknowledged. Other pharmacokinetic parameters and characteristics, such as plasma protein binding, the differential binding to various tissues, volume of distribution, and extrahepatic routes of clearance, should also be considered for the estimation of *in vivo* kinetics on the basis of *in vitro* studies.[19,20]

INDUCTION OF CYP ENZYMES

Induction of drug metabolism is an adaptive cellular response that usually leads to enhanced metabolism and termination of the pharmacological action of drugs.[21] Induction of a CYP isoform can sometimes have harmful consequences if the metabolite produced by the induced CYP is reactive. It is noteworthy that a clear majority of CYP enzymes that are involved in drug metabolism are also inducible—including the important CYP3A4 isoform (see Table 20.1). Advances in molecular and cell biology during the past decade have helped elucidate the major mechanisms by which drugs and xenobiotics induce the expression of CYP genes. This occurs mainly via receptor proteins that upon inducer binding are transformed into transcriptionally active DNA-binding forms. These active receptors then bind to their DNA response elements present in the regulatory region of the CYP genes and greatly speed up the production of CYP mRNA, up to 50-fold or more.[22] In a few instances, there are alternative mechanisms for CYP induction. The best-characterized example is the induction of CYP2E1 by ethanol and other small solvent molecules, which appears to involve stabilization of CYP2E1 protein through binding of the inducer to the CYP2E1-active site.

ASSAY SYSTEMS FOR INDUCTION

Induction of human CYP enzymes is difficult to study because human liver cell lines neither express the full complement of CYP enzymes nor reproduce the induction observed *in vivo*, apart from the Ah receptor-mediated induction of the CYP1A1 gene. Problems in tissue availability, interindividual differences, reproducibility, and ethical issues preclude the efficient large-scale use of human primary hepatocytes for induction screening. Therefore, there is an increasing interest in the development of mechanism-based test systems for CYP induction—systems based on the characterized nuclear receptors AHR, CAR, PXR, and PPARα.[23–25] The interaction of inducers with the particular receptor is exploited in an attempt to predict CYP induction at least in the following ways:

1. Direct binding assays rely on the ability of the putative inducer to displace a radioactively labeled receptor ligand.
2. Indirect binding assays measure the association of fluorescently labeled coactivator peptide with the inducer-bound receptor. The receptor can also be linked to another fluorophor to utilize the fluorescence resonance energy transfer (FRET) phenomenon in the assay.
3. Cell-based reporter gene assays measure the ability of inducer-bound receptor to interact with endogenous cellular coactivators after binding to its DNA response element that drives the expression of an easily assayable reporter gene such as luciferase or CAT enzyme.

MODELING OF CYP ENZYMES

The following are two important questions concerning the metabolism of an NCE: (1) What kinds of metabolites are formed from the NCE? and (2) which enzymes catalyze their formation? Modeling of CYP enzymes will most probably be of great help in answering these questions.[26] Although at the moment the various approaches used to predict CYP-mediated metabolism of substances have been explanatory rather than predictive, the situation is improving. Prediction of metabolism and inhibitory properties of human CYPs is becoming more accurate because more resources are invested in three types of research efforts: (1) crystallization of mammalian CYPs, (2) homology modeling based on sequences and on crystallized CYPs, and (3) structure–activity relationship of CYP substrates or inhibitors.

ROLE OF ANIMAL METABOLISM STUDIES DURING DRUG DEVELOPMENT

Due to species differences in metabolic pathways,[27] previously unknown metabolites are often found in humans, meaning that the human subjects are exposed to compounds with unknown toxic potential. This may lead to a delay in clinical testing until an appropriate animal species can be tested with the synthesized metabolite. Metabolic stability assays employing different test species and human liver make it possible to select the species that best represent the human metabolic fate of an NCE. These results can be utilized in selecting most appropriate test species for further toxicological tests.

CONCLUSIONS

Metabolism is a major determinant governing both pharmacokinetics and clinical response (pharmacodynamics) of most small-molecule drugs, and a great deal of effort is now directed at assessing key metabolic parameters at early stages of drug development. Several *in silico* and *in vitro* methods are now available for determination of metabolic features, often yielding data that reasonably well predict *in vivo* behavior of the studied drug molecules. Further refining of these methods will provide us with methods having increasing precision and speed, allowing for truly high-throughput analysis of metabolic features. In addition, analogous methodology will be employed on predicting absorption, organ uptake, and efflux mediated by various transporter systems, plasma protein binding, and cellular determinants of intrinsic clearance. As an outcome of this development, it is likely that we will witness a diminishing number of drug candidates being withdrawn from clinical studies (or the market) due to major kinetic problems, such as strong metabolism induction or interaction potential.

REFERENCES

1. Boobis, A.R., Kremers, P., Pelkonen, O., and Pithan, K., Eds., *EUR 18569—European Symposium on the Prediction of Drug Metabolism in Man: Progress and Problems,* Office for Official Publications of the European Communities, Luxembourg, 1999.
2. White, R.E., Anthony Y.H. Lu commemorative issue. Short- and long-term projections about the use of drug metabolism in drug discovery and development, *Drug Metab. Disp.,* 26, 1213, 1998.
3. Pelkonen, O. et al., Inhibition and induction of human cytochrome P450 (CYP) enzymes, *Xenobiotica,* 28, 1203, 1998.
4. Ito, K. et al., Quantitative prediction of *in vivo* drug clearance and drug interactions from *in vitro* data on metabolism, together with binding and transport, *Annu. Rev. Pharmacol. Toxicol.,* 38, 461, 1998.
5. Andersson, T.B. et al., An assessment of human liver-derived *in vitro* systems to predict the *in vivo* metabolism and clearance of almokalant, *Drug Metab. Dispos.,* 29, 712, 2001.
6. Pelkonen, O. et al., Carbamazepine: a "blind" assessment of CYP-associated metabolism and interactions in human liver-derived *in vitro* systems, *Xenobiotica,* 31, 321, 2001.

7. Salonen, J.S. et al., Comparative studies on the CYP-associated metabolism and interaction potential of selegiline between human liver-derived *in vitro* systems, *Drug Metab. Dispos.*, 31, 1093, 2003.

8. Rendic, S. and Di Carlo, F.J., Human cytochrome P450 enzymes: a status report summarizing their reactions, substrates, inducers, and inhibitors, *Drug Metab. Rev.*, 29, 413, 1997.

9. Pacifici, G.M. and Pelkonen, O., Eds., *Variation of Drug Metabolism in Humans*, Taylor & Francis, London, 2001.

10. Ingelman-Sundberg, M., Genetic variability in susceptibility and response to toxicants, *Toxicol. Lett.*, 120, 259, 2001.

11. Kremers, P., *In vitro* tests for predicting drug-drug interactions: the need for validated procedures, *Pharmacol. Toxicol.*, 91, 209, 2002.

12. Donato, M.T. and Castell, J.V., Strategies and molecular probes to investigate the role of cytochrome P450 in drug metabolism: focus on *in vitro* studies, *Clin. Pharmacokinet.*, 42, 153, 2003.

13. Rodrigues, A.D., Integrated cytochrome P450 reaction phenotyping. Attempting to bridge the gap between cDNA-expressed cytochromes P450 and native human liver microsomes, *Biochem. Pharmacol.*, 57, 465, 1999.

14. Obach, R.S. et al., The prediction of human pharmacokinetic parameters from preclinical and *in vitro* metabolism data, *J. Pharmacol. Exp. Ther.*, 283, 46, 1997.

15. Kostiainen, R. et al., Liquid chromatography/atmospheric pressure ionization-mass spectrometry in drug metabolism studies, *J. Mass Spectrom.*, 38, 357, 2003.

16. Lee, M.S. and Kerns, E.H., LC/MS applications in drug development, *Mass Spectrom. Rev.*, 18, 187, 1999.

17. White, R.E., High-throughput screening in drug metabolism and pharmacokinetic support of drug discovery, *Annu. Rev. Pharmacol. Toxicol.*, 40, 133, 2000.

18. Houston, J,B. and Carlile, D.J., Prediction of hepatic clearance from microsomes, hepatocytes and liver slices, *Drug Metab. Rev.*, 29, 891, 1997.

19. Lin, J.H., Applications and limitations of interspecies scaling and *in vitro* extrapolation in pharmacokinetics, *Drug Metab. Dispos.*, 26, 1202, 1998.

20. Theil, F.P. et al., Utility of physiologically based pharmacokinetic models to drug development and rational drug discovery candidate selection, *Toxicol. Lett.*, 138, 29, 2003.

21. Okey, A.B., Enzyme induction in the cytochrome P-450 system, *Pharmacol. Ther.*, 45, 241, 1990.

22. Honkakoski, P. and Negishi, M., Regulation of cytochrome P450 (CYP) genes by nuclear receptors,. *Biochem. J.*, 347, 321, 2000.

23. Waxman, D.J., P450 gene induction by structurally diverse xenochemicals: central role of nuclear receptors CAR, PXR, and PPAR, *Arch. Biochem. Biophys.*, 369, 11, 1999.

24. Pelkonen, O. et al., *In vitro* screening of cytochrome P450 induction potential, in *Ernst Schering Research Foundation Workshop 37: Pharmacokinetics Challenges in Drug Discovery*, Pelkonen, O., Baumann, A., and Reichel, A., Eds., Springer, Berlin, 2002, pp. 105–137.

25. Honkakoski, P., Nuclear receptors CAR and PXR in metabolism and elimination of drugs, *Curr. Pharmacogenomics*, 1, 75, 2003.

26. Boobis, A. et al., *In silico* prediction of ADME and pharmacokinetics. Report of an expert meeting organised by COST B15, *Eur. J. Pharm. Sci.*, 17, 183, 2002.

27. Pelkonen, O. and Breimer, D.D., Role of environmental factors in the pharmacokinetics of drugs: considerations with respect to animal models, P-450 enzymes, and probe drugs, in *Handbook of Experimental Pharmacology*, Vol. 110, Welling, P.G. and Balant, L.P., Eds., Springer-Verlag, Berlin, 1994, pp. 289–332.

28. Nebert, D.W. and Russell, D.W., Clinical importance of the cytochromes P450, *Lancet*, 360, 1155, 2002.

29. Kremers, P., Liver microsomes: a convenient tool for metabolism studies but …, in *European Symposium on the Prediction of Drug Metabolism in Man: Progress and Problems*, Boobis, A.R., Kremers, P., Pelkonen, O., and Pithan, K., Eds., Office for Official Publications of the European Communities, Luxembourg, 1999, pp. 38–52.

30. Guillouzo, A. et al., The isolated human cell as a tool to predict *in vivo* metabolism of drugs, in *Advances in Drug Metabolism in Man*, Pacifici, G.M. and Fracchia, G.N., Eds., Office for the Official Publications of the European Communities, Luxembourg, 1995, pp. 756–782.

31. Gomez-Lechon, M.J. et al., Human hepatocytes as a tool for studying toxicity and drug metabolism, *Curr. Drug Metab.*, 4, 292, 2003.

32. Beamand, J.A. et al., Culture of precision-cut liver slices: effect of some peroxisome proliferators, *Food Chem. Toxicol.*, 31, 137, 1993.

33. Boobis, A.R., Prediction of inhibitory drug-drug interactions by studies *in vitro*, in *Advances in Drug Metabolism in Man*, Pacifici, G.M. and Fracchia, G.N., Eds., Office for the Official Publications of the European Communities, Luxembourg, 1995, pp. 513–539.

34. von Moltke, L.L. et al., *In vitro* approaches to predicting drug interactions *in vivo*, *Biochem. Pharmacol.*, 55, 113, 1998.

Part V

Physical Organic and Theoretical
Medicinal Chemistry

21 How to Probe the Sites of Action of Drug Molecules

A. Makriyannis and F. Bitter

CONTENTS

INTRODUCTION

To better understand the molecular basis for activity of a certain group of pharmacologically active agents, it is important to understand how these drug molecules
interact with their sites of action. In our laboratories, such studies have involved the
combined use of biophysical and biochemical methods and could only be realized
with the successful design and synthesis of appropriate molecular probes.

The various methods we use are aimed at answering the following questions:

What are the conformational properties of the drug molecule in solution and
  in the membrane?
What orientation does the drug molecule assume when it partitions in the
  membrane?
What is the location of the drug molecule in the membrane bilayer?
What is the distribution of the receptors associated with the pharmacological
  activity of the drug molecule in the various tissues, and how can the receptor-mediated biochemical events be followed?

How can the receptors be covalently labeled as an aid to their isolation and characterization?

This chapter will briefly review this multifaceted approach for studying drug-active site interactions by focusing on the cannabinoids, a research area with which we have considerable familiarity. Examples will be used to illustrate each of these methods and to demonstrate how the results from the different experiments can be integrated in order to provide a more unified picture of the mechanism of action of these drug molecules.

MECHANISM OF CANNABINOID ACTIVITY

Cannabinoids produce a complex pattern of pharmacological actions, some of which are believed to be related to their effects on cellular membranes,[1-3] whereas others are thought to be produced through an interaction with one or more cannabinoid receptors.[4-7] Existing evidence indicates that the membrane effects involve cannabinoid interactions with noncatalytic amphipathic sites, resulting in perturbation of the membrane and a modification of the functions of membrane-associated proteins. There is also an evidence that the cannabinoid–membrane interactions are governed by stereoelectronic requirements of a varied degree of specificity.

Regarding the receptor-mediated effects, cannabinoids are believed to first partition into the membrane where they assume a specific orientation and location, and then laterally diffuse across the bilayer to reach their receptor sites. Thus, for a productive interaction with its active sites, a cannabinoid must fulfill two criteria. First, it must possess certain pharmacophores in a specific stereochemical arrangement (stereoelectronic requirements). Second, it must assume an appropriate orientation and location in the bilayer, which would allow an optimal alignment of its pharmacophores with the corresponding binding components at the active site (drug–membrane interaction requirements).

CANNABINOID RECEPTORS

Although it had been generally recognized that at least some of the cannabinoid effects are due to interactions with specific receptor sites, only recently has direct evidence for a cannabinoid receptor become available. This evidence comes from experiments in which the tritiated cannabinoid-like ligand [3H] CP-55940, synthesized at Pfizer, was shown to bind stereospecifically to a brain membrane preparation.[4] The relative abilities of cannabinoid analogues to inhibit [3H] CP-55490 binding were found to parallel their potencies for producing behavioral effects in animals,[5] as well as for regulating adenylate cyclase *in vitro*.[8] The same tritiated ligand was also used in autoradiographic experiments for cannabinoid-receptor distribution in brain sections from several mammalian species including man.[6] The study revealed a unique and conserved distribution with the highest receptor density in the basal ganglia, hippocampus, and cerebellum. Subsequently, the receptor was cloned; the cDNA was then injected in CHO-K_4 cells, and a cannabinoid-responsive G-protein-coupled receptor was expressed.[9-11] More recently, a second cannabinoid receptor was identified in the periphery, thus opening the door to the development of cannabimimetic analogues possessing the highest degree of pharmacological specificity.[12]

CANNABINOID–LIPID INTERACTIONS

Interactions of lipophilic drug molecules, including cannabinoids, with membrane lipids are generally discussed in terms of changes in the bulk properties of the lipid component, which are often described as changes in membrane fluidity. However, there is an increasing evidence from a variety of sources for specific drug–membrane interactions. For example, evidence has been presented for the formation of Δ^9-THC-phospholipid complexes, the stoichiometry and stability of which varied depending on the nature of the phospholipids in the membrane.[13] More detailed evidence comes from our own work showing that cannabinoids assume specific orientations in the phospholipid bilayers.[14]

Because of their high lipid solubility, cannabinoids partition preferentially into the lipid component of the membrane. The cannabinoid–lipid interactions can play two roles in determining drug activity: (1) cannabinoids may affect the functions of membrane-associated proteins by perturbing their lipid environment; and (2) cannabinoids may diffuse laterally across the membrane–lipid bilayer to reach a specific site of action on the cannabinoid receptor. It has been argued for other groups of lipophilic drugs,[15] and evidence was provided, that the location and orientation of the drug molecule in the membrane is critical in determining its ability to reach its sites of action in the proper orientation and conformation required for a productive interaction. We have studied the location and orientation of a number of cannabinoids in model and biological membranes and have developed arguments correlating these properties with their abilities to interact with the receptor.

CANNABINOID CONFORMATIONS

To explore the conformational requirements for interaction of cannabinoids with their sites of action on the receptor, we have compared pairs of analogues that differ only in one structural feature but exhibit different pharmacological potencies and affinities for the receptor.

The conformational properties of the drug molecules in solution and in membrane-like environments can be studied using high-resolution nuclear magnetic resonance (NMR) spectroscopy or computational methods. For example, we have compared the two isomeric cannabinoids (–)-Δ^9-THC and (–)-$\Delta^{9,11}$-THC using two-dimensional NMR. [16] Our results led us to suggest that a slight deviation from planarity in the carbocyclic ring C of the cannabinoid, as is the case with the pharmacologically potent Δ^9-THC, is a requirement for activity (Figure 21.1). Conversely, the much less potent $\Delta^{9,11}$-analog has all three rings coplanar. The same conclusions were reached by Reggio and co-workers, who compared the theoretically obtained conformations of the previous two analogues.[17] More recently, we have used a combination of solution NMR and theoretical computational methods to study the conformational properties of CP-47497, a bicyclic nonclassical cannabinoid synthesized at Pfizer, which resembles many of the structural features of the tricyclic classical cannabinoids but

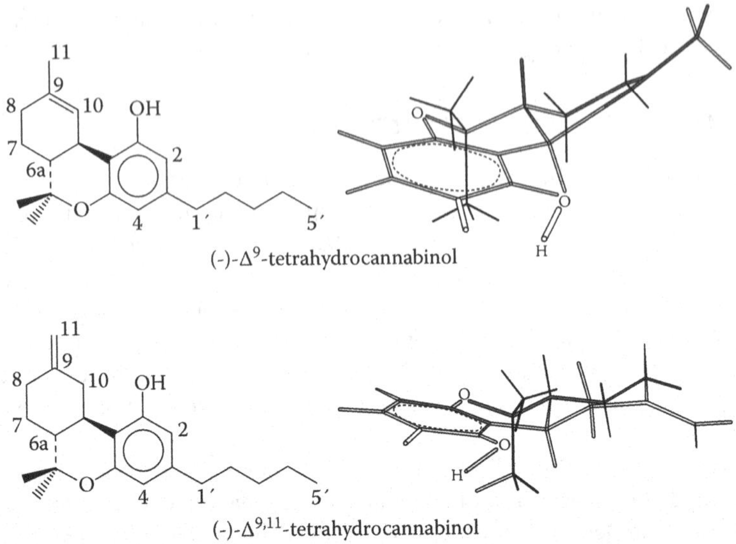

FIGURE 21.1   Left: Structures of $(-)$-Δ^9-tetrahydrocannabinol and $(-)$-$\Delta^{9,11}$-tetrahydrocannabinol. Right: Drawings showing the experimentally determined conformations of Δ^9-THC and $\Delta^{9,11}$-THC.

lacks their tetrahydropyran ring. The conformational features of the nonclassical cannabinoid were shown to closely resemble those of a classical counterpart (Figure 21.2). The only apparent conformational difference between the two analogues is the dihedral angle between the planes of the two rings in the nonclassical analog when compared to that of a classical hexahydrocannabinol. CP-47497 can be superimposed over the more rigid hexahydrocannabinol only after rotation around the phenyl-cyclohexyl bond. The molecule then assumes a conformation that is energetically higher than the preferred one by approximately 3.0 kcal/mol. This may explain the slightly lower potencies of nonclassical cannabinoids when compared to their structurally equivalent classical congeners.

CANNABINOID ORIENTATION IN THE MEMBRANE

As with many other membrane-active lipophilic molecules, cannabinoids preferentially partition in the membrane where they assume a thermodynamically favorable orientation and location. In this preferred orientation and location, the cannabinoid experiences fast lateral diffusion within the membrane bilayer. As mentioned earlier, these features of the drug–membrane interaction are believed to play an important role in determining the drug's ability to interact with its site of action on the receptor.

    The orientation of the cannabinoid molecule in the membrane can be calculated from the solid-state ^{2}H NMR spectra of membrane preparations into which the appropriate ^{2}H-labeled analog is introduced. The ^{2}H-labels are introduced in strategic positions of the tricyclic ring system. Each experiment involves the singly ^{2}H-labeled

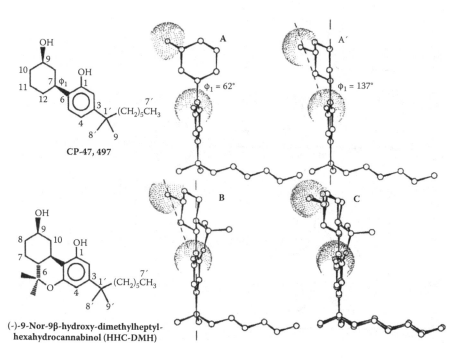

FIGURE 21.2 Graphic representation of the stereochemical correlation of CP-47,497 with HHC-DMH using a superimposition method. The preferred conformer (A) of CP-47,497 ($\varphi 1=$ 62°) was allowed to rotate to a confomration in which $\varphi 1= 137°$ (A'). In this conformation, CP-47,497 is superimposable with the ring systems of the classical cannabinoid HHC-DMH (B). The superimposed structures of CP-47,497 and HHC-DMH are displayed (C).

drug molecule. However, preparations in which the drug molecule is labeled in more than one position can be used if each ^{2}H-subspectrum in the composite ^{2}H-spectrum can be positively assigned to an individual label. This approach was initially used to study the orientation of cholesterol in the membrane and subsequently expanded in our laboratory into a more general method.[14]

Studies with (–)-Δ^9-THC required the stereospecific introduction of six individual deuterium atoms in the tricyclic component of the molecule (Figure 21.3). We found that the cannabinoid assumes an awkward orientation in the bilayer with the long axis of its tricyclic system perpendicular to the bilayer chains.[14,18] This can be accounted for by assuming that this amphipathic molecule uses its phenolic hydroxyl group to anchor itself at the interface of the amphipathic membrane bilayer; small-angle x-ray results confirmed this interpretation. We also found that cannabinoids with two hydroxyl groups orient in a manner that allows both hydroxyls to be near the interface facing the polar surface of the membrane.[14,19] Conversely, if the phenolic hydrogen of THC is replaced with a methyl group, the cannabinoid orients with its natural long axis parallel to the chains.[20]

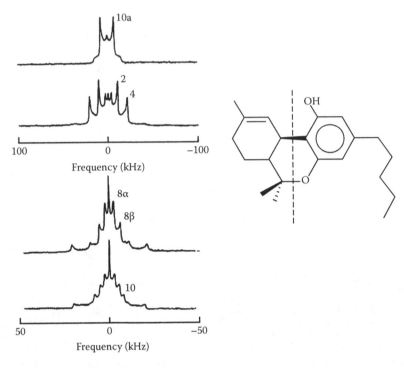

FIGURE 21.3   Left: Solid-state ^{2}H-NMR spectra of model membrane samples containing specifically ^{2}H-labeled Δ^9-THC. Four deuterated Δ^9-THC analogues have ^{2}H-labels in the 2, 4, 8α, 8β, 10, 10α-positions. Right: Orientation of Δ^9-THC in the membrane as determined from the solid-state ^{2}H-NMR results. The dashed line represents the direction of the lipid acyl chains in the membrane.

CANNABINOID TOPOGRAPHY IN THE MEMBRANE

Information on the location of the drug in the bilayer can be obtained from small-angle x-ray or neutron diffraction experiments using carefully hydrated membrane samples into which the cannabinoid analog is introduced. The well-layered membrane preparation is exposed to a small-angle x-ray beam, from which several diffraction orders are obtained. After transformation, these orders can provide an electron density profile of the membrane bilayer (Figure 21.4) in which the highest electron density corresponds to the phosphate head groups and the lowest to the ends of the chains in the center of the bilayer. In such an experiment, three membrane preparations are compared. The first of these contains no drug molecule. The second preparation contains the cannabinoid molecule under study, and the third contains the same cannabinoid analog into which a high electron density substituent (e.g., Br, I) is introduced as a marker. The difference between the electron density profiles of the previous three preparations reveals the positions of the cannabinoid tricyclic ring system and that of the heavy atom, and thus provides direct information on the location of the drug in the membrane. If the high electron density atom is placed at the end of the cannabinoid chain, then the experiment also provides information on the conformation of this flexible portion of the drug molecule in the membrane.

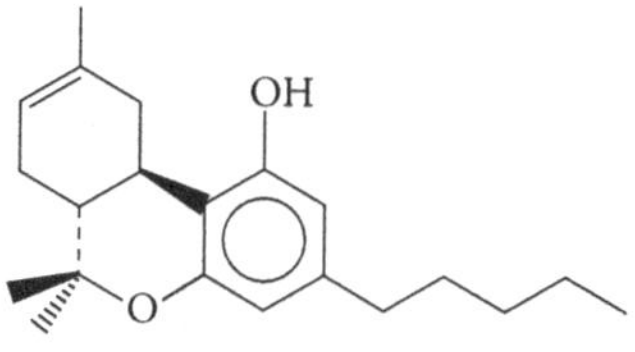

(-)-Δ⁸-tetrahydrocannabinol

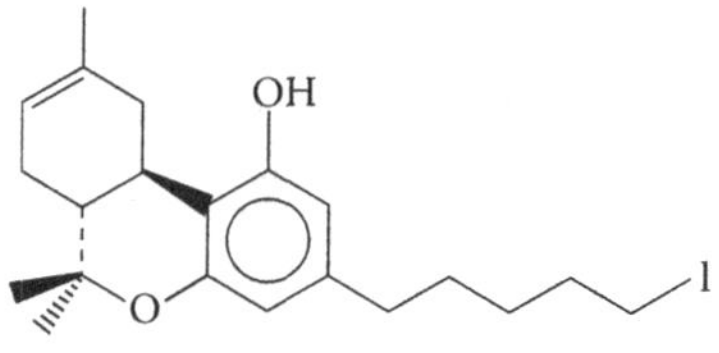

(-)-5′-iodo-Δ⁸-tetrahydrocannabinol

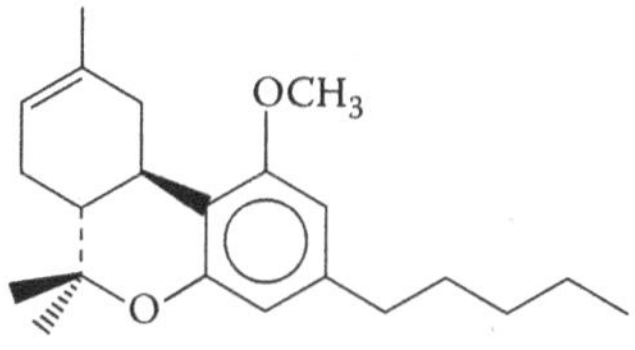

(-)-O-methyl-Δ⁸-tetrahydrocannabinol

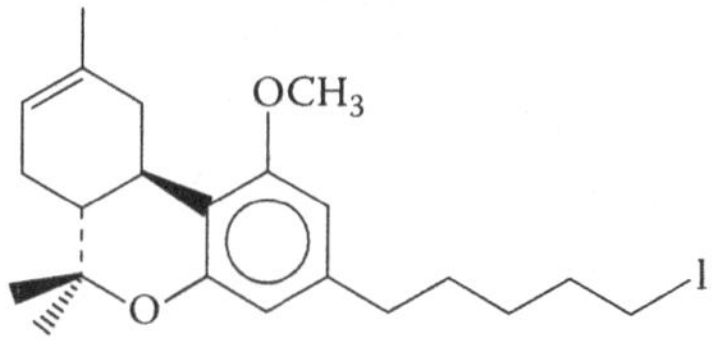

(-)-5′-iodo-O-methyl-Δ⁸-tetrahydrocannabinol

FIGURE 21.4 Electron density profile obtained by Fourier transformation of small-angle diffraction intensities from a DMPC model membrane. A molecular graphical representation above the profile shows the space correlation in the dimension across the bilayer.

Our experiments with (−)-Δ⁹-THC, (−)-Δ⁸-THC, and their inactive O-methyl analog provided us with data that complemented the ²H-NMR results. Such experiments required the synthesis of these cannabinoids as well as analogues having a halogen (I) substituent in the 5′-position of the cannabinoid side chain (Figure 21.5).

The study, which made use of membrane preparations from partially hydrated phosphatidyicholine, showed that the biologically active (−)-Δ⁹-and (−)-Δ⁸-THC analogues are located in the upper portion of the phospholipid bilayer with the phenol hydroxyl anchored at the bilayer interface.[20] However, when the polar phenolic hydroxyl is replaced with the more lipophilic methoxyl group, the cannabinoid now sinks deeper into the bilayer. The study also revealed that the location of the iodine in the bilayer implies that the cannabinoid side chain is parallel with the bilayer chains and in a fully extended *all-trans* conformation.

When the studies were extended from model to biological membranes obtained from rat or calf brain synaptosomes, there was no significant change in the manner in which the cannabinoid analogues orient in the membrane.[21,22] However, the x-ray data revealed a small but significant variation in the location of the drug in the bilayer. Indeed, we found that (−)-Δ⁸-THC and its halogenated analog are located 2–3Å deeper in the membrane preparations obtained from calf brain synaptosomes when compared with the model membrane preparations. The data from the synaptosomal plasma membrane preparations also showed that the THC side chain assumes a more compact *gauche-trans-gauche* conformation unlike the fully extended conformation observed in the phosphatidylcholine preparations. When molecular modeling was used to represent our data from the biological membranes, we found that the phenolic hydroxyl in the cannabinoid can still interact with the bilayer interface through hydrogen bonding with either one of the phospholipid carbonyls or with a water molecule located slightly below the interface (Figure 21.6).

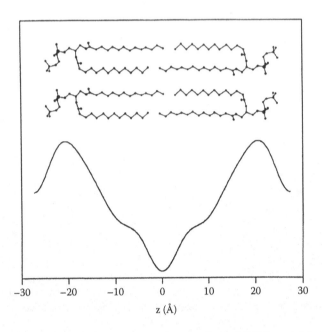

FIGURE 21.5 Structures of (−)-Δ^8-tetrahydrocannabinol (Δ^8-THC) and (−)-O-methyl-Δ^8-tetrahydrocannabinol (Me-Δ^8-THC), along with the respective iodinated analogues 5′-I-Δ^8-THC and 5′-I-Me-Δ^8-THC.

A biophysical method that can be used in an analogous manner with the x-ray method described earlier is neutron diffraction. In this method, the high electron density halogen marker used in the x-ray experiment is now replaced with deuterium atoms. The two membrane preparations in this method contain, respectively, unlabeled and specifically ^{2}H-labeled cannabinoids. The location of the ^{2}H-label is then revealed by the difference between the data from the two preparations. One advantage of this method is that the two membrane samples being compared are almost identical. This avoids the potential pitfall encountered in the x-ray experiment where the samples being compared are not totally identical and may have somewhat different phase properties, thus requiring very rigorous controls. Data using ^{2}H-labeled (−)-Δ-THC in neutron diffraction experiments[23] have confirmed, to a large extent, the earlier x-ray diffraction data described previously.

The combined data from the solid-state ^{2}H-NMR and small-angle x-ray diffraction experiments support our hypothesis that the ability of a cannabinoid to orient "properly" in an amphipathic membrane can seriously influence its ability to bind to its active sites on the receptor. This "proper" orientation is one in which all the cannabinoid's polar hydroxyl groups face the polar side of the membrane bilayer and, in turn, is determined by the position and relative stereochemistry of these hydroxyl groups in the cannabinoid molecule.

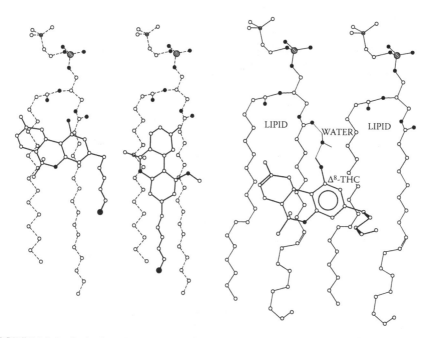

FIGURE 21.6 Left: Graphical representation of the locations and orientations of 5′-I-Δ8-THC and 5′-I-Me-Δ8-THC in a DMPC model membrane based on results from a small-angle x-ray diffraction and solid-state 2H-NMR. Right: Graphical representation of a Δ8-THC molecule in a synaptosomal plasma membrane.

CANNABINOID BINDING SITES IN MAMMALIAN BRAINS

The distribution of a cannabinoid in the brains of live animals can be studied using positron emission tomography (PET). This method requires the labeling of the drug probe with a short-lived positron-emitting isotope. To carry out such an experiment, we synthesized an 18F-containing analog of the parent (–)-Δ8-THC, namely (–)-5′-18F-Δ8-THC. The probe was administered to live baboons, and the presence and distribution of the drug in the brain was followed using computerized tomography. Among our findings[24] was that the drug reaches its maximum concentration in the brain (approximately 0.02% dose/cc) during the first 5 min and persists (0.005% dose/cc) up to 2 h after injection (Figure 21.7). High drug concentrations were found in the striatum, thalamus, and cerebellum. A shortcoming in these early experiments was the high lipophilicity of the probe as well as its relatively low affinity for the receptor sites. Both of these factors resulted in relatively high levels of nonspecific binding by the ligand in the brain. Currently under way is the synthesis of novel cannabinoid PET probes possessing higher affinities for the receptor and more favorable partitioning properties.

CANNABINOID AFFINITY PROBES

The task of obtaining direct information on the molecular features of the cannabinoid receptor-binding sites can be successfully pursued with the help of cannabinoid

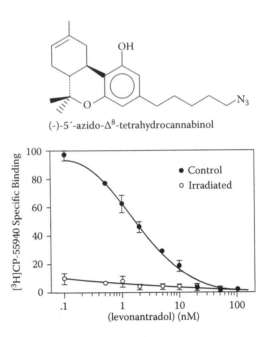

(-)-5′-azido-Δ⁸-tetrahydrocannabinol

FIGURE 21.7 Time courses for (–)-5′-¹⁸F-Δ⁸-THC distribution in baboon brain, showing the rates of accumulation and clearance of the radioactivity in the various areas of the brain.

affinity labels. The ligands, after interacting with the cannabinoid-binding sites, can covalently attach themselves to the receptor. The affinity probes should be especially useful for the isolation and characterization of the cannabinoid receptor. The most successful of the first-generation cannabinoid probes was (–)-5′-azido-Δ⁸-THC (Figure 21.8), the first photoaffinity label for the cannabinoid receptor. This molecule was shown to specifically bind and inactivate the receptor after irradiation with ultraviolet light.[25] Furthermore, its ¹²⁵I-radiolabeled analog, when incubated with two different membrane preparations, was shown to covalently label the receptor.[26] The use of an aliphatic azide as a photoaffinity label represents a departure from the generally used aromatic azides which, upon photoactivation, are transformed into reactive aromatic nitrenes. The aliphatic azido group in the photoaffinity probe described here is also expected to be transformed to a nitrene when subjected to ultraviolet irradiation. However, it is still unclear whether the originating highly unstable aliphatic nitrene directly reacts with the receptor or whether it first rearranges to a more stable intermediate that, in turn, covalently attaches itself at or near the active sites on the receptor. The mechanism of this photoactivation reaction is under study.

We have developed second-generation photoaffinity probes that possess considerably higher affinity and reactivity for the cannabinoid receptor, as exemplified by the one depicted in Figure 21.9 (IC$_{50}$ about 0.25 nM).[25,27] These probes are ideally suited for pursuing the important but difficult task of labeling and characterizing the cannabinoid receptor. In addition to the photoactivatable receptor probes, we have also developed a series of electrophilic probes that react covalently with the receptor without requiring an intervening activation step. In this new family of probes,

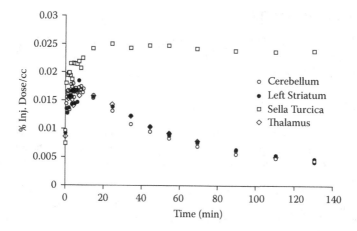

FIGURE 21.8 Photoirradiation of (-)-5'-N$_3$-Δ^8-THC with forebrain membranes inhibits specific binding of [^{3}H]-CP-55,940 to the cannabinoid receptor. Heterologous displacement by levonantradol is shown for membranes irradiated in the absence or after equilibration with 5'-azido-THC.

the azido group is substituted with an isothiocyanate, a group that has low reactivity with thiol and amino groups. Because of their relative stability in water, these analogues may prove to also act *in vivo* as irreversible antagonists.

The previously described affinity labels should prove to be useful biochemical tools in our efforts to characterize the active sites of the cannabinoid receptor. Currently, this effort is being pursued using cloned cannabinoid-receptor preparations. The availability of molecular probes that covalently react at or near the receptor active site, hopefully, will allow us to obtain detailed structural information on this active site and, perhaps, help in elucidating the molecular events leading to receptor activation or inactivation. Such data can also be combined with the results obtained from the biophysical experiments so that a more integrated view can be obtained of the molecular basis for cannabinoid activity.

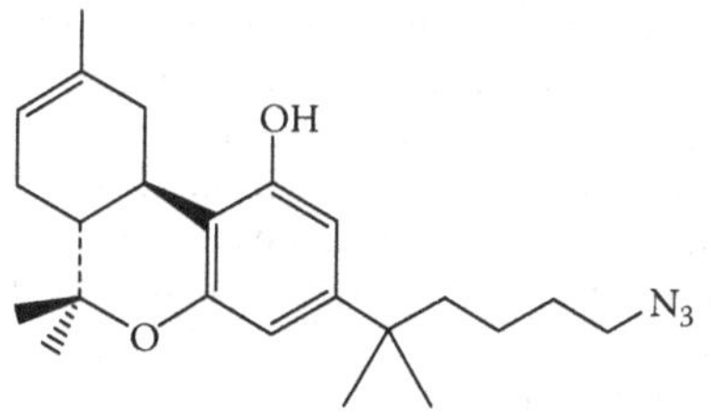

(-)-1',1'-dimethyl-5'-azido-Δ^8-tetrahydrocannabinol

FIGURE 21.9 Structure of (–)-1', 1'-dimethyl-5'-N$_3$-Δ^8-THC.

ACKNOWLEDGMENTS

We would like to thank D.P. Yang for useful discussions and R. Galligan for help in the preparation of this manuscript. This work was supported by grants DA-3801, DA-00152, from NTDA.

REFERENCES

1. Martin, B.R., Cellular effects of cannabinoids, *Pharmacol. Rev.* 38, 45, 1986.
2. Hilliard, C.J. et al., Studies of the role of membrane lipid order in the effects of delta 9-tetrahydrocannabinol on adenylate cyclase activation in heart, *J. Pharmacol. Exp. Therap.*, 252, 1075, 1990.
3. Leuschner, J.T.A. et al., The partitioning of delta 1-tetrahydrocannabinol into erythrocyte membranes *in vivo* and its effect on membrane fluidity, *Experientia*, 40, 866, 1984.
4. Devane, W.A. et al., Determination and characterization of a cannabinoid receptor in rat brain, *Mol. Pharmacol.*, 34, 605, 1989.
5. Devane, W.A. et al., In *An International Research Report*, Chesher, G., Consore, P., and Musty, R., Eds., Australian Government Publishing Service, Canberra, 1988, p. 141.
6. Herkenham, M. et al., Cannabinoid receptor localization in brain, *Proc. Natl. Acad. Sci. U.S.A.*, 87, 1932, 1990.
7. Bidaut-Russel, M., Devane, W.A., and Howlett, A.C., Cannabinoid receptors and modulation of cyclic AMP accumulation in the rat brain, *J. Neurochem.*, 55, 21, 1990.
8. Howlett, A.C. and Fleming, R.M., Cannabinoid inhibition of adenylate cyclase biochemistry of the response in neuroblastoma cell membranes, *Mol. Pharmacol.*, 27, 429, 1985.
9. Matsuda L.A. et al., Structure of a cannabinoid receptor and functional expression of the cloned cDNA, *Nature*, 346, 561, 1990.
10. Gerard, G.M. et al., Nucleotide sequence of a human cannabinoid receptor cDNA, *Nucl. Acid. Res.*, 18, 7142, 1990.
11. Gerard, G.M. et al., Molecular cloning of a human cannabinoid receptor which is also expressed in testis, *Biochern. J.*, 279, 129, 1991.
12. Munro, S., Thomas, K.L., and Abu-Shaar, M., Molecular characterization of a peripheral receptor for cannabinoids, *Nature*, 365, 61, 1993.
13. Bruggemann, E.P. and Meichior, D.L., Alterations in the organization of phosphatidylcholine/cholesterol bilayers by tetrahydrocannabinol, *J. Biol. Chem.*, 258, 8298, 1983.
14. Makriyannis, A. et al., The orientation of (–)-delta 9-tetrahydrocannabinol in DPPC bilayers as determined from solid-state 2H-NMR, *Biochim. Biophys. Acta.*, 986, 141, 1989.
15. Herbette, L.G., Chester, D.W., and Rhodes, D.G., Structural analysis of drug molecules in biological membranes, *Biophys. J.*, 49, 91, 1986.
16. Kriwacki, R. and Makriyannis, A., The conformational analysis of delta 9- and delta 9,11-tetrahydrocannabinols in solution using high resolution nuclear magnetic resonance spectroscopy, *Mol. Pharmacol.*, 35, 495, 1989.
17. Reggio, P.H., Greer, K.V., and Cox, S.M., The importance of the orientation of the C9 substituent to cannabinoid activity, *J. Med. Chem.*, 32, 1630, 1989.
18. Makriyannis, A., DiMeglio, C., and Fesik, S.W., Anesthetic steroid mobility in model membrane preparations as examined by high resolution 1H and 2H NMR spectroscopy, *J. Med. Chem.*, 34, 1700, 1991.
19. Yang, D.P. et al., Solid state 2H-NMR as a method for determining the orientation of tetrahydro- and hexahydro-cannabinols in membranes, *Pharmacol. Biochem. Behav.*, 40, 553, 1991.
20. Yang, D.P. et al., Amphipathic interactions of cannabinoids with membranes. A comparison between delta 8-THC and its *O*-methyl analog using differential scanning calorimetry, X-ray diffraction and solid state 2H-NMR, *Biochim. Biophys. Acta.*, 1103, 25, 1992.

21. Mavromoustakos, T. et al., Study of the topography and thermotropic properties of cannabinoids using small angle x-ray diffraction and differential scanning calorimetry, *Biochim. Biophys. Acta.* 1024, 336, 1990.

22. Mavromoustakos, T. et al., Small angle x-ray diffraction studies on the topography of cannabinoids in synaptic plasma membranes, *Pharmacol. Biochem. Behav.*, 40, 547, 1991.

23. Martel, P. and Makriyannis, A., Topography of tetrahydrocannabinol in model membranes using neutron diffraction, *Biochim. Biophys. Acta.*, 1151, 51, 1993.

24. Charalambous, A. et al., PET studies in the primate brain and biodistribution in mice using (–)-5′-18F-delta 8-THC, *Pharmacol. Biochem. Behav.*, 40, 503, 1991.

25. Charalambous, A. et al., 5′-Azido-delta 8-THC: a novel photoaffinity label for the cannabinoid receptor, *J. Med. Chem.*, 35, 3076, 1992.

26. Burstein, S. et al., Detection of cannabinoid receptors by photoaffinity labeling, *Biochem. Biophys. Res. Commun.*, 176, 492, 1991.

27. Charlambous, A., Grzybowska, Y., and Makriyannis, A., *Problems of Drug Dependence: Proceedings, 54th Annu. Sci. Meet.*, 1992.

22 Physicochemical Profiling in Early Drug Discovery: *New Challenges at the Age of High-Throughput Screen and Combinatorial Chemistry*

Bernard Faller

CONTENTS

INTRODUCTION

Early physicochemical profiling has gained considerable importance over the last years as inappropriate pharmacokinetics in human volunteers has been shown to be the primary cause of compound withdrawal of new chemical entities. The paradigm in today's drug discovery is that the majority of research programs are intended for oral therapy, although most high-throughput screening techniques tend to shift leads toward compounds with unfavorable properties for absorption as hits tend to be (1) more lipophilic and therefore less soluble compounds, and (2) compounds with a higher number of hydrogen bond donors and acceptors and a larger molecular volume.

In the past, drug candidates were first optimized for potency (binding affinity, selectivity). Absorption, distribution, metabolism, and excretion (ADME) were usually considered at a later stage. In this scenario, when development was hindered by

	High Solubility	Low Solubility
High Permeability	**Class I** Dissolution rate limits absorption Diltiazem Labetalol Enalapril Propranolol	**Class II** Solubility limits absorption Flurbiprofen Naproxen
Low Permeability	**Class III** Permeability limits absorption Acyclovir Famotidine Nadolol	**Class IV** Significant problems for oral drug delivery are expected Terfenadine Furosemide

FIGURE 22.1 The Biopharmaceutical Classification Scheme (BCS).

ADME issues, the only rescue option available was to enter an extensive formulation program to minimize factors that prevent absorption. This approach is, however, always expensive and not always successful. In addition, it basically only applies to the absorption part, with distribution, metabolism, and excretion being closely associated with the intrinsic properties of the active ingredient.

Nowadays, strategies have been developed that allow parallel optimization of potency and ADME properties at a time when chemical modifications of the molecule are still possible. This new, parallel optimization approach triggers the development of new, innovative methods able to cope with the particular constraints of early profiling: high-throughput, low sample requirements, low sensitivity to impurities, ability to digest noncongeneric sets of molecules, and good predictive value. This chapter will focus on past, present, and future approaches to predict drug absorption based on solubility and permeability, the two components of the Biopharmaceutical Classification Scheme (BCS)[1] (see Figure 22.1).

PHYSICOCHEMICAL PARAMETERS GOVERNING DRUG ABSORPTION

As a first step, only absorption via the transcellular passive diffusion route is considered. This is the most common absorption pathway for generic drugs. Transcellular passive diffusion is mainly controlled by solubility and permeability, which together define the flux of the drug through the gastrointestinal barrier. In most absorption processes, the first step is the dissolution of the active ingredient (Figure 22.2). The link between solubility and permeability is given by the following relation, which is simply Fick's first law:

$$F = AC \times Pe \tag{22.1}$$

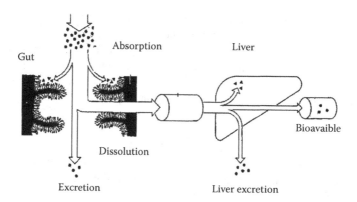

FIGURE 22.2 Drug absorption pathway.

where F is the flux, AC is the concentration gradient through the membrane, and Pe is the effective permeability. Both AC and Pe are pH dependent if the drug is ionizable with pKa's within the pH gradient encountered in the GI (3–8).

APPROACHES TO SOLUBILITY

Solubility is probably the most simple parameter to consider and, at the same time, the most difficult to accurately predict. We recently looked at different software programs to predict water solubility and came to the conclusion that thus far none was really significantly better than simple correlations based on CLOGP.[2] The difficulty lies in the fact that aqueous solubility is not only influenced by physicochemical properties of an isolated molecule such as molecular volume and hydrogen bond donor and acceptor strengths but also involves the energy necessary to disrupt the crystal lattice of the solid to bring it into solution. A more promising approach to predicting solubility may be to rely on structure similarity with compounds for which experimental data are available. This approach, however, requires a large database of experimental data with drug-like molecules. The use of kinetic solubility data obtained via nephelometric titration[2–6] appears to be the fastest way to build such a database.[7] Other approaches to measure solubility include shake-flask/HLPC and dissolution template potentiometric titration.[8] These methods provide high-quality data but are limited with respect to the number of samples that can be analyzed. They can be used to characterize more advanced compounds and serve as a test set to check the predictive value of the methods based on kinetic solubility data.

Solubility is particularly important for immediate release of class II drugs for which absorption is limited by solubility (thermodynamic barrier) or dissolution rate (kinetic barrier).[1,9] In early drug discovery, it is very difficult to measure dissolution rate because accepted methods require large amounts of material and are labor-intensive. Instead, one measures solubility as it correlates with dissolution rate as shown by the Noyes–Whitney equation:[10]

$$J = KA(Cs-C) \tag{22.2}$$

where J is the dissolution rate, K is a constant, A is the surface area of the dissolving solid, Cs is the saturation solubility, and C is the concentration in the medium. There are, however, a number of outliers in the solubility/dissolution rate correlation, particularly for medium and highly soluble ionizable compounds. The reason is that pH and viscosity in the diffusion layer might differ substantially from conditions in the bulk.[11] To take these two factors into account, Equation 22.2 can be rewritten as follows:

$$J = D \; A/h \; (Cs-C) \hspace{4cm} (22.3)$$

where D is the diffusion coefficient of the drug and h is the effective diffusion boundary layer thickness.

Inspection of Equation 22.3 shows what are—besides saturation solubility—the additional parameters influencing dissolution rate. These are surface area of the drug (A), diffusivity of the drug (D), and boundary layer thickness (h). In drug substance terms, this means molecular size (D parameter), particle size and wettability (A parameter), and stirring conditions (h parameter).

SOLUBILITY CLASSES: HIGH–MEDIUM–LOW

The next question related to drug absorption and solubility to be answered is: What is meant by low, medium, and high solubility?

If one follows the FDA guidelines for bioavailability studies[12] and assumes an average potency of 1 mg/kg, a drug is classified soluble when 60–70 mg of the active substance can be dissolved in 250 mL of water. Soluble therefore means that water solubility is higher than 0.2 mg/mL. Subsequently, by medium solubility is usually meant 0.01–0.2 mg/mL, and low when solubility drops below 0.01 mg/mL. It is important to point out that the solubility class needs to be put in perspective with the therapeutic dose (or potency). The boundaries just given can therefore move up or down by a factor of 5 to 10. It has been reported that less than 10% of the marketed drugs have a low kinetic (turbidimetric) solubility[6] according to the definition just given. In contrast, 30–40% of the compounds synthesized in medicinal chemistry programs typically fall in that category.[7]

APPROACHES TO PERMEABILITY

Permeability is a measure of how fast a given molecule is able to cross a barrier. Most of the times, the barrier of interest is the gastrointestinal wall, or the blood–brain barrier. Other barriers are also of interest, such as the cornea or the skin for eye and dermal drug delivery, respectively. Molecules can cross biological membranes through three routes: transcellular diffusion (the most common for xenobiotics), paracellular diffusion (small, hydrophilic compounds), and active transport systems. From a physicochemical viewpoint, permeability is mainly dictated by the hydrogen-bond acceptor, donor strength, and molecular size. These parameters are traditionally obtained through lipophilicity determinations that are obtained by distribution studies in a biphasic system (i.e., octanol/water).

This approach is, however, rather labor-intensive because (1) distribution behavior itself can be complicated to measure, particularly for ionizable substances, and (2) several partition solvents are necessary to deconvolute the molecular forces encoded in lipophilicity (i.e., octanol–alkane–chloroform).

More recently, assays based on artificial model membranes were introduced, in which true permeability rather than distribution behavior can be measured.[13–15] The main advantage of these assays is that they (1) can be carried out in microtiterplate (high-throughput), (2) are highly reproducible and easy to standardize, and (3) allow for the measurement of permeability pH-profiles (>50% of drug candidates ionize within the pH range 3–8). Indeed, ionizable compounds must be characterized by their solubility/permeability pH-profile rather than single-point measurements; otherwise, the predictive value of the assay can be lost. This point is illustrated with examples later in the text. These assays also have limitations because some features of the GI tract are not modeled by artificial membranes. For instance, the paracellular route by which a number of water-soluble drugs are taken up is not taken into account. In a first attempt to circumvent this limitation, one can simply flag water-soluble compounds with a molecular weight lower than 250 as these are potential candidates for paracellular transport. This is, however, not totally satisfactory because the shape of the molecule and the surface polarity distribution are not taken into account. Proper prediction of paracellular transported compounds remains an issue even when cellular models such as Caco-2 monolayers are used because cell junctions are usually tighter than in the GI epithelium. It should also be noticed that paracellular transported compounds are prone to high interspecies variations, presumably due to some structural differences in the GI tract. The second kind of limitation lies in the absence of active transport or efflux systems that can play a major role in GI and blood–brain barrier penetration. When active transport systems are involved more sophisticated tools are required, such as Caco-2 monolayers (GI-absorption)[16] or cocultures of astrocytes and endothelial cells (BBB penetration).[17]

CASE STUDIES

The following section is aimed at illustrating some of the points raised previously with concrete examples taken from drug discovery.

Permeability of Ionizable Compounds is pH-Dependent

We have previously shown that permeability is pH-dependent for ionizable compounds. This is illustrated in Figure 22.3 in which Caco-2 cell monolayers were used to measure the permeability of CGP75254A, an orally active iron chelator. Although the compound is well permeable at pH 6–7, Papp strongly decreased at pH 8 to values where poor absorption is anticipated. Physicochemical factors responsible for this variation of permeability around pH 7.4 were discussed by Lowther et al.[18] Generally, when compounds with pKa values in the range of 6–8 (as well as acids with pKa values below 7) are measured in Caco-2 cells at a single pH (usually 7.4), the results must be interpreted carefully, and possibly call for additional investigations. More recently, we have measured permeability

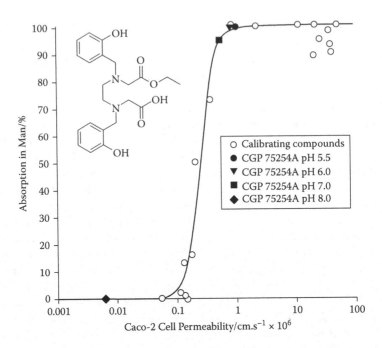

FIGURE 22.3 Permeability pH-profile measured with CaCo-2 cells.

of the angiotensin II antagonist Valsartan with the same system: very poor permeability values were found using the standard assay (pH 7.4), whereas in animal studies the fraction absorbed was 30–40%. The reason is that Valsartan is a di-acid with pKa values at 4.7 and 3.9; therefore, it is highly charged at pH 7.4. In contrast, permeability pH-profile showed that permeability was becoming significant at pH <5, thus allowing for absorption in the upper GI. Artificial model membranes are better suited to measure permeability pH-profiles than Caco-2 monolayers because (1) the latter are difficult to maintain at pH <6.5, and (2) it is easier to reduce the unstirred layer thickness with artificial membranes.

Solubility-Limited Absorption (BSC Class II)

This is to illustrate how the flux (F = Pe × AC) is influenced by ionization constants, permeability, and dose. The example illustrated in Figure 22.4 is a dibasic compound with pKa's at 7.8 and 3.3. Figure 22.4A shows the solubility pH-profile measured using the pH-metric technique.[8] The permeability pH-profile of the compound was measured using the artificial membrane technique[15] at two different concentrations in the donor compartment (Figure 22.4B, C): 0.05 mg/mL and 0.001 mg/mL (MS detection). Figure 22.4B shows that logPe versus pH is bell-shaped with a maximum at pH 6.5. At low pH, the amount of compound that reaches the acceptor compartment is limited by a low permeability because the compound is largely ionized at pH <7. At pH >7, the flux between donor and acceptor compartments is limited by solubility. In fact, Figure 22.4B represents

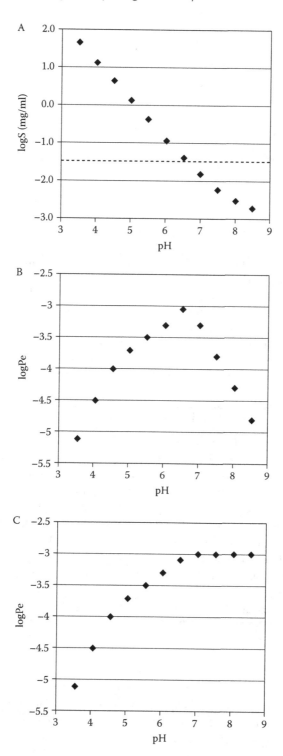

FIGURE 22.4 Artificial membranes to identify BCS class II compounds.

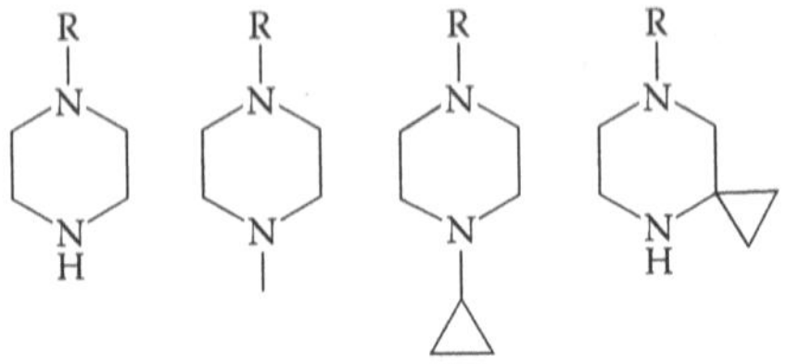

Binding (nM)	2.5	7.5	210	19
pKa	8.5	7.5	6.5	7.6
BAV (%)	4	28	–	–
cpKa (ACD)	8.4	6.6	6.8	9.4

FIGURE 22.5 The ionization/binding/bioavailability triad.

the flux pH-profile rather than the permeability pH-profile. The maximum flux is reached at the pH where solubility becomes rate limiting at the concentration in the donor compartment (see dotted line in Figure 22.4A). When the experiment is repeated at a lower concentration (Figure 22.4C), the compound flux measured with the artificial membranes corresponds again to the permeability pH-profile, with permeability increasing with a fraction of neutral substance. The plateau of Figure 22.4C corresponds to the diffusion limit by the unstirred layers.

The Ionization/Affinity/Permeability Triad

In some cases, ionic bonds participate in the binding to the receptor or ligand. Hence, permeability and binding affinity will be strongly intercorrelated, and only a subtle balance between the two parameters will lead to good binding/ selectivity and acceptable absorption. This is illustrated in Figure 22.5 with a real example from one of our lead optimization programs. In this series, binding was lost when pKa was lower than 7.5 (presumably because the ionic bond was destroyed), whereas bioavailability was lost when the pKa was higher than 7.5 because the fraction of neutral (permeable) species became too small. It is also worthy of mention that, in this case, experimental values are better suited than calculated values because subtle chemical modifications introduced to modulate ionization are not always properly taken into account by the software.

CONCLUSION AND OUTLOOK

Physicochemical profiling has changed quite dramatically in the last decade. On the one hand, a number of new experimental and theoretical approaches have been explored, whereas on the other, considerable information has been generated related to the *in vivo* significance of the parameters measured or calculated. Numerous software programs are now available to predict octanol/water partition coefficients, ionization constants, and solubility. Although the two former have probably reached maturity, significant improvements are still needed for the latter. Solvents other than octanol have been explored (hexane, hexadecane, chloroform, and dichloroethane) and have given additional insight into the understanding of the underlying forces governing drug disposition. Hydrogen-bond potential has emerged as perhaps the most critical parameter governing solubility/permeability properties. H-bond acidity and basicity can be estimated using software tools such as HYBOT (Raevsky approach) or ABSOLV (Abraham approach). As an alternative, the Polar Surface Area, a less computer-demanding parameter, was proposed. New experimental concepts have emerged, such as dissolution template potentiometric titration for solubility pH-profile measurement, or the use of artificial membranes to measure permeability in the microtiterplate format. The most striking development from the end of the millennium is probably the focus on physicochemical screening methods to rapidly assign BCS classification to large numbers of compounds, rather than the determination of true physicochemical parameters. This trend will probably continue and develop due to the pressure on early ADME profiling. New technological developments can be anticipated to cope with the increasing number of targets (through proteomics/genomics approaches) and the increasing number of compounds to characterize. Challenges for the next generation of profiling tools will be to reduce the constraints on compound requirements for analysis (amounts, purity), whereas the predictive value should increase through improved data quality, parameter diversity, and data mining.

REFERENCES

1. Amidon, G.L. et al., A theoretical basis for a biopharmaceutic drug classification: the correlation between *in vitro* drug product dissolution and *in vivo* bioavailability, *Pharm. Res.*, 12, 413, 1995.
2. Faller, B. and Wohnsland, F., Physicochemical parameters as tools in drug discovery and lead optimization, in *Pharmacokinetic Optimization in Drug Research: Biological, Physicochemical and Computational Strategies,* Testa, B., Van de Waterbeemd, H., Folkers, G., and Guy, E., Eds., Verlag HCA, Basel, Switzerland, 2001, pp. 257–273.
3. Brooker, P.J. and Ellison, M., The determination of water solubility of organic compounds by a rapid turbidimetric method, *Chem. Ind.*, 1974, 785.
4. Davis, W.W. and Parke, T.V., A nephelometric method for the determination of solubilities of extremely low order, *Am. Chem. Soc.*, 64, 101, 1942.
5. Bevan, C.D. and Lloyd, R.S., A high-throughput screening method for the determination of aqueous drug solubility using laser nephelometry in microtiter plates, *Anal. Chem.*, 72, 1781, 2000.
6. Lipinski, C.A. et al., Experimental and computational approaches to estimate solubility and permeability in drug discovery and development settings, *Adv. Drug. Del. Rev.*, 23, 3, 1997.
7. Lipinski, C.A., Avoiding investment in doomed drugs, *Curr. Drug Discov.*, 1, 17, 2001.
8. Avdeef, A., pH-Metric solubility. 1. Solubility-pH profiles from Bjerrum plots. Gibbs buffer and pKa in the solid state, *Pharm. Pharmacol. Commun.*, 4, 165, 1998.
9. Dressman, J.B. et al., Dissolution testing as a prognostic tool for oral drug absorption: immediate release dosage forms, *Pharm. Res.*, 15, 11, 1998.
10. Noyes, A. and Whitney, W.R., The rate of solution of solid substances in their own solutions, *Am. Chem. Soc.*,19, 930, 1897.

11. Serajuddin, A.T. and Jarowski, C.I., Effect of diffusion layer pH and solubility on the dissolution rate of pharmaceutical acids and their sodium salts II: salicylic acid, theophylline and benzoic acid, *Pharm. Sci.*, 74, 148, 1985.

12. Guidance for Industry Waiver of *In Vivo* Bioavailability and Bioequivalence Studies for Immediate Release Solid Oral Dosage Forms, www.fda.gov.cder/guidance/index.htm.

13. Camenisch, G., Folkers, G., and van de Waterbeemd, H., Comparison of passive drug transport through Caco-2 cells and artificial membranes, *Int. J. Pharm.*, 147, 61, 1997.

14. Kansy, M., Senner, F., and Gubernator, K., Physicochemical high-throughput screening: parallel artificial membrane permeation assay in the description of passive absorption processes, *Med. Chem.*, 41, 1007, 1998.

15. Wohnsland, F. and Faller, B., High-throughput permeability pH profile and high-throughput alkane/water logP with artificial membranes, *Med. Chem.*, 44, 923, 2001.

16. Artursson, P. and Karlsson, J., Correlation between oral drug absorption in humans and apparent drug permeability coefficients in human intestinal epithelial (Caco-2) cells, *Biochem. Biophys. Res. Commun.*, 175, 880, 1991.

17. Gaillard, P.J. et al., Establishment and functional characterization of the blood-brain barrier, comprising a co-culture of brain capillary endothelial cells and astrocytes, *Eur. J. Pharm. Sci.*, 12, 215, 2001.

18. Lowther, N. et al., Caco-2 cell permeability and physicochemical properties of a new oral iron chelator, *J. Pharm. Sci.*, 87, 1041, 1998.

23 Drug–Membrane Interaction and Its Importance for Drug Efficacy

J.K Seydel, E.A. Coats, K. Visser, and M. Wiese

CONTENTS

INTRODUCTION

In this review, the possible role of membranes on drug transport, distribution, accumulation, efficacy, and resistance is discussed.

The interest in drug development, research, and QSAR is especially focused on the interaction of ligand (drug) molecules and the proteins of the specific receptor. It is widely assumed that the biological activity of drugs (such as tranquilizers, anesthetics, β-blockers, antidepressants, etc.) arises as a result of binding to active sites in membrane-bound proteins, whereas the lipid background is considered to play a more passive role. In the last decade, however, evidence is increasing that we may have underestimated the influence of drug–membrane interactions on drug action. These membranes do not consist of lipids only but possess polarized phosphate groups and neutral, positively or negatively charged head groups, and are highly structured and chiral. The consequences of drug–membrane interaction on transport, distribution, accumulation, efficacy, and resistance cannot always be explained sufficiently by mere partitioning processes, i.e., by log $P_{octanol/water}$.

In the case of chemotherapeutics, the interaction of the drug molecules with membranes of, for example, bacteria or fungi must be considered additionally. These membranes have a much more complex construction than mammalian membranes.

TABLE 23.1

Drug Membrane Interaction

I. Action of the membrane on drug molecules

   Diffusion through membranes may become rate limiting

   Membrane may completely prevent diffusion to the active site

   Solvation of the drug in the membrane may lead to conformational changes of the drug molecule

II. Drug action on membrane properties

   Drug may change conformation of acyl groups (trans-gauche)

   Drug may increase membrane surface

   Drug may increase membrane fluidity

   Drug may change membrane potential and hydration of head groups

   Drug may change membrane fusion

In many cases they are asymmetric, as for example, the membrane of Gram-negative bacteria.

This membrane is characterized by an outer hydrophilic core rich in polysaccharides, and a hydrophobic phospholipid bilayer. Therefore, a balance in hydrophobic/hydrophilic properties is indicated for an effective inhibitor.

Experimental work on artificial lipid layers or model membranes has demonstrated that the structural properties of these membranes may be strongly affected by the presence of membrane-associated molecules. Some of the possible events that can arrive from drug–membrane interaction are summarized in Table 23.1.

Examples of these events will be given during this review. A general model has been discussed by Herbette and co-workers.[1] Because binding sites of some receptors are embedded in the membrane, drug molecules need to interact favorably with membrane components in order to get access to the receptor. In addition to the influence of the membrane on drug distribution and rate of diffusion, the change induced in the organization of the bilayer by drug molecules can affect the conformation of membrane-associated receptors, thus modifying agonist–receptor interaction. Alternatively, the conformation of the drug may be changed in the membrane, or equilibrium between conformers is changed.

How can we measure and quantify drug–membrane interaction? Because of the physicochemical properties of phospholipids, they readily form bilayers (artificial membranes, liposomes) that can be used to study drug–membrane interaction and diffusion. The importance of neutral and charged liposomes for modeling such interactions in various tissues has been shown on numerous examples. Methods to measure and quantify such interactions are summarized in nuclear magnetic resonance (NMR) techniques, differential scanning calorimetry, x-ray diffraction, Ca^{++}-displacement, fluorescence, and many other techniques that are excellent tools that allow sensitive determinations of the interaction of drugs with artificial membranes.[2]

Amphiphilic drugs, in particular, can interact strongly with membranes. A large number of drug molecules possess amphiphilic properties for producing a wide variety of pharmacological effects. These effects may be direct results of interaction with proteins embedded in the membrane, or the indirect result of changes induced in the organization of the phospholipids. The latter effect can, for example, lead to changes in phase transition temperature at which transition from the gel to liquid crystalline phase occurs, which is characteristic for each phospholipid. Such drugs can displace Ca^{++} from phospholipid monolayers.[3,4]

NMR techniques offer even more detailed information. NMR spectra can be characterized by the magnetic field at which resonance occurs, depending on the surrounding of the nucleus considered, the degree of spin–spin coupling produced by neighboring effects, the spin lattice relaxation rate $1/T_1$, and the spin–spin relaxation rate $1/T_2$, the latter expressed as line width of the resonance signal.

The interaction between drug and phospholipid can change one or several of these parameters in a characteristic manner in both the drug and the receptor molecules.

Additional information on molecular conformation can be obtained by using the nuclear Overhauser effect (NOE) or transfer NOE, used to measure the distance between nuclei.[5]

These methods allow the changes in membrane organization to be localized and quantified, and they also provide information on substructures of the drug molecule that are involved in the interaction. Proton and ^{13}C NMR spectra in the presence of phospholipids can lead to changes in $1/T_1$ and $1/T_2$ relaxation rates. The change in $1/T_2$ is related to a decrease in rotational freedom of drug molecules in the presence of "receptor". These changes, expressed as function of the drug–lipid ratio, can be used to determine K_D values to quantify the degree of interaction. This parameter can then be used for the derivation of structure–activity (binding) relationships.[6]

DRUG TRANSPORT

Drug transport may occur by passive diffusion processes through bilayers or pores. In the first case, we can expect to describe this by the octanol/water partition coefficient, and in the second case by molecular weight or surface of the drug molecule. This has been exemplified by the diffusion-controlled onset of inhibition of *E. coli* cultures by rifampicin derivatives. The delay in onset of inhibition was a function of lipophilicity and concentration.[7]

On a series of sulfonamides, it can be shown that molecular weight becomes the rate-limiting step for the inhibitory activity if it becomes larger than 300. The pore size of *E. coli* membranes is about 600 kDa. If the activity is determined against whole cell bacteria, the variation in inhibitory activity (MIC) can be described by the steric effect of *o*-substituents and the molecular weight or the surface of the substituents, respectively.

In contrast, the inhibitory activity of the same sulfonamides against the isolated target enzyme depends solely on the steric effect of the substituents, whereas molecular weight is no longer a limiting factor.[8]

In case of mammalian membranes, it has been shown that the placental transfer ratio (TR) of a heterogeneous set of compounds through the placental membrane can significantly be described by the octanol/water partition coefficient, i.e., only the overall partition coefficient of the molecule in octanol mimics drug transport.[9]

$$\log TR = 0.39 \ (0.068) \log P - 0.478 \ (0.054) \log (0.53 \ P + 1) - 0.03 \ (0.11) \quad (23.1)$$

$$n = 20 \qquad r = 0.96 \qquad s = 0.09 \quad F = 54$$

If other than bulk effects, as for example hydrogen-binding capacity, are involved in the diffusion process, the octanol/water partition coefficient alone is not a sufficient parameter. This is true for the transport across the more complicated blood–brain barrier. In this case, R.C. Young et al.[10] have introduced $\Delta \log P$, the difference between the partition coefficient in the system octanol/water and cyclohexane–water, as a measure of overall hydrogen-binding capacity. A satisfying relation between $\Delta \log P$ and the rate of brain penetration for a heterogeneous series of compounds has been found.

DRUG DISTRIBUTION AND ACCUMULATION

There is another important aspect of drug–membrane interaction, the distribution or accumulation of drug molecules in membranes. When one ligand interacts with two different biological samples, one containing the receptor subtype I and the other the receptor subtype II, *apparent* dissociation constants of the ligand-receptor complex can be obtained from log dose–response curves. The usual interpretation is that the stronger affinity, e.g., at side II, is a consequence of the ligand structure meeting the receptor requirements of site II better than of I.

An alternative interpretation, however, is that the ligand is present near site II in a greater concentration or in a more suitable conformation and orientation than near site I.[11] This could be the consequence of a ligand–membrane interaction near the membrane-bound receptor. In that case, the membrane requirements of site I and II are different.

This can be shown by the partition behavior of dihydropyridine type compounds into biological membranes. The partitioning into sarcoplasmic reticulum for some derivatives has been determined by Herbette and co-workers and compared with partition coefficients obtained for the system octanol/water[12] (Table 23.2).

Two facts are becoming immediately obvious:

1. The partitioning into biological membranes is much stronger as compared to the octanol/water system.
2. The ranking is different.

The first point is of great importance. It means that the concentration within the membrane is considerably higher as compared to the surrounding water phase. This means that K_D values using concentrations in the water phase are certainly not correct. It is obvious that the partition coefficients do not only depend on the structure of the drug molecules but also on the structure and composition of the biological membranes. There are data available from the literature, which strongly support these arguments. Validiva and co-workers[13] have used a calcium channel reconstituted into planar bilayers of different phospholipid constitution. They were able to show that the affinity of a positively charged dihydropyridine is increasing by a factor of 10

TABLE 23.2
Partition Coefficients of Dihydropyridine-Type Drugs into Biological Membranes and into *n*-Octanol

Drug	Partition Coefficient into Biological Membranes[a]	Partition Coefficient into Octanol/Buffer
Bay P 8857	125,000	40
Iodinine	26,000	—
Amlodipine	19,000	30
Nisoldipine	13,000	40
Bay K 8644	11,000	290
Nimodipine	6,300	730
Nifedipine	3,000	—

[a] The values shown were obtained using sarcoplasmic reticulum, but similar values were obtained with cardiac sarcolemmal lipid extracts. The results indicate a primary interaction of the drug with the membrane bilayer component of these biological membranes.

when the ratio of charged-to-uncharged phospholipids was increased by a factor of 2. As cholesterol is also changing membrane properties and, by this, drug distribution into membranes, this effect also has to be considered for disease states such as hypercholesterolemia.

Scheufler et al.[14] have published the partition coefficients using biological membranes for another series of compounds (Table 23.3). The partition coefficients of these drugs into biological membranes correlate—as shown in previous tables— only on a statistically, not highly, significant level with log $P_{octanol/water}$. The prediction power is very poor ($r^2 = 0.27$). In contrast, a highly significant correlation is obtained if the change in spin–spin relaxation rate ($\Delta 1/T_2$) of protons of the ligands in the presence of phospholipids is used as a measure of drug–membrane interaction (atria/medium AT/M).

$$\log AT/M = 0.861 \ (0.33) \log 1/T_2 + 1.428 \ (0.34) \qquad (23.2)$$

$$n = 5 \qquad r^2 = 0.96 \qquad s = 0.21 \qquad r^2_{cross\ val.} = 0.89 \qquad F = 68.8$$

The predicitive power is on the 90% level using the leave-one-out procedure.

Additional information on the substructure involved in the interaction can be obtained by NMR measurements. In the case of verapamil, a clear-cut difference in the involvement of substructures can be derived from the NMR measurements. In contrast, amiodarone shows the involvement of all substructures of the molecule in the drug–lipid interaction. This finding is supported by results from x-ray diffraction measurements showing amiodarone deeply buried into the lipid hydrocarbon chain.[14,15] Simulation of the binding mode of the energy-minimized structures of these two drugs into a complex of four lipid PPC molecules are in agreement with the x-ray and NMR studies. A high-binding energy is found for amiodarone buried between the hydrocarbon chains, whereas the calculated binding-energy for

TABLE 23.3

Erythrocyte (E/M) and Tissue (T/M) Medium Ratios of the Drugs in Human Erythrocytes, Rat Aortas, and Left Atria at pH 7.2 and Medium Concentration 10^{-6} mol/L

Drugs	E/M	T/M (Aortas)	T/M (Left Atria)
Flunarizine	196 ± 3	682 ± 52	784 ± 155
R 56865	37 ± 2	230 ± 47	402 ± 37
Nitrendipine	18 ± 1	77 ± 2	96 ± 4
Verapamil	6 ± 1	23 ± 2	57 ± 6
Diltiazem	3 ± 0.3	11 ± 0.8	28 ± 2
Lidocaine	1.5 ± 0.5	4.5 ± 1.1	5.0 ± 0.5

Note: The values represent mean ± S.D. (n = 6).

verapamil is significantly larger in the extended conformation where that part of the molecule for which no change in $1/T_2$ has been observed is not interacting with the phosphor head group but reaches into the water phase.

An example where even the preferred conformation of the drug molecule is changed in lipid environment is N-alkyl substituted benzylamines. These compounds produce negative inotropic effects (beating rate and contractile force, guinea-pig atria), which increases strongly at chain length larger than 4. We were able to show that this nonlinear increase in activity with chain length was due to a change in preferred conformation.[3,5] Up to a chain length of *n*-butyl, the extended molecular structure exists in the absence and presence of phospholipids. With longer chain length, however, a folded conformation is preferred in the presence of lipids.[6] Thus, the active conformation is created in lipid environment.

Next, some examples are presented where drug–membrane interaction seems to be directly related to drug efficacy. Neuroleptics act on the sodium/potassium channels, which are embedded on phospholipid-containing membranes. In a cooperative study with Astra Medica with the aim to reduce animal experiments, the interaction of flupirtine derivatives with phospholipid vesicles using NMR technique has been analyzed. The interaction was quantified, as discussed before, by the change in $1/T_2$ for the methylene protons as a function of increasing phospholipid concentration. Linear dependence was observed for a large range of lipid concentrations. The slope of such plots was used to characterize the degree of interaction. A highly significant correlation between the degree of interaction and the observed *in vivo* neuroleptic effect, measured in the metrazole electroshock test (MES), was observed.[6]

$$\log I/I_{50.\,MES} = 0.722 \, (0.12) \log I/T_2 + 3.00 \, (0.19) \qquad (23.2a)$$

$$n = 13 \qquad r^2 = 0.94 \qquad s = 0.086 \qquad r^2_{cross\,val.} = 0.92 \qquad F = 178$$

The increase in interaction was paralleled by an increase in the pharmacological effect over a large range of activity. It is also obvious that the correlation obtained using octanol/water partition coefficients of the compounds instead of the degree of interaction with phospholipids $(1/T_2)$ is inferior $(r^2_{c.v.} = 0.678)$.[6]

$$\log I/I_{50.\ MES} = 0.411\ (0.134)\ \log k'_{oct.} + 3.482\ (0.23) \qquad (23.2b)$$

$$n = 13 \qquad r^2 = 0.79 \qquad s = 0.158 \qquad r^2_{cross\ val.} = 0.678 \qquad F = 45$$

Especially, those derivatives bearing a polar substituent at the benzene ring deviate from the regression. This indicates again that the octanol/water partition coefficient is not a sufficient system if other than bulk effects are involved.

A change in lipophilicity at the ethyl carbamate group of the pyridine ring leads to a decrease in activity. Probably, the distribution of hydrophobic surface with respect to the cationic center of the flupirtine molecule is of importance. It could influence the orientation of the drug molecule in the membrane.

Another example, where the biological effect could be related to the partitioning into membranes, is taken from a paper of Choi and Rogers.[16] These authors have studied the partition behavior of adrenergic compounds using differently charged liposomes and also octanol. They found an excellent linear correlation between the hypotensive effect (PC_{25}) and the partition coefficient (k_{m2}) determined with neutral and negatively charged liposomes but not with $\log P_{octanol}$ $(r = 0.69)$.

$$PC_{25} = -2.562\ (1.13)\ \log k_{m2} + 6.33 \qquad (23.3)$$

$$n = 5 \qquad r^2 = 0.94 \qquad s = 0.228 \qquad r^2_{cross\ val.} = 0.84 \qquad F = 51.6$$

We have determined the retention time for these derivatives on a phosphatidylserine-covered column and found the same excellent correlation with the hypotensive effect.

$$PC_{25} = -2.449\ (0.79)\ \log k_{rPPC} + 3.44\ (0.79) \qquad (23.4)$$

$$n = 6 \qquad r^2 = 0.95 \qquad s = 0.21 \qquad r^2_{cross\ val.} = 0.91 \qquad F = 81.4$$

An example in which the inflammatory effect is correlated with the broadening of the transition peak of DPPC vesicles in the presence of a sulindac metabolite has also been published.[17] Unlike the active sulfite metabolite—produced *in vivo*—neither sulindac nor the sulfone derivative shows any effect on the sharpness of the transition peak, and both are pharmacologically inactive. The change in cooperativity of the lipid chains seems to be correlated with activity, i.e., important for the functioning of anti-inflammatory agents.

DRUG–MEMBRANE INTERACTION AND DRUG RESISTANCE

Drug resistance is an increasing problem in the therapy of diseases caused by bacteria as well as in tumor therapy. Various mechanisms are responsible for the occurrence of drug resistance, for example:

- Mutation or selection under therapy
- Change in cell wall of target cells
- Overproduction of target enzymes or proteins
- Change in conformation of binding site

Again, we would like to discuss two examples with respect to the possible role of drug–membrane interactions for resistance—one on resistant bacterial cells, and the other on resistant tumor cells.

The first example is on *E. coli*, highly resistant toward inhibitors of dihydrofolic acid reductase (DHFR), of the benzylpyrimidine type. The results are at the same time a warning to be careful in interpreting QSAR equations. From QSAR analysis of a series of benzylpyrimidines tested against drug-sensitive and drug-resistant *E. coli*, Hansch and co-workers concluded that one can design more effective drugs against resistant bacteria and cancer cells by making more lipophilic congeners.[18] We have repeated such studies using resistant and sensitive *E. coli* and DHFR isolated from such bacteria. The resistant strain used, *E. coli* RT 500, is a DHFR overproducer. In whole-cell bacteria, a steady decrease, and in resistant cells, an increase in inhibitory activity was noticed with increasing chain length, i.e., lipophilicity. Thus far, this is in agreement with the results by Hansch.

The results from the cell-free systems, using DHFR extracted from those two *E. coli* strains, show, however, that there is no change in the conformation of the active site. The inhibitory activity toward DHFR derived from sensitive and resistant cells follows the same ranking, trimethoprim being about 10 times more effective than the other derivatives, with no influence on chain length (Table 23.4).

At first glance, this could support the hypothesis of Hansch that more lipophilic derivatives are more easily transported through the cell wall of resistant cells. However, comparing the dose–response curves, it became obvious that there was a dramatic difference, indicating a change in mechanism of action with chain length in the case of resistant cells.

This was further supported by the observation that these derivatives show no synergism with sulfonamides in the case of resistant strains and acting antagonistically in combination with trimethoprim.[6] The structure–activity relationship found for the sensitive strain can be described by Equation 23.5 for the resistant strain by Equation 23.6 using the capacity factor $\log k_1$ from HPLC experiments on an octanol-covered reversed-phase column.

$$\log 1/I_{50 \, E. \, coli \, sens.} = -0.939 \log k'_r + 0.121 (\log k'_r)^2 + 296 \qquad (23.5)$$

$$n = 7 \qquad r = 0.99 \qquad s = 0.10 \qquad F = 107$$

$$\log 1/I_{50 \, E. \, coli \, res.} = -0.276 \log k'_r + 1.164 \log (0.051 \, k' + 1) - 3.02 \qquad (23.6)$$

$$n = 9 \qquad r = 0.99 \qquad s = 0.167 \qquad F = 83.9$$

TABLE 23.4

Inhibitory Activity (I_{50}, µM) of 3-Methoxy-4-Alkoxybenzylpyrimidines in Cell-Free and Whole-Cell Systems of *E. coli* ATCC (TMP Sensitive) and *E. coli* RT 500 (TMP Resistant) and the Lipophilicity Descriptor logk$'_r$

NH$_2$ OCH$_3$ CH$_2$ O(CH$_2$)$_x$CH$_3$ H$_2$N

GH3OX	I_{50} (µM) Cell-Free *E. coli*		I_{50} (µM) Whole Cells *E. coli*		
	ATCC 11775	RT 500	ATCC 11775	RT 500	logk$'_r$
Trimethoprim, TMP	0.0018	0.0023	0.97	1147	0.211
GH O1	—	—	1.46	1168	0.706
GH O2	0.02	0.017	4.99	885	1.299
GH O3	0.014	0.013	13.43	738.2	1.849
GH O4	0.015	0.015	17.52	163	2.427
GH O5	0.018	0.018	29.20	55.6	3.001
GH O6	0.022	0.021	31.20	18.6	3.575
Brodimoprim	0.0018	0.0011	0.81	118.4	1.382

One can conclude that these benzylpyrimidines are no longer DHFR inhibitors in the case of the resistant strain but inhibit bacterial growth by interacting with the membrane of this resistant strain as a detergent, a cationic soap. The resistant strain is not only an overproducer of DHFR but also has defects in its membrane. It has partially lost its hydrophilic outer core, so that these compounds can partition into lipids and may form micelle structures.

Finally, some of our results are presented: the reversal of multidrug resistance in tumor and plasmodial cells by amphiphilic drugs. This property was first reported for the Ca^{++}-antagonist, verapamil.

In several papers, it is argued that these amphiphilic compounds prevent binding of antitumor and antimalarial drugs to a *P*-glycoprotein, which transports the drug out of the cells, thus preventing accumulation of the drug in the infected resistant cells.[19] The *P*-glycoprotein is overproduced in several resistant cell types. We remain unconvinced of the general applicability of this hypothesis because:

1. Resistance of very different cells (tumor, bacteria, plasmodia) can be reversed, and also in cases where no overproduction of *P*-glycoprotein is observed.
2. Amphiphilic drugs with very different structures and conformation can reverse multidrug resistance.

Among potential mechanisms under investigation in our laboratories, the ability to reverse drug resistance by changes in membrane fluidity (permeability) remains most attractive.

Support for this mechanism can be achieved from results obtained from studies on tetracycline-resistant *E. coli*. This resistance was effectively reversed upon addition of known membrane-active drugs: desipramine and maprotiline.[20]

TABLE 23.5

Reversal of Chloroquine Resistance with Antidepressant Drugs

Drug	Concentration to Reverse Chloroquine Resistance to 50% West African Strain FCR-3 (μmol/l)*	Concentration to Reverse Chlorphentermine Binding to the Phospholipid (μM)
Imipramine	107	0.877
Desipramine	150	0.530
Doxepin	250	1.400
Maprotiline	595	0.270
Verapamil	1300	2.07
Trazodone	>1345	>>2.00

* Bitonti, A.J. et al., Reversal of chloroquine resistance in malaria parasite *Plasmodium falciparum* by desipramine, *Science*, 242, 1301, 1988. With permission.

For a series of catamphiphilic drugs, the interaction with phospholipids was studied using NMR relaxation measurements. It was found that the ranking in interaction strength for various drugs was similar to the reported ranking in the concentration required to reverse chloroquine resistance in plasmodia (Table 23.5).

An interesting example comes from reports comparing reversal of doxorubicin resistance in MCF-7 tumor cells by tricyclic drugs related to chloropromazine. Particularly intriguing was the finding that *trans*-flupentixol (a neuroleptic drug) was about three times more effective in reversal of resistance as compared to the cis-isomer. This suggests stereospecific interaction with some biomolecule. Both flupentixol isomers were subjected to NMR-binding studies with the result that *trans*-flupentixol binds more than two times stronger to liposomes of lecithin than the *cis*-isomer.[6]

In this respect, it is interesting to note that it has been shown by Aftab et al.[21] that *trans*-flupentixol is also about three times more potent in the inhibition of protein kinase C. The latter enzyme needs phosphatidylserine for its activation and is responsible for the phosphorylation of the glycoprotein and its drug-pumping activity. As phosphorylation is in equilibrium with dephosphorylation by phosphatase 1 and 2A, the inhibition of protein kinase C by the interaction of catamphiphilic drugs with the activating phosphatidylserine could be the mechanism as proposed by us. It could explain why compounds possessing very deviating structure, but similar physicochemical properties, are active in reversal of multidrug resistance.

Before proceeding to make some concluding remarks, it is noteworthy to point to the pharmacological effects of lipids themselves, as for instance their platelet activation activity, the histamine release by phosphatidylserine, the various effects of lysophosphatidylcholine, and last but not least, the anticancer activity of several lipid ethers.

CONCLUSION AND SUMMARY

Some aspects of drug–membrane interaction and its influence on drug transport, accumulation, efficacy, and resistance have been discussed. The interactions manifest themselves macroscopically in changes in the physical and thermodynamic properties of "pure membranes" or bilayers. As various amounts of foreign molecules enter the membrane, in particular, the main gel-to-liquid crystalline phase transition can be dramatically changed. This may change permeability, cell fusion, and cell resistance, and may also lead to changes in conformation of the embedded receptor proteins. Furthermore, specific interactions with lipids may lead to drug accumulation in membranes, and thus to much larger concentrations at the active site, than present in the surrounding water phase. The lipid environment may also lead to changes in the preferred conformation of drug molecules. These events are directly related to drug efficacy.

The determination of essential molecular criteria for the interaction could be used to design new and more selective therapeutics. This excursion in some aspects of drug–membrane interaction underlines the importance of lipids and their interaction with drug molecules for our understanding of drug action. This is not really a new thought but was formulated in 1884 by THUDICUM:

"Phospholipids are the centre, life and chemical soul of all bioplasm whatsoever, that of plants as well as animals."

For further information and more details on drug–membrane interaction, see Reference 22.

REFERENCES

1. Mason, R.P., Rhodes, D.G., and Herbette, L.G., Reevaluating equilibrium and kinetic binding parameters for lipophilic drugs based on a structural model for drug interaction with biological membranes, *J. Med. Chem.* 34, 869–877, 1991.
2. Seydel, J.K., Nuclear magnetic resonance and differential scanning calorimetry as tools for studying drug-membrane interactions, *Trends Pharmacol. Sci.*, 12, 368, 1991.
3. Seydel, J.K. et al., QSAR and multivariate data analysis of amphiphilic benzylamines and their interaction with various phospholipids determined by different methods, *Quant.-Struct.-Act. Relat.*, 8, 266, 1989.
4. Scheufler, E. and Peters, T., Phosphatidylserine monolayers as models for drug uptake into membranes and tissue, *Cell Biol. Int. Rep.*, 14, 381, 1990.
5. Jardetzky, O. and Roberts, G.C.K. Eds., *NMR in Molecular Biology*, Academic Press, 1981.
6. Seydel, J.K. et al., The importance of drug-membrane interaction in drug research and development, *Quant. Struct.-Act. Relat.*, 11, 205, 1992.
7. Seydel, J.K. et al., In *QSAR: Rational Approaches to the Design of Bioactive Compounds*, Silipo, C. and Vittoria, A., Eds., Elsevier Science, Amsterdam, 1991, pp. 367–376.
8. Koch, A. et al., QSAR and molecular modelling for a series of isomeric X-sulfanil-amido-1-phenylpyrazoles, *Quant.-Struct.-Act. Relat.*, 12, 373, 1993.

9. Akbaraly, J.P. et al., In *QSAR and Strategies in the Design of Biocative Compounds,* Seydel, J.K., Ed., VCH, Weinheim, Germany, 1985, p. 313.
10. Young, R.C. et al., Development of a new physicochemical model for brain penetration and its application to the design of centrally acting H2 receptor histamine antagonists, *J. Med. Chem.,* 31, 656, 1988.
11. Schwyzer, R., Molecular mechanism of opioid receptor selection, *Biochemistry,* 25, 4281, 1986.
12. Herbette, L.G., Rhodes, D.G., and Mason, R.P., New approaches to drug design and delivery based on drug-membrane interactions, *Drug Des. Deliv.,* 7, 75, 1991.
13. Valdivia, H. Dubinski, W.P., and Coronado, R., Reconstitution and phosphorylation of chloride channels from airway epithelium membranes, *Science,* 242, 1441, 1988.
14. Scheufler, E. et al., Uptake of cataraphiphilic drugs into erythrocytes and muscular tissue correlates to membrane enrichment and to 45Ca displacement from phosphatidylserine monolayers, *J. Pharmacol. Exper. Ther.,* 252, 333, 1990.
15. Herbette, L.G. et al., Possible molecular basis for the pharmacokinetics and pharmacodynamics of three membrane-active drugs: propranolol, nimodipine and amiodarone, *J. Mol. Cell. Cardiol.,* 20, 373, 1988.
16. Choi, Y.W. and Rogers, J.A., The liposome as a model membrane in correlations of partitioning with alpha-adrenoceptor agonist activities, *Pharm. Res.,* 7, 508, 1990.
17. Hwang, S.B. and Shen, T.Y., Membrane effects of antiinflammatory agents. 2. Interaction of nonsteroidal antiinflammatory drugs with liposome and purple membranes, *J. Med. Chem.,* 24, 1202, 1981.
18. Coats, E.A. et al., Quantitative structure-activity relationship of antifolate inhibition of bacteria cell cultures resistant and sensitive to methotrexate, *J. Med. Chem.,* 28, 1910, 1985.
19. Martin, S.K., Odoula, A.M.J., and Milhous, W. K., Reversal of chloroquine resistance in plasmodium falciparum by verapamil, *Science,* 235, 899, 901, 1987.
20. Seydel, J.K. et al., in *Trends in Medicinal Chemistry,* Sarel, G., Mechoulam, R., and Agranat, I., Eds., Blackwell Scientific, Oxford, 1992, p. 397.
21. Aftab, D.T., Yang, J.M., and Hait, W.N., Functional role of phosphorylation of the multidrug transporter (P-glycoprotein) by protein kinase C in multidrug-resistant MCF-7 cells, *Oncol. Res.,* 6, 59, 1994.
22. Seydel, J.K. and Wiese, M., Drug-membrane interactions, analysis, drug distribution, modeling, in *Methods and Principles in Medicinal Chemistry* (Vol. 15), Mannhold, R., Kubinyi, H., and Folkers, G., Eds., Wiley-VCH, Weinheim, Germany, 2002.

24 The Fight Against AIDS: *New Avenues for Inhibiting Reverse Transcriptase (RT), an Old Target*

*Maurizio Botta, Lucilla Angeli,
Marco Radi, and Giovanni Maga*

CONTENTS

ABSTRACT

The acquired immunodeficiency syndrome (AIDS) related to HIV-1 infection is one of the most serious threats to human health, and it has been estimated that more than 25 million people have died since it was first recognized. In the fight against AIDS, first- and second-generation non-nucleoside reverse transcriptase inhibitors (NNRTIs) are now established as part of highly active antiretroviral therapy (HAART) for treating HIV infection. However, the efficacy of currently available NNRTIs, e.g., nevirapine (NVP, Viramune®), delavirdine (DLV, Rescriptor®), and efavirenz (EFV, Sustiva®, Stocrin®), is impaired by rapid emergence of drug resistance. On the other hand, as patients live longer on HAART therapy and the pool of NNRTI-resistant viruses increases, so does the need for the development of new NNRTIs with antiviral activity against clinically relevant mutant strains. Our research group was recently involved in a multitarget approach to defeat the HIV virus, focusing on the inhibition of HIV-1 reverse transcriptase according to both classical and nonclassical approaches.

Here, we will report an efficient methodology for the parallel solution-phase synthesis of a series of thiouracils, in turn selectively S-benzylated under microwave irradiation to give new S-DABOs. S-DABO derivatives, endowed with subnanomolar anti-HIV-1 activity, were subjected to docking and molecular dynamic studies, with the aim of rationalizing their activity both within the wild-type:RT and K103N: RT nonnucleoside binding pocket (NNBP).

A combinatorial approach led instead to the identification of a new class of compounds (namely 6-vinylpyrimidines) endowed with an unprecedented mechanism of action: these compounds are the first NNRTIs competing with the nucleotide substrate. An enzymological and computational study has been conducted to elucidate their unique mechanism of action.

INTRODUCTION

The current therapy against the human immunodeficiency virus type 1 (HIV-1), the causative agent of AIDS, is based on four classes of drugs: the nucleoside reverse transcriptase inhibitors (NRTIs), the NNRTIs, the protease inhibitors (PIs), and the fusion inhibitor enfuvirtide. The United States Food and Drug Administration (FDA) has to date approved 22 individual monotherapies to directly treat the HIV infection. Initial binding of the virus to host CD4[+] cells is blocked by one of these inhibitors, and ten of these agents target the viral protease, the key enzyme involved in protein maturation and viral assembly. The remaining inhibitors are directed against the viral reverse transcriptase (RT) and represent the cornerstone of anti-HIV therapy.[1] Although NRTIs are equally active against HIV-1 and HIV-2 RT, acting at the catalytic site as DNA chain terminators,[2] NNRTIs are highly specific for HIV-1 and include more than 30 structurally different classes of molecules, such as nevirapine,[3] TIBO,[4] BHAP,[5] α-APA,[6] PETT,[7] HEPT,[8] TNK-651,[9] ITU,[10] DATA,[11] and DAPY[12–14] (Chart 24.1).

The HIV-1 RT is a multifunctional enzyme, consisting of a p66 and a p51 subunits (Figure 24.1), responsible for the conversion of single-stranded RNA viral genome into double-stranded DNA. Lacking a biological counterpart in the eukaryotic systems, RT is an attractive target for development of selective inhibitors.

Studies carried out on crystal structures of different RT/NNRTI complexes suggest that NNRTIs share a common mode of action, binding the RT enzyme at an allosteric site corresponding to a hydrophobic pocket in p66, called nonnucleoside inhibitor binding pocket (NNIBP). NNIBP is located approximately 10 Å from the catalytic site in the p66 subunit, and only a small portion of it is formed by amino acid residues belonging to the p51 subunit.[9,15–22] During the process of inhibitor binding, significant conformational changes occur in the orientation of the side chain of some residues (particularly Tyr-181 and Tyr-188), leading to the formation of the NNIBP to accommodate the inhibitors.

It is evident, from a comparison of the various RT structures, that the NNIBP has a very flexible structure, and this characteristic allows the enzyme to accommodate structurally diverse inhibitors having different shapes and sizes,[23] such as

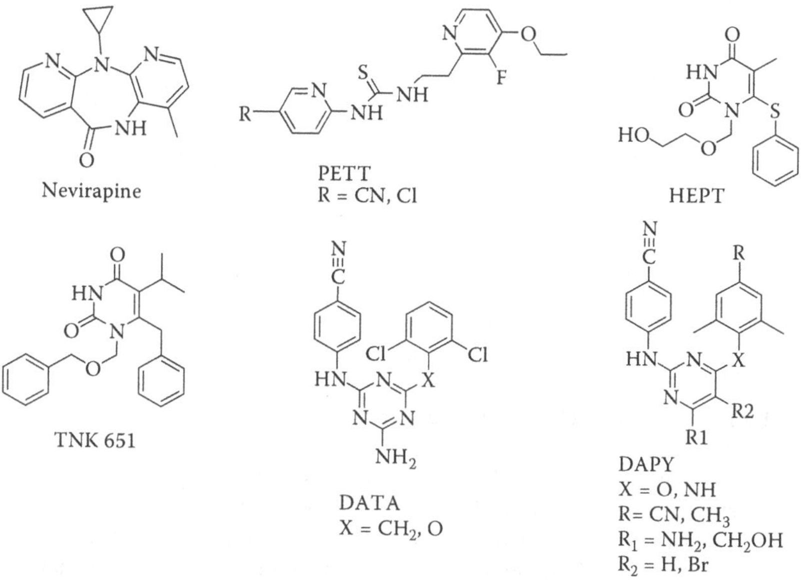

Nevirapine

PETT
R = CN, Cl

HEPT

TNK 651

DATA
X = CH₂, O

DAPY
X = O, NH
R = CN, CH₃
R₁ = NH₂, CH₂OH
R₂ = H, Br

CHART 24.1 Examples of known NNRTIs.

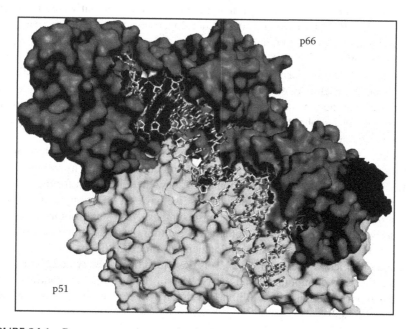

FIGURE 24.1 Reverse transcriptase subunits in complex with DNA.

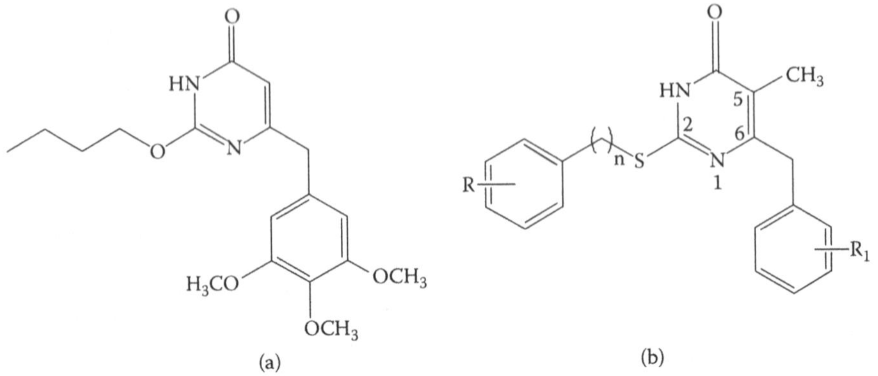

CHART 24.2   (a) First disclosed DABO and (b) S-DABOs compounds.

the approved drugs nevirapine, delavirdine, and, efavirenz, which all bind at a similar site distal to the active site within RT.[15,16,24] Although NNRTIs generally exhibit low toxicity and favorable pharmacokinetic properties, it has been observed that the rapid replication of HIV and its inherent variability often lead to the generation of drug-resistant variants responsible for the clinical NNRTI treatment failure.[25,26]

Therefore, the development of novel NNRTIs with improved pharmacological, pharmacokinetic, and drug resistance profiles is critical for a more successful application of NNRTIs in combination therapy.

Among NNRTIs, dihydro-alkoxy-benzyl-oxopyrimidines (DABOs) are an interesting class of compounds active at nanomolar concentration that were first discovered in 1992[27] (Chart 24.2a) and further developed during the following years into S-DABO (Chart 24.2b) and related analogues.

Here, we will describe the microwave-assisted synthesis and biological evaluation of a series of S-DABO derivatives characterized by the presence of an arylalkylthio substituent of variable length at position 2, a CH_3 group at C-5, and a halogenated benzyl group at the 6 position.[28] Results from in vitro anti-HIV RT assays and anti-HIV activity in lymphoid cells showed activity values in the nanomolar and sub-nanomolar range, respectively.

Moreover, with the aim of identifying new lead compounds for the development of novel NNRTIs capable of overcoming the effects of resistant mutations, and in continuous efforts toward the development of new methodologies for the synthesis of pyrimidine and pyrimidinone derivatives, we have set up a simple and efficient methodology for the parallel solution-phase synthesis of a small library of pyrimidine derivatives. A preliminary biological screening of these compounds on wild-type (wt) HIV-1 RT allowed the development of a focused library of derivatives, a selection of which (a–e; see Chart 24.3) is described here.[29] Enzymological studies revealed that such compounds bind the NNIBP of the enzyme but, contrary to the NNRTIs reported to date, they inhibit HIV-1 RT by a competitive mechanism with the nucleotide substrate.

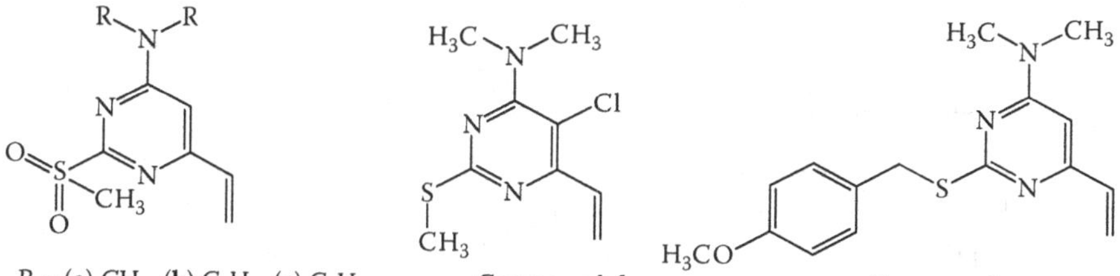

R = (a) CH₃, (b) C₂H₅, (c) C₃H₇   Compound **d**   Compound **e**

CHART 24.3   6-Vinylpyrimidine inhibitors **a–e**.

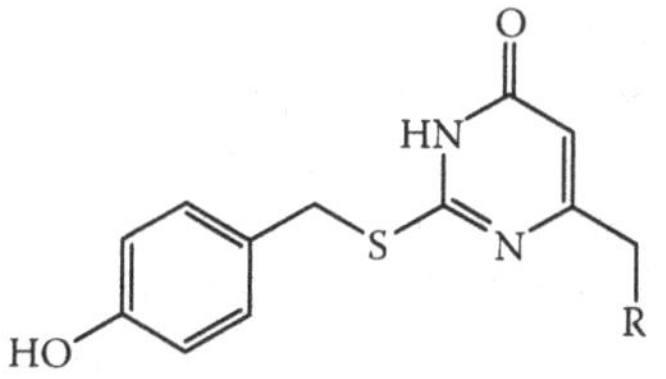

6 R = H
7 R = Ph

CHART 24.4   Compounds **6** and **7**.

S-DABO

Recently, compounds **6** and **7** were identified by us as by-products of cleavage of thiopyrimidinones from Wang resin (Chart 24.4).[30] Considering their structural similarity to known *S*-DABOs, these compounds were tested as possible NNRTIs and, despite their moderate activity (IC$_{50}$ 400 µM), we became interested in investigating the influence of the arylalkyl moiety on anti-HIV activity because only scattered examples of *S*-DABOs having this kind of substitution at position 2 had been reported in the literature. Therefore, we planned to design a number of *S*-DABO derivatives characterized by the presence of the following:

1. An arylalkylthio substituent at position 2 that represents the focus of our investigations. To expand the SAR of *S*-DABOs, the substitution pattern on the phenyl ring was broadly varied, as well as the length of the alkyl spacer, which was increased from one to three carbon atoms.
2. A 5-methyl group that was kept fixed in all the molecules, as a conformational constraint with the aim of enhancing the affinity for the enzyme.
3. A halogenated benzyl group at the 6 position, which was expected to improve a putative π-stacking interaction between the electron-deficient benzene ring of the ligand and the electron-rich benzene ring of Tyr-188 located in the allosteric site of the enzyme.[31]

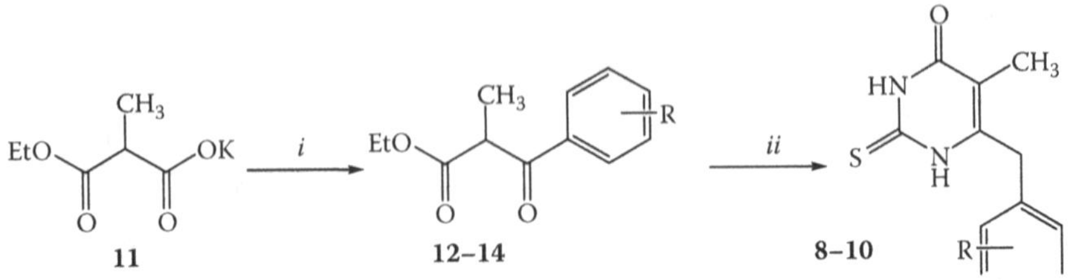

SCHEME 24.1 (*i*) a. MgCl$_2$, Et$_3$N, CH$_3$CN, rt, 360 rpm, 2 h; b. substituted phenylacetic acid, *N, N*-carbonyldiimidazole, rt, 360 rpm, overnight then reflux, 360 rpm, 2 h; c. 13% HCl, rt, 360 rpm, 10 min; (*ii*) thiourea, EtONa, reflux, 360 rpm, overnight.

CHEMISTRY

In view of a more extensive functionalization of the positions 5 and 6 of the thiopyrimidinone scaffold, we devised a simple and efficient methodology for the parallel solution-phase synthesis of thiouracils **8–10** using a Büchi Syncore® synthesizer (Scheme 24.1). Potassium ethyl 2-methylmalonate **11** was partitioned into three reaction vessels and reacted with three substituted phenylacetyl imidazolides in the presence of magnesium dichloride/triethylamin system in acetonitrile, according to the Clay procedure,[32] to give, after a simple liquid-phase extractive purification, the β-keto esters **12–14**. Subsequent condensation of **12–14** with thiourea in the presence of sodium ethoxide in refluxing ethanol afforded 6-benzyl-2,3-dihydro-5-methyl-2-thioxopyrimidin-4(1*H*)-ones (**8–10**).

Exploitation of this procedure might allow the rapid synthesis of a large number of *S*-DABO derivatives variously substituted at positions 2, 5, and 6 for further SAR studies in this class of compounds.

5,6-Disubstituted thiouracils **8–10** were selectively S-benzylated under microwave irradiation with the appropriate substituted benzyl halide in dry DMF in the presence or absence of potassium carbonate (method A) (Scheme 24.2).[33]

Alternatively, **8–10** were S-benzylated using the appropriate benzyl alcohol via a microwave-assisted Mitsunobu reaction in the presence of trimethylphosphine and diisopropylazodicarboxylate (DIAD) in dry DMF (method B). In both cases, the use of microwaves allowed us to obtain, in a few minutes, the desired products in high yield and good purity.

Finally, sulfides **18, 20, 25** were oxidized to sulfones **41–43** with *m*-chloroperbenzoic acid in dichloromethane at room temperature.[34]

All the compounds were obtained with more than 95% purity, as shown by HPLC-MS analysis (Table 24.1)

RESULTS AND DISCUSSION

Compounds **15–40** and **41–43** constitute a new family of *S*-DABO characterized by the presence of an arylalkylthio substitution at C-2, a methyl group at C-5, and a halogenated benzyl group at C-6 of the pyrimidinone nucleus, which were prepared in a straightforward fashion by alkylation of three different 6-substituted 2-thiouracils (**8–10**) and subsequent oxidation of the sulfur atom (**41–43**).

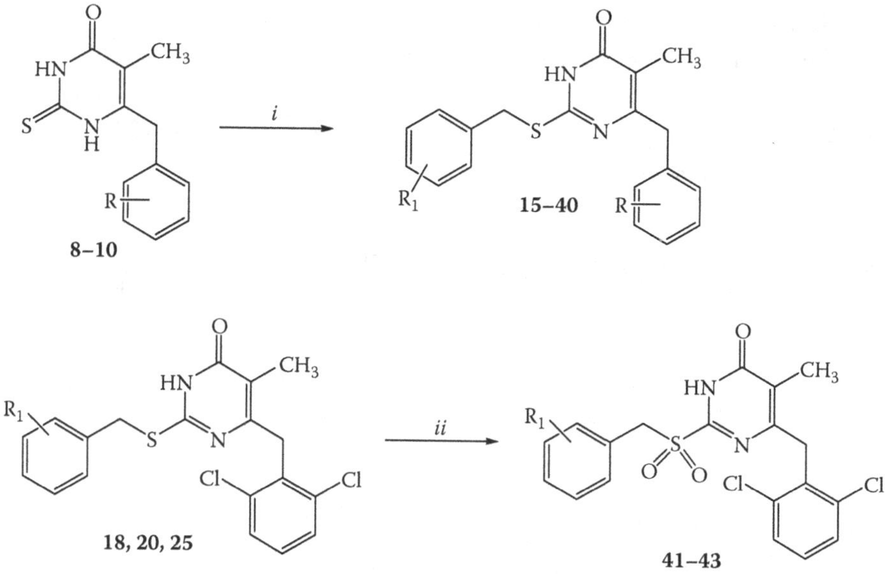

SCHEME 24.2 (*i*) substituted benzyl halide, DMF, MW, 130°C, 5 min (method A) or substituted benzyl alcohol, trimethylphosphine, DIAD, DMF, MW, 40°C, 10 min (method B); (*ii*) mCPBA; CH_2Cl_2, rt, 360 rpm, overnight.

The new compounds were evaluated in enzymatic tests for their ability to inhibit either wt or mutated RTs as well as on MT-4 cells for cytotoxicity and anti-HIV-activity, in comparison with nevirapine and efavirenz used as reference drugs. In particular, the following mutants were used: K103N and Y181I for enzymatic tests, K103N, Y188L, and IRLL98 (bearing the K101Q, Y181C, and G190A mutations conferring resistance to nevirapine, delavirdine, and efavirenz) for tests on cell lines. The results of these assays for selected compounds are reported in Table 24.2. Although previous findings obtained with other *S*-DABO series highlighted the importance for optimal activity of a 2,6-difluorobenzyl substituent at position 6,[35] in our case this type of substitution proved to be less profitable with respect to the corresponding 2,6-dichlorobenzyl group. In fact, although compound **20** was highly active both in enzymatic and cell tests, showing appreciable activity also against IRLL98, compound **34** (i.e., the analogue of compound **20** with 2,6-difluorobenzyl group) retained activity only against the wt RT. Moreover, a striking difference in activity can be observed between **25** (which displays a full range of activity at low concentrations, and is thus one of the most interesting compounds of the series) and **37**, whose activity was limited to wild-type virus and unrelated to RT inhibiting properties (see Chart 24.5).

In line with these observations, the presence of a 6-(4-fluorobenzyl) group led to the substantially inactive compounds **39** and **40** (data not shown).

On the basis of these results, we kept fixed the 6-(2,6-dichlorobenzyl) substituent and systematically modified the C-2 position. The 4-hydroxybenzyl group, found in our lead compounds **6** and **7** (Chart 24.4) and also present in compound **19**, did not positively contribute to the antiviral activity. Replacement of the OH group with

TABLE 24.1
Chemical and Physical Data of Compounds 15–43

Compound	R	R$_1$	n	mp (°C)	Yield (%)	Method of Synthesis
15	2,6-diCl	H	1	236–237	50	A
16	2,6-diCl	4-F	1	205–206	50	A
17	2,6-diCl	4-Cl	1	214	65	A
18	2,6-diCl	4-Br	1	219	67	A
19	2,6-diCl	4-OH	1	204–206	64	B
20	2,6-diCl	4-OCH	1	197–199	72	A
21	2,6-diCl	4-OCH$_2$CH$_3$	1	218–221	73	B
22	2,6-diCl	4-OBu	1	201–205	79	B
23	2,6-diCl	4-iPr	1	221–223	60	A
24	2,6-diCl	4-CN	1	232–235	78	A
25	2,6-diCl	4-NO$_2$	1	231–232	55	A
26	2,6-diCl	3-F	1	238–239	50	A
27	2,6-diCl	3-Cl	1	232	51	A
28	2,6-diCl	3-CN	1	203	77	A
29	2,6-diCl	2,4-diOCH$_3$	1	217–219	72	B
30	2,6-diCl	3,4-diOCH$_3$	1	195–196	65	B
31	2,6-diCl	3,5-diOCH$_3$	1	246–247	50	A
32	2,6-diCl	4-OCH$_3$	2	182–183	53	B
33	2,6-diCl	4-OCH$_3$	3	180–182	67	B
34	2,6-diF	4-OCH$_3$	1	170–171	52	A
35	2,6-diF	4-iPr	1	179–181	49	A
36	2,6-diF	4-CN	1	247–248	57	A
37	2,6-diF	4-NO$_2$	1	248–250	40	A
38	2,6-diF	3,5-diOCH$_3$	1	228–230	59	A
39	4-F	4-NO$_2$	1	246	56	A
40	4-F	4-OCH$_3$	1	190	63	A
41	2,6-diCl	4-OCH$_3$	1	171–174	71	
42	2,6-diCl	4-NO$_2$	1	175–180	55	
43	2,6-diCl	4-Br	1	241–244	50	

H, F, Cl, Br, *i*Pr, and CN improved the general biological profile (compounds **15–18**, and **24**), with activity ranging from 0.39 to 103 μM in enzymatic tests, and from 0.11 to 0.83 μM on cells infected with wild-type virus. No interesting activity was observed on mutant strains, with the exception of **24**, which retained activity on clinically relevant mutants at a micromolar or submicromolar concentration. Moving the substituent (F, Cl, CN) from the para to meta position on the aromatic ring did not modify the biological response to the molecules (**26–28**). On the other hand, replacement of the OH group with NO$_2$ led to the very interesting compound **25** (*vide supra*), whereas substitution of the OH group with alkoxy groups gave compounds **20–22**, whose activity was strongly dependent on the length of the alkyl moiety. Thus, whereas **21** and **22** (*p*-ethoxy and *p*-butoxy group, respectively) were only

TABLE 24.2
Anti-HIV-1 Activity Data of a Selection of S-DABOs and of Three Reference Compounds

Compound	ID_{50} (μM)[a,b]			EC_{50} (μM)[a,c]			
	wt	K103N	Y181I	NL4-3 wt	IRLL98	K103N	Y188L
20	0.004	102	150	0.007	0.047	4.7	>32
21	2	na[d]	na	0.25	1.9	>58	>58
22	2.8	na	na	>3.1	34% at 11[e]	>3.1	>3.1
24	103	na	na	0.17	1.0	0.10	>60
25	0.009	0.12	0.22	0.79	3.3	11	31%
29	25	na	na	1.33	>55	>55	>55
30	1.2	na	na	0.20	0.93	10	>55
32	0.026	3.2	1.5	<0.00014	0.00022	0.87	6.9
33	0.037	0.5	0.3	0.00044	0.11	>1.0	>1.0
34	0.007	3.5	na	0.026	0.69	0.69	10
36	0.76	na	na	0.052	>65	>65	>65
37	na	na	na	0.074	>62	>62	>62
41	0.006	10	27	>2.8	>2.8	>2.8	>2.8
42	na	na	na	>53	>53	>53	>53
43	na	na	na	>50	>50	>50	>50
Nevirapine	0.4	8	20	0.052	>7.5	3.9	>7.5
Efavirenz	0.04	0.4	0.1	0.001	0.2	0.057	0.3
AZT				0.003	0.008	0.003	0.003

[a] Data represent mean values of at least two experiments.
[b] ID_{50}: Inhibiting dose 50 or needed dose to inhibit 50% of enzyme.
[c] EC_{50}: Effective concentration 50 or needed concentration to inhibit 50% HIV-induced cell death, evaluated with MTT method in MT-4 cells.
[d] na: Not active at 400 μM (the highest concentration tested).
[e] Percentage inhibition of HIV-induced cell death at the reported μM concentration.

moderately active on the wt, compound **20** (*p*-methoxy) was endowed with very significant activity in the nanomolar range toward both wild-type and IRLL98 strains, as well as with low toxicity (SI > 4600). The simultaneous presence on the aromatic ring of more than one methoxy group negatively affected the biological activity. In particular, whereas a modest effect on wt RT can still be ascribed to compounds **29** and **30** (having a substituent at the para position), no interesting activity was demonstrated by **31** and **38** (R = 3,5-dimethoxyphenyl). Even more detrimental to antiviral activity was the presence of a bulky group on the aryl moiety, as for compounds **23** and **35** (data not shown).

A very marked improvement of the biological profile was obtained by increasing the length of the linker connecting the aromatic ring to the sulfur atom. Compounds **32** and **33** (*n* = 2 and 3, respectively) proved to be the most interesting among the new derivatives. In particular, **32** exhibited anti-HIV activity on cells infected with either

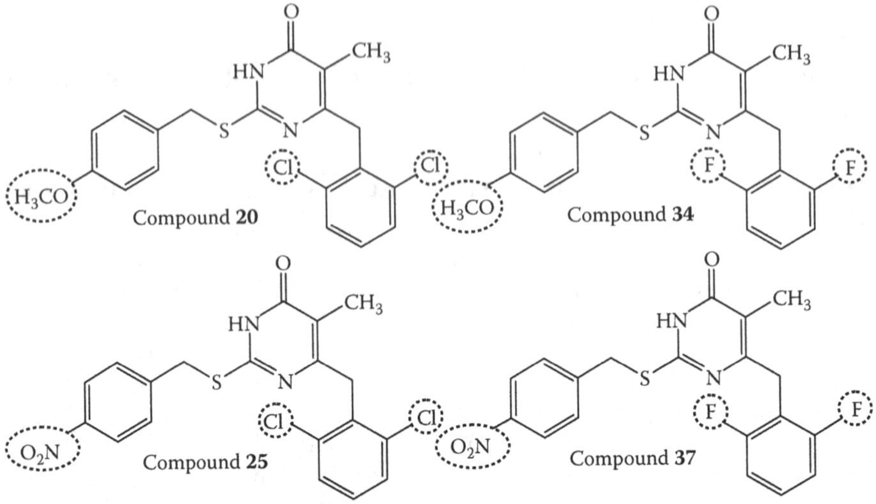

CHART 24.5 Structure of compounds **20**, **25**, **34**, and **37**.

wild-type or pluriresistant virus (IRLL98) at subnanomolar concentrations, together with low cytotoxicity (SI > 410000). On the whole, this compound showed an anti-HIV profile superior to that of nevirapine and comparable to that of efavirenz, thus emerging as the most active *S*-DABO analogue reported thus far. Further lengthening of the linker led to compound **33**, which, though retaining very good anti-HIV activity, was endowed with higher cytotoxicity.

Finally, basically inactive compounds (**41–43**) were obtained by oxidation of the sulfur atom. Biological data showed some contradictions between ID_{50} and EC_{50} values for several compounds, probably because cellular tests reflect the interference of a compound in viral steps not shown in the pure RT enzymatic tests (i.e., enzymatic tests account only for the DNA- and RNA-dependent synthesis inhibition, whereas cellular test results represent the inhibition of all viral replication phases). Moreover, we cannot exclude that the low cellular activity or the inactivity of some of our compounds could be the consequence of poor adsorption through the T-cell membrane, in addition to a poor affinity for the corresponding receptor counterpart.[28]

MOLECULAR MODELING CALCULATIONS

To investigate the binding mode of the new RT inhibitors, molecular docking simulations were performed by means of the software Autodock 3.0,[35] and the HIV-1 RT/ligand complexes were minimized with Macromodel 8.5[36]/Maestro.[37] Three-dimensional coordinates of the HIV-1 RT/MKC-442 (emivirine) complex (Brookhaven Protein Data Bank entry 1RT1) were used as the input structure for docking calculations. All cocrystallized water molecules were deleted, and all polar hydrogens were added using the appropriate tool in the builder module of Maestro. The structures of the new *S*-DABO derivatives were built using Maestro 3D-sketcher and fully minimized (Polak-Ribiere conjugate gradient, of 0.05 kJ/Å·mol convergence). Atom charges assigned to compounds during the minimization step were retained for the following docking calculation.

For molecular docking purposes, a box of 64 × 50 × 62 points was set that comprised all residues constituting the NNRTI binding pocket. Starting structures of the selected compounds were randomly defined to obtain totally unbiased results. The GA-LS method was used with the default settings while retrieving 100 docked conformations from each compound. Results from Autodock calculations were clustered using an RMSD tolerance of 2 Å, and the lowest energy conformer of the most populated cluster (the lowest energy cluster in most cases) was selected as the most probable binding conformer. HIV-1 RT/ligand complexes were submitted to a full minimization of the whole structure to a 0.1 kJ/Å·mol gradient, using Amber force field as implemented in the Macromodel software package.

Results showed **20** assuming an orientation very similar to that of the cocrystallized inhibitor (MKC-442, emivirine). In particular, the pyrimidinone ring of **20** was superposable to that of MKC-442, allowing for a hydrogen bond contact between its NH group at position 3 with the carbonyl moiety of Lys-101 (1.9 Å), suggested to be a crucial interaction key for inhibitors belonging to the DABO and S-DABO classes of compounds.[38] Moreover, the side chains of both Lys-103 and Val-106 were also found in close contact with the pyrimidinone nucleus. The extended side chains at position 2 of both **20** and MKC-442 were lined up in a parallel way and pointed toward the solvent accessible surface defined by Pro-225 and Pro-236. In further detail, although the sulfur atom of **20** was located at a distance of about 3.7 Å from the backbone-NH group of Lys-101, the benzyl moiety was accommodated into a large pocket mainly defined by Val-106, Pro-225, Pro-236, and Tyr-318. Good hydrophobic interactions were found between the alkyl side chain of Val-106 and the phenyl ring of the inhibitor as well as between the terminal methyl group with both Pro-225 and Pro-236. Finally, the benzyl substituent at position 6 of **20** was embedded into an extended hydrophobic region defined by the aromatic side chains of Tyr-181, Tyr-188, Phe-227, and Trp-229 as well as by Leu-100 and Leu-234. The major difference between the experimental and calculated complexes involved the interaction between the 6-benzyl side chain of the inhibitors and the side chain of Trp-229. In fact, although a T-tilted interaction was shown in the complex with MKC-442, a π–π interaction was found in the complex with **20**. This was mainly because a significant conformational rearrangement occurred within the aromatic cage accommodating the substituent at position 6. In fact, the aromatic side chains of Tyr-183, Tyr-188, and Trp-229 approached the inhibitor, leading to a structural system in which the indole nucleus was sandwiched between the 6-benzyl group of the inhibitor and the phenyl ring of Tyr-183 (allowing π–π interactions) and interacted with a T-tilted contact with the aromatic moiety of Tyr-188. On the other hand, although Phe-227 retained its original position, a rotation of the Tyr-181 side chain occurred that pushed it about 1.8 Å away from the NNIBP.

Similar results were found for the nitro derivative **25**, with a major difference involving the benzylthio side chain that was located in a region of space between the corresponding side chain of **20** and the N1 side chain of the crystallized inhibitor. In particular, owing to a rotation of about 20° around the C2-S bond, the nitro group was inserted between Pro-225 and the side chain of Phe-227, and one of its electron-rich

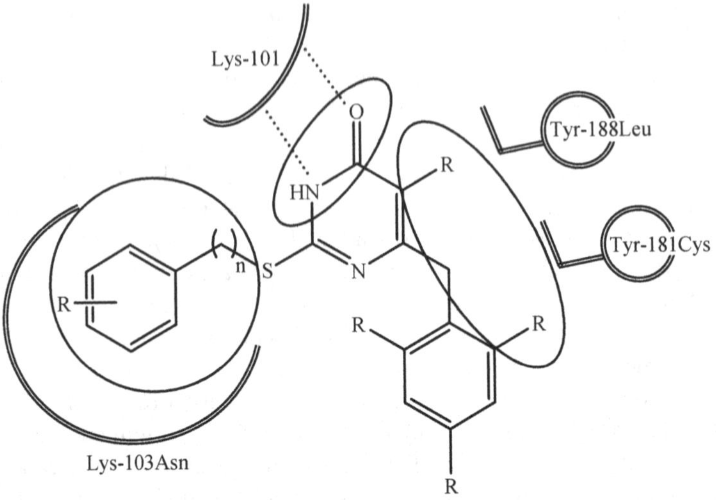

FIGURE 24.2 Schematic representation of the *S*-DABOs and of their interactions in three different RT mutants: Lys-103Asn, Tyr-181Cys, and Tyr-188Leu.

oxygen atoms was found at the proper distance (1.6 Å) to make a hydrogen-bond contact with the backbone-NH group of Phe-227. Lengthening the benzyl moiety of **20** to a phenylethyl (**32**) and phenylpropyl (**33**) chain led their *p*-methoxy group to go beyond the opening defined by Pro-225 and Pro-236 and, thus, to be exposed to the solvent. Moreover, such a shift of the phenyl ring caused the lack of its hydrophobic interactions with the side chain of Val-106. The pyrimidinone nucleus of **33** maintained an orientation comparable to that of **20**, including the hydrogen-bond contact with Lys-101 as well as the hydrophobic interactions with a part of the Val-106 side chain. Similarly, no substantial difference was found for the location of the benzyl substituent at position 6, in comparison to that of **20** and the crystallized inhibitor. On the other hand, the 6-benzylpyrimidinone system of **32** underwent a significant conformational rearrangement, allowing for a hydrogen-bond contact involving the carbonyl oxygen and the backbone-NH group of Lys-101, in addition to the hydrogen-bond contact between the 3-NH group of the inhibitor and the carbonyl of the same amino acid. Moreover, because the benzyl group at position 6 was characterized by an alternative orientation into the previously described hydrophobic cage, the profitable aromatic–aromatic interaction involving the side chain of Trp-229 was lost. In summary, docking calculations suggested that the *p*-methoxybenzylthio group at position 2 of **20** seems to be characterized by the optimal length to fulfill the tunnel, which is delimited at the surface of the RT structure by Pro-225 and Pro-236. Moreover, the lower activity of **32** and **33** toward the wt RT, with respect to that of their shorter analogue **20**, was in part due to reduced hydrophobic interactions mainly involving the benzyl substituents at position 2 as well as to unfavorable contacts between the terminal methyl substituent and the solvent. In a similar way, analogues of **20**, bearing at the para position of the 2-benzyl substituent a hydrophobic group larger than a methoxy moiety (exceeding the surface of the RT and contacting the solvent), showed lower activity than **20**. In fact, the ethoxy, butoxy, and isopropyl derivatives (**21**, **22**, and **23**) showed activity in the micromolar range or lower.[28]

REVERSE TRANSCRIPTASE MUTANTS

However, resistance, which has emerged against all available antiretroviral drugs, represents a major challenge in the therapy of HIV infection, and it has become the main obstacle in the management of the infection.

Drug resistance can quickly emerge as a result of mutations at one or more locations in the NNIBP. Common NNRTI resistance mutations include L100I, K103N, V106A, Y181C, Y188L, and G190A, which contribute to drug resistance through different mechanisms (See Figure 24.2). The structures of early NNRTIs were described as existing in a "butterfly" conformation, in which two chemical side group "wings" are attached to a central linker/backbone moiety in a relatively rigid manner. Resistance to these compounds can occur easily by amino acid substitutions that generate steric hindrance and prevent binding of the compounds to the NNIBP. For example, steric hindrance is created when leucine is substituted for the β-branched isoleucine in the L100I mutation, or if a methyl group is introduced in a G190A mutant. Another mechanism of NNRTI resistance involves mutations leading to the loss of aromatic amino acids, such as Y181C and Y188L. This results in the loss of hydrophobic interactions necessary to stabilize the binding of NNRTIs to the NNIBP. A third mechanism of NNRTI resistance involves K103N. This mutation has little effect on inhibitor binding. Instead, the asparagine side chain of the K103N mutant forms a hydrogen bond with the phenoxyl group of tyrosine located at position Y188 of the NNIBP as it exists in the unbound state. The result of this interaction is the formation of a molecular "gate" that prevents the access of the NNRTI class of drugs into the NNIBP. Thus, K103N results in cross-resistance against all three clinically approved NNRTIs.[39]

Among compounds **15–40**, **25** and **33** have emerged as the most potent inhibitors against this mutation. From the K103N activity data, it appears that different substituents in *p*-position of benzyl moiety of *S*-DABOs deeply modify the activity against the K103N mutant (see Table 24.1 and Table 24.2): compounds **16, 17, 19,** and **20**, bearing, respectively, an F, Cl, OH, and OCH_3 at the para position of aromatic ring are inactive, whereas compounds **18** and **20**, with, respectively, Br and NO_2 at the same position showed activity in the low micromolar range.

For compounds **20**, **32**, and **33**, with a *p*-OCH3 benzylthio side chain, lengthening of the spacer (from phenylethyl up to phenylpropyl) induces an increase of activity. However, our docking studies, performed on both wt and K103N mutant with all the aforementioned compounds, did not clarify their different activities because the binding mode into the NNIBP was essentially the same, and none of the molecules interacted with the mutated residue (see as an example compounds **20** and **25** docked within the K103N NNBP in Figure 24.3). In agreement with this observation, the crystallographic complexes of several inhibitors with wt and K103N mutant RT also have revealed comparable binding modes and similar interactions with the enzyme in the bound state.[40]

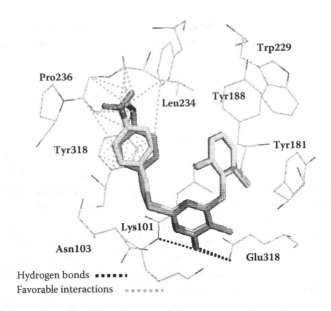

FIGURE 24.3 Docking studies on K103N did not clarify the different activity profiles of compounds **20** (shown in stick notation) and **25** (shown in stick notation) on this specific mutant. For the sake of clarity, only representative amino acids are shown.

Therefore, another approach had to be employed for this mutation, to consider both the flexibility of NNIBP and the events that promote the opening of the allosteric pocket and the entry of inhibitors.

To take into account the flexibility of the NNIBP, a study of molecular dynamic simulations was performed on both **20**:K103N and **25**:K103N complexes, starting from the conformations obtained by docking studies. The molecular dynamic simulations were performed using the software NAMD,[41] with a simulation time of 400 ps and a time step of 1fs with a water box of 32,970 water molecules. The trajectory analysis of the dynamic simulations showed that compound **20** had an electrostatic energy of interaction with the protein of −15.59 Kcal/mol, whereas compound **25** had an electrostatic energy of −32.77 Kcal/mol. These first results are in agreement with the biological data because compound **25**, with a much more favorable electrostatic energy of interaction compared to compound **20**, is about 3 orders of magnitude more active than the other.

These results are encouraging enough for us to persist with the design of novel inhibitors having a good activity profile against common drug-resistant HIV-1 RT strains.

6-VINYLPYRIMIDINES

Besides *S*-DABOs, which act as classical noncompetitive NNRTIs, our interest was also focused on synthesis and biological evaluation of 6-vinylpyrimidine derivatives, whose structure may be related to that of TNK-651 more than to that of any other known NNRTIs.[42] Enzymological studies revealed that such compounds bind the

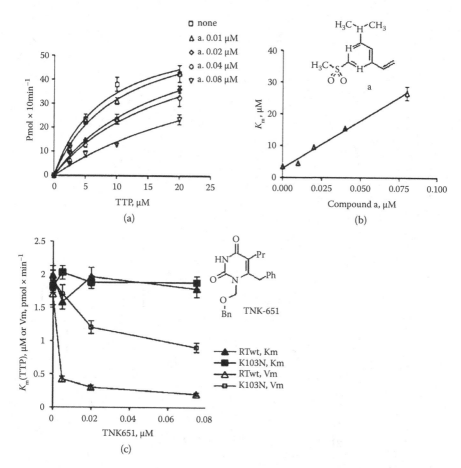

FIGURE 24.4 (a) Plot of the incorporation rates of wt HIV-1 RT, reporting the variation of the reaction rate as a function of the dTTP substrate concentration in the absence or presence of increasing amounts of compound **a**. Curves were fitted to a Briggs–Haldane mechanism. Error bars represent ±S.D. of three independent replicates. (b) Variation of the apparent affinity (K_m) for the nucleotide substrate as a function of the concentration of **a**. K_m values were determined as described in the Supporting Information of Reference 29 from the curves shown in panel a. (c) Variations of the apparent affinity (K_m; filled symbols) and maximal velocity (V_m; open symbols) for dTTP incorporation catalyzed by HIV-1 RT wild-type (triangles) or K103N (squares) as a function of TNK-651 concentration. Data are the means of three independent replicates. Bars represent ±S.D. TNK-651 was synthesized following a literature procedure; physical and spectroscopic data were consistent with those reported (*J. Med. Chem.*, 39, 1589–1600, 1996.)

NNIBP of the enzyme but, contrary to the NNRTIs reported to date, inhibit HIV-1 RT by a competitive mechanism with the nucleotide substrate. The most potent analogue, 2-methylsulfonyl-4-dimethylamino-6-vinylpyrimidine (a; see Chart 24.3), has high activity toward both wt RT and drug-resistant mutants.

To determine the mechanism of inhibition of 6-vinylpyrimidines, they were titrated in reverse-transcription assays *in vitro* in the presence of various

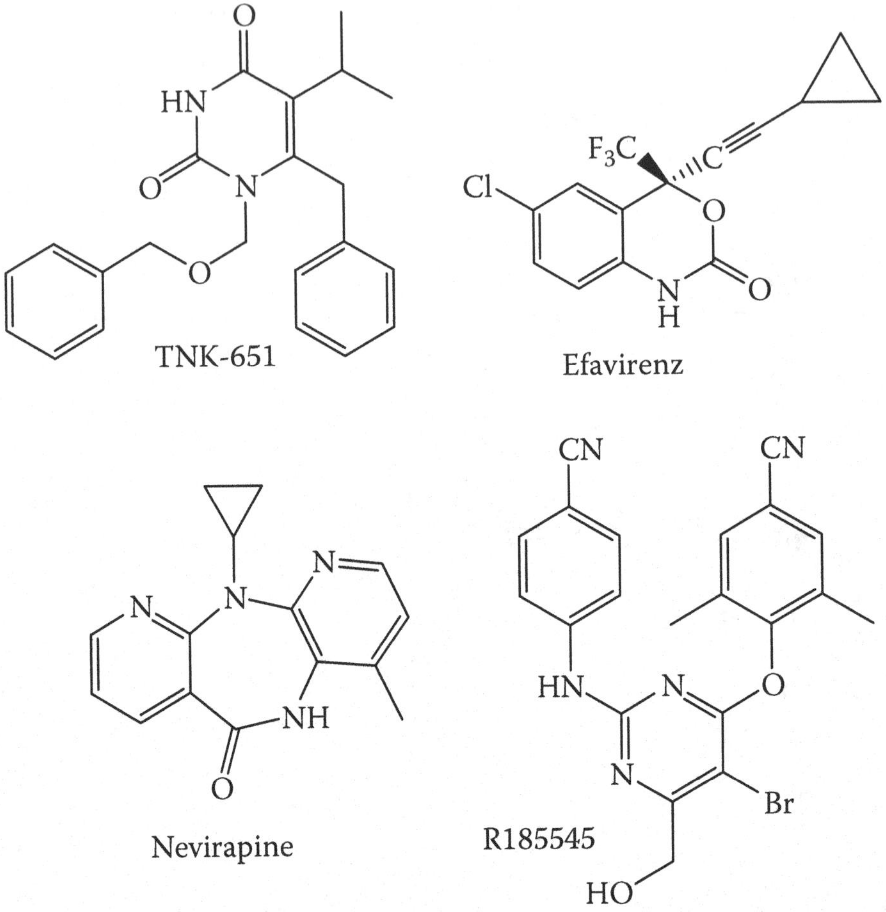

CHART 24.6 Examples of first- and second-generation NNRTIs.

concentrations of either the nucleic acid or the nucleotide substrates. As a result, inhibition exerted by the tested compounds was sensitive to changes in the nucleotide concentration (Figure 24.4a), resulting in an increase in the apparent K_m for 2′-deoxythymidine-5′-triphosphate (dTTP; Figure 24.4b). On the other hand, no effect on RT inhibition was observed when the nucleic acid concentration was varied (data not shown). These results clearly indicate that the tested 6-vinylpyrimidines **a-e** are competitive inhibitors of RT with respect to the nucleotide.[29] As a comparison, the structurally related 1-[(2-hydroxyethoxy)methyl]-6-(phenylthio)thymine (HEPT) analogue TNK-651 (see Chart 24.6) showed a purely noncompetitive mechanism of inhibition with respect to the nucleotide substrate, as shown by the lack of significant variations in the apparent K_m value for dTTP, both with the RT wt and the K103N mutant. In the absence of cocrystals between RT and our compounds, their binding site was established by testing their sensitivity to known NNRTI-resistant mutations localized in the NNIBP.[43-45]

TABLE 24.3

Inhibitory Activity of 6-Vinylpyrimidines a–e Toward wt and Mutants HIV-1 RT

Compound	K$_i$ µM				
	Wild type	K103N	Y181I	L100I	V179D
a	0.008	3.2	39	0.012	0.015
b	3.5	>400	>400	n.d.	n.d.
c	>400	>400	>400	n.d.	n.d.
d	90	>400	>400	n.d.	n.d.
e	325	>400	>400	n.d.	n.d.

Note: n.d. = not determined.

As shown in Table 24.3, inhibitory activity of the tested compounds was strongly affected by these mutations, which supports the hypothesis that 6-vinylpyrimidines bind to the NNIBP and therefore act as nonclassical competitive inhibitors.

Molecular docking and dynamics simulations were finally performed to investigate the interaction mode of these ligands with the NNIBP (both of wt and mutated RT) and to suggest a possible explanation for their unique mechanism of action. Compound **a** was docked into the wt HIV-1 RT NNIBP starting from the x-ray coordinates of the TNK-651:HIV-1 RT complex (PDB code: 1RT2),[9] which were chosen on the basis of the similarity between compounds **a-e** and TNK-651. The reliability of the docking protocol was tested on the prediction of the binding geometries of the reference compound TNK-651 into the NNIBP. As a result, the experimental binding conformation of the reference drug was successfully reproduced with acceptable root-mean square deviation (0.966 Å) of atom coordinates. The energetically preferred docked conformation of **a** revealed interactions that may contribute to the stability of the resulting inhibitor:RT complex. The heterocyclic ring of the ligand was found in close contact with Tyr-188, Tyr-181, and Phe-227 (allowing π–π interactions), whereas the vinyl group interacted with Tyr-318. Moreover, additional profitable hydrophobic contacts between the methyl groups of the amine moiety and Trp-229 were noted. The very polar sulfone group is oriented toward the water-exposed surface, in proximity to the positive charge of the Lys-101 ammonium group. To shed light on the peculiar mechanism of action of the 6-vinylpyrimidines **a–e**, the progression of the conformational changes in the side chain of Met-230, Asp-110, Asp-185, and Asp-186 (key residues for the polymerization process) was monitored by means of molecular dynamics (MD) simulations. In this regard, it was clearly seen that, as the simulation progresses, the side chain of Met-230 achieves an extended conformation. This new conformation could affect the positioning of the growing viral DNA (forcing the growing nucleic acid chain to reorient) and the subsequent polymerization process. By monitoring the conformational changes of the aspartic acid triad, it was interesting to note that, as the simulation progresses, the catalytic site opened so that the Asp-185 side chain was gradually shifted 5.59 Å away from its initial position ($t = 0$ in the MD simulation). In case of the **b**:RT (wt) complex, the catalytic

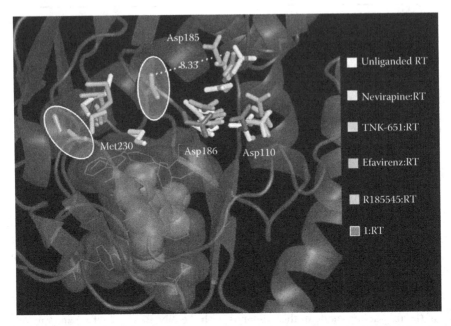

FIGURE 24.5 Superimposition of the structures corresponding to the **a**:RT complex at *t* = 1000 ps, TNK-651:RT complex (PDB code 1RT2), efavirenz:RT complex (PDB code 1IKW), nevirapine:RT complex (PDB code 3HVT), R185545:RT complex (PDB code 1SUQ), unligated RT (PDB code 1HMV). Circles evidence conformational rearrangements of Met-230 and Asp-185 for the **a**:RT complex in comparison to the other inhibitor:RT complexes. For reasons of clarity, only the fundamental residues (Met-230, Asp-110, Asp-185, and Asp-186) of unligated RT and RT complexed with reference inhibitors are shown as sticks. Compound **a** bound to the NNIBP is shown as spheres.

site triad also assumed an open conformation during the MD simulations even if this distortion was smaller than that observed for the most active compound (**a**). In fact, the Asp-185 side chain extended only 3.52 Å away from its initial position, and this could account for the lower inhibitory activity of **b**. Whatever their mechanism of inhibition, all the known NNRTIs bind the NNIBP. It is therefore reasonable to expect that a comparison of the crystal structures of non-ligand-bound HIV-1 RT with those of HIV-1 RT complexed with common NNRTIs and with the output of our MD simulations should yield valuable insights into the peculiar mechanism of action of the 6-vinylpyrimidines described here. Superposition analysis shows that the binding of common first- and second-generation NNRTIs (TNK-651,[9] efavirenz,[46] nevirapine,[47] and R185545[12]) (Chart 24.6) to the NNIBP determines only a major shift of the primer grip (Met-230), whereas the aspartic acid triad does not seem to experience substantial modifications compared with the non-ligand-bound enzyme. By contrast, it was interesting to note that although the binding of compound **a** to the NNIBP (**a**:RT complex at *t* = 1000 ps) determined only a small change to the side chain of Met-230 (white circle on the left in Figure 24.5), a significant and peculiar shift was noted for Asp-185 (white circle on the right in Figure 24.5), which was shifted far away from the aspartic acid triad (i.e., 7.66 Å away from its position in the TNK-651:RT complex). Furthermore, when the same MD simulation was performed on the non-ligand-bound RT and the RT:TNK-651 complex, no significant shift for Asp-185 was observed. On the basis of these observations, it is reasonable to argue that the unusual shift of Asp-185 could ultimately be responsible for the competitive mechanism of action exerted by the 6-vinylpyrimidines as, when **a** is bound to the allosteric site, the magnesium ions in the catalytic site may lie so apart that no phosphodiester bond formation can take place. From a visual inspection of the polymerase active site, we moreover speculated that the conformational rearrangements responsible for the competitive mechanism of action may originate from the disruption of the typical type II geometry of the turn formed by the conserved Tyr-Met-Asp-Asp sequence (residues 183–186), which are responsible for the correct positioning of the aspartate residues of the catalytic site.[48] In fact, it is well known that the formation of a hydrogen bond between Gln-182 and Met-184 is required for the stabilization of the otherwise-strained type II conformation of this turn in the wt RT. In a similar way, the complex RT:TNK-651 retained the Gln182:Met184 hydrogen bond and, consequently, displayed the same strained type II geometry of the β turn. On the contrary, in the **a**:RT complex, a conformational rearrangement of the turn occurred so that the side chains of Asp-185 and Glu-182 were shifted away from their original position. As a consequence, the unfavorable steric interaction between the Cb atom of Met-184 and the amide group of Asp-185 disappeared and the hydrogen bond between Gln-182 and Met-184 was lost. This result further supports the peculiar behavior of compound **a**, which is able to induce conformational modifications otherwise not found in complexes between RT and common NNRTIs. In summary, the present work reports the identification of a new class of NNRTIs with a 6-vinylpyrimidine scaffold found to exhibit a peculiar behavior: contrary to the NNRTIs reported to date, enzymological studies reveal that such compounds inhibit HIV-1 RT by a competitive mechanism with the nucleotide substrate after binding to the NNIBP of the enzyme. To the best of our knowledge, these compounds represent the first example of NNRTIs found to exhibit such a behavior.

ACKNOWLEDGMENT

The authors thank Fabrizio Manetti, Luca Bellucci, Cesare Bernardini, Montserrat Terrazas, Jaume Vilarrasa, Maddalena Alongi, Chiara Falciani, Harsukh Gevariya, Claudia Mugnaini, Andrea Togninelli, and José A. Esté.

REFERENCES

1. Furman, P.A., Painter, G.R., and Anderson, K.S., An analysis of the catalytic cycle of HIV-1 reverse transcriptase: opportunities for chemotherapeutic intervention based on enzyme inhibition, *Curr. Pharm. Des.*, 6, 547, 2000.
2. St. Clair, M.H. et al., 3'-Azido-3'-deoxythymidinetriphosphate as an inhibitor and substrate of purified human immunodeficiency virus reverse transcriptase, *Antimicrob. Agents Chemother.*, 31, 1972, 1987.
3. Merluzzi, V.J. et al., Inhibition of HIV-1 replication by a nonnucleoside reverse transcriptase inhibitor, *Science*, 250, 1411, 1990.

4. Pauwels, R. et al., Potent and selective inhibition of HIV-1 replication *in vitro* by a novel series of TIBO derivatives, *Nature*, 343, 470, 1990.

5. Romero, D.L. et al., Nonnucleoside reverse transcriptase inhibitors that potently and specifically block human immunodeficiency virus type 1 replication, *Proc. Natl. Acad. Sci. U.S.A.*, 88, 8806, 1991.

6. Pauwels, R. et al., Potent and high selective HIV-1 inhibition by a new series of R-anilinophenylacetamide R-APA derivatives targeted at HIV-1 reverse transcriptase, *Proc. Natl. Acad. Sci. U.S.A.*, 90, 1711, 1993.

7. Cantrell, A.S. et al., Phenethylthiazolylthiourea (PETT) compounds as a new class of HIV-1 reverse transcriptase inhibitors. 2. Synthesis and further structure-activity relationship studies of PETT analogues, *J. Med. Chem.*, 39, 4261, 1996.

8. Baba, M. et al., Highly specific inhibition of human immunodeficiency virus type-1 by a novel 6-substituted acyclouridine derivative, *Biochem. Biophys. Res. Commun.*, 165, 1375, 1989.

9. Hopkins, A.L. et al., Complexes of HIV-1 reverse transcriptase with inhibitors of the HEPT series reveal conformational changes relevant to the design of potent non-nucleoside inhibitors, *J. Med. Chem.*, 39, 1589, 1996.

10. Ludovici, D.W. et al., Evolution of anti-HIV drug candidates. Part 1: From R-anilinophenylacetamide (R-APA) to imidoyl thiourea (ITU), *Bioorg. Med. Chem. Lett.*, 11, 2225, 2001.

11. Ludovici, D.W. et al., Evolution of anti-HIV drug candidates. Part 2: Diaryltriazine (DATA) analogues, *Bioorg. Med. Chem. Lett.*, 11, 2229, 2001.

12. Das, K. et al., Roles of conformational and positional adaptability in structure-based design of TMC125-R165335 (etravirine) and related non-nucleoside reverse transcriptase inhibitors that are highly potent and effective against wild-type and drug-resistant HIV-1 variants, *J. Med. Chem.*, 47, 2550, 2004.

13. Guillemont, J. et al., Synthesis of novel diarylpyrimidine analogues and their antiviral activity against human immunodeficiency virus type 1, *J. Med. Chem.*, 48, 2072, 2005.

14. Janssen, P.A.J. et al., In search of a novel anti-HIV drug: Multidisciplinary coordination in the discovery of 4-[[4-[[4-[(1E)-2-cyanoethenyl]-2,6-dimethylphenyl]amino]-2-pyrimidinyl]amino]benzonitrile (R278474, rilpivirine), *J. Med. Chem.*, 48, 1901, 2005.

15. Ren, J. et al., High resolution structures of HIV-1 RT from four RT-inhibitor complexes, *Nat. Struct. Biol.*, 2, 293, 1995.

16. Smerdon, S.J. et al., Structure of the binding site for nonnucleoside inhibitors of the reverse transcriptase of human immunodeficiency virus type 1, *Proc. Natl. Acad. Sci. U.S.A.*, 91, 3911, 1994.

17. Ding, J. et al., Structure of HIV-1 reverse transcriptase in a complex with the nonnucleoside inhibitor α-APA R 95845 at 2.8 Å resolution, *Structure*, 3, 365, 1995.

18. Ren, J. et al., The structure of HIV-1 reverse transcriptase complexed with 9-chloro-TIBO: lessons for inhibitor design, *Structure*, 3, 915, 1995.

19. Das, K. et al., Crystal structures of 8-Cl and 9-Cl TIBO complexed with wild-type HIV-1 RT and 8-Cl TIBO complexed with the Tyr181Cys HIV-1 RT drug-resistant mutant, *J. Mol. Biol.*, 264, 1085, 1996.

20. Esnouf, R.M. et al., Unique features of the complex between HIV-1 reverse transcriptase and the bis-(heteroaryl)piperazine (BHAP) U-90152 explain resistance mutations for this nonnucleoside inhibitor, *Proc. Natl. Acad. Sci. U.S.A.*, 94, 3984, 1997.

21. Ren, J. et al., Crystal structures of HIV-1 reverse transcriptase in complex with carboxanilide derivatives, *Biochemistry*, 37, 14394, 1998.

22. Högberg, M. et al., Urea-PETT compounds as a new class of HIV-1 reverse transcriptase inhibitors. 3. Synthesis and further structure-activity relationship studies of PETT analogues, *J. Med. Chem.*, 42, 4150, 1999.

23. Arnold, E. et al., Targeting HIV reverse transcriptase for anti-AIDS drug design: structural and biological considerations for chemotherapeutic strategies, *Drug. Des. Discov.*, 13, 29, 1996.

24. De Clercq, E., Non-nucleoside reverse transcriptase inhibitors (NNRTIs): past, present, and future, *Chem. Biodiv.*, 1, 44, 2004.

25. De Clercq, E., Toward improved anti-HIV chemotherapy: therapeutic strategies for intervention with HIV infections, *J. Med. Chem.*, 38, 2491, 1995.

26. De Clercq, E., Antiviral therapy for human immunodeficiency virus infections, *Clin. Microbiol. Rev.*, 8, 200, 1995.

27. Botta, M. et al., Synthesis, antimicrobial and antiviral activities of isotrimethoprim and some related derivatives, *Eur. J. Med. Chem.*, 27, 251, 1992.

28. Manetti, F. et al., Parallel solution-phase and microwave-assisted synthesis of new S-DABO derivatives endowed with subnanomolar anti-HIV-1 activity, *J. Med. Chem.*, 48, 8000, 2005.

29. Maga, G. et al., Discovery of non-nucleoside inhibitors of HIV-1 reverse transcriptase competing with the nucleotide substrate, *Angew. Chem. Int. Ed.*, 46, 1810, 2007.

30. Parlato, M.C. et al., Solid-phase synthesis of 5,6 disubstituted pyrimidinone and pyrimidindione derivatives, *ARKIVOC*, v, 349, 2004.

31. Mai, A. et al., 5-Alkyl-2-(alkylthio)-6-(2,6-dihalophenylmethyl)-3,4-dihydropyrimidin-4(3H)-ones: novel potent and selective dihydro-alkoxybenxyl-oxopyrimidine derivatives, *J. Med. Chem.*, 42, 619, 1999.

32. Clay, R.J. et al., Economical method for the preparation of β-oxo esters, *Synthesis*, 290, 1993.

33. Petricci, E. et al., Microwave-assisted acylation of amines, alcohols, and phenols by the use of solid-supported reagents (SSRs), *J. Org. Chem.*, 69, 7880, 2004.

34. Obrecht, D. et al., A novel and efficient approach for the combinatorial synthesis of structurally diverse pyrimidines on solid support, *Helv. Chim. Acta.*, 80, 65, 1997.

35. Goodsell, D.S., Morris, G.M., and Olson, A.J., Automated docking of flexible ligands: applications of AutoDock, *J. Mol. Recognit.*, 9, 1, 1996.

36. Mohamadi, F. et al., MacroModel—An integrated software system for modeling organic and bioorganic molecules using molecular mechanics, *J. Comput. Chem.*, 11, 440, 1990.

37. Maestro, version 6.0, is distributed by Schroedinger.

38. Mai, A. et al., 5-Alkyl-2-alkylamino-6-(2,6-difluorophenylalkyl)-3,4-dihydropyrimidin-4(3H)-ones, a new series of potent, broad-spectrum nonnucleoside reverse transcriptase inhibitors belonging to the DABO family, *Bioorg. Med. Chem.*, 13, 2065, 2005.

39. Yin, P.D., Das, D., and Mitsuya, H., Overcoming HIV drug resistance through rational drug design based on molecular, biochemical and structural profiles of HIV resistance, *Cell. Mol. Life Sci.*, 63, 1706, 2006.

40. Rodriguez-Barrios, F. and Gago, F., Understanding the basis of resistance in the irksome Lys103Asn HIV-1 reverse transcriptase mutant through targeted molecular dynamics simulations, *J. Am. Chem. Soc.*, 126, 15386, 2004.

41. Phillips, J.C. et al., Scalable molecular dynamics with NAMD, *J. Comput. Chem.*, 26, 1781, 2005.

42. Radi, M. et al., Parallel liquid-phase synthesis of 2-methylsulfonyl-4-dialkylamino-6-vinyl pyrimidines, *J. Comb. Chem.*, 7, 117, 2005.

43. Balzarini, J. et al., HIV-1 specific reverse transcriptase inhibitors show differential activity against mutant strains containing different amino acid substitutions in the reverse transcriptase, *Virology*, 192, 246, 1993.

44. Mellors, J.W. et al., *In vitro* selection and molecular characterization of human immunodeficiency virus-1 resistant to non-nucleoside inhibitors of reverse transcriptase, *Mol. Pharmacol.*, 41, 446, 1992.

45. Nunberg, J.H. et al., Viral resistance to human immunodeficiency virus type 1-specific pyridinone reverse transcriptase inhibitors, *J. Virol.*, 65, 4887, 1991.
46. Lindberg, J. et al., Structural basis for the inhibitory efficacy of efavirenz (DMP-266), MSC194 and PNU142721 towards the HIV-1 RT K103N mutant, *Eur. J. Biochem.*, 269, 1670, 2002.
47. Wang, J. et al., Structural basis of asymmetry in the immunodeficiency virus type 1 reverse transcriptase heterodimer, *Proc. Natl. Acad. Sci. U.S.A.*, 91, 7242, 1994.
48. Rodgers, D.W. et al., The structure of unligated reverse transcriptase from human immunodeficiency virus type 1, *Proc. Natl. Acad. Sci. U.S.A.*, 92, 1222, 1995.

Index

Printed in the United States
by Baker & Taylor Publisher Services